i

Linear Integrated Circuits as Sensor Signal Conditioning Amplifiers:
Analog Sensors and Operational Amplification for Data Acquisition and Experimental
Testing

Table of Contents

Strain Gauge Conditioning

Example Stress, Strain and Strain Gauge Computation

Dummy Gauges, Leadwire Resistance and Thermal Compensations

Bridge Gain and Offset

Display Scaling

Stress/Strain Application Notes

    Ductile Materials and Strain Gauges

    Stress coat

Stress/Strain Problems

Lab Experiments

Total pages of text w/o labs w/ ToC = 448
Total pages of labs = 93
Total pages including labs = 541

Chapter 1 - Introduction to Sensors, Signal Conditioning and Data Acquisition

-This chapter provides the necessary framework to begin a study of experimental measurements, signal conditioning and data analysis. Sensors, measurement, signal conditioning and data acquisition systems of an analog nature are presented. Linear, Y=MX+B relations are reviewed. Typical specifications of measurement components are investigated, terms and characteristics are defined, and fundamental operating characteristics are investigated. Chapter topics are included following.

Objectives:
Upon completion of this chapter, you should be able to:
- Explain in general terms the purpose and operation of common linear analog sensors
- Explain the operational characteristics of gain and offset
- Interpret the important operational specifications of common sensors
- Suggest an appropriate signal conditioner for a given sensor
- Suggest calibration techniques for common sensors

Measurement Variables and Sensors

As defined in mathematics, any quantity capable of change is a variable. A measurement variable is a quantity associated with a product or a process, usually manufacturing, capable of change. Measurement variables need not be only associated with manufacturing however. Nor will measurement variables be used for purely measurement or analysis purposes.

Sensors encompass our daily lives. From the remote control of one's television to the unseen complexities under the hood of an automobile, automation of one form or another will continue to influence the quality of life that all people experience. At the center of an automated system is the ability to detect the state or condition of a process. To perform the detection function, sensors have been developed. Sensors can assume many forms with most today providing an electronic output to be easily adapted for computer interfacing.

Many sensors are *transducers*. Transducers convert energy forms. A temperature transducer in a chemical plant converts thermal energy into a proportional electronic signal, ultrasonic transducers convert sound into electronic impulses to scan the human anatomy, or open a door at a grocery store. Pressure transducers are used to convert fluid pressure into a voltage signal to measure human blood pressure or to interpret liquid level within a tank. Transducers provide a means of converting real world physical signals of pressure, level, flow, velocity, acceleration, temperature, position and other variables into electrical quantities which can be introduced to electronic components for analytical, control, display and other purposes. As such, transducers convert physical phenomena for electronic use in a manner similar to the performance of an analog to digital converter processing sensory information for compatibility with computers.

Experimental Measurement Systems

In recent years manufacturers have become quality conscious. A product that is not reliable, or that is not perceived as being worth the price will not sell in the contemporary marketplace. To assist manufacturers in developing a quality product, experimental testing facilities have been developed. Products and their sub-assemblies are tested in experimental facilities for reliability, durability, safety, long-life and other characteristics. All of which contribute to product quality. Sensors performing in the measurement and signal conversion capacity are necessary components of the experimental testing process. Such testing is common in automotive, aircraft, agricultural and most industries developing products for human use where safety or comfort is a primary concern.

Yet many physical variables need not be tested for quality but merely measured and displayed or analyzed for informational purposes. Many biophysical processes as in hospitals are measured, analyzed and displayed for purely informational reasons. Similar sensing, measurement, display and analysis systems are found in scientific research and development laboratories where the information provided by sensors ultimately enhance products and processes that humans contact routinely. Regardless of the purpose of the measurement, sensors from the same group of manufacturers find a plethora of applications with products and processes that surround our daily lives.

This text will examine common experimental measurement sensors and associated conditioning, calibration, display and data analysis functions. As an introductory text all measurement sensors, amplifiers, display devices and data analysis topics will be of a linear, proportional nature. Linear components and systems are described by straight-line characteristic functions. A linear characteristic is one whose input/output plot is a straight line. A review of linear, Y = MX + B characteristics follow.

Linear Review

Without question, a majority of components in measurement systems today are naturally, or have been made linear. Proportional components and systems have a number of benefits including ease of understanding, simplistic operation, straightforward analysis, design and repair. An intuitive comprehension of proportional concepts will be very useful when checking, calibrating, designing, modifying or operating sensing and measurement apparatus. Before proceeding, a review of basic Y=MX+B, linear relations is in order.

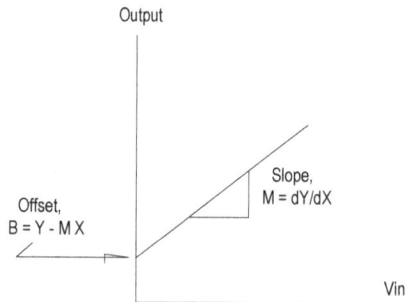

Figure 1 – The linear, Y=MX+B characteristic is representative of the majority of sensing, measurement, conditioning and control components.

The Y=MX+B model allows an investigation of proportional input/output relations. For our purposes Y=MX+B is representative of an amplifier performing a *scaling* function. Scaling involves amplification (input signal voltage span or amplitude change) and offset (output is shifted from an initial value of zero) of a voltage signal span. Signal spans shifted (offset) from zero volts are commonly provided in the output of sensors measuring physical variables such as pressure, temperature, flow, stress, strain, acceleration, position and so forth. The signals available to amplifier inputs are supplied by a sensor's output. The following diagram is representative of a common sensor-based measurement system.

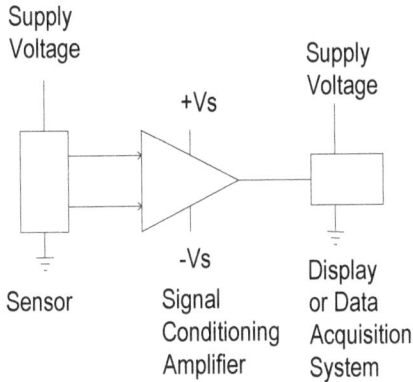

Figure 2 - A typical measurement system is composed of a sensor, signal conditioner and display or data acquisition system.

Where Y=MX+B is a general and universal model used in any linear (proportional) application, applications are best suited by substituting quantities appropriate to the application of experimental testing and sensor signal conditioning. Assume input (X) and output (Y) voltages, gain (M) and offset (B) quantities are substituted into the Y=MX+B equation.

Given,

$$Y = (M*X) + B$$

or more specifically using electronic terms,

$$Vout = (Gain * Vin) \pm Offset$$

Where,   The 'Y' quantity is Vout,

The slope of the plot, 'M', is Gain,

The 'X' quantity is Vin,

The 'B' of the plot is the Output Offset Voltage, V_B,

The following figure helps to demonstrate these quantities. The Slope (M) is the dynamic characteristic or "rise over run" of the resulting input/output plot. The Offset (B) is the where the plot passes (or would pass if extended) through the 'Y' axis. The offset, or input offset, is properly (mathematically) referred to as the Y-intercept value of the plot, and represents the 'Y' (output) value at an 'X' (input) value of zero. In electronic amplification terms, the slope (M) is the amplifiers gain (A_V) and the Y-intercept value (B) is the required amount of Output Offset, also referred to as V_B.

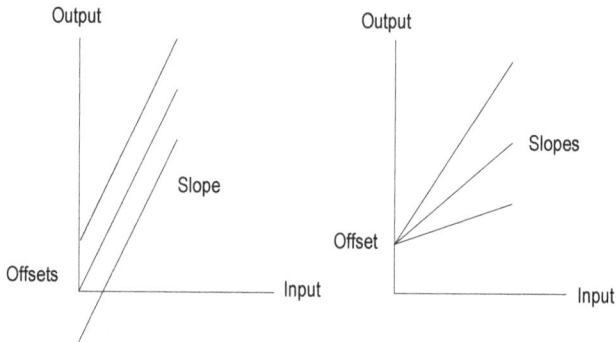

Figure 3 - Examples of same slope, differing offset values (left) and same offset, differing slopes (right).

As will be described later, the ratio of two resistors used to assemble the amplifier circuit will determine the amplifier circuit's gain, and the graphs' slope. The amount of DC voltage introduced to the amplifier's input (in conjunction with the signal) will determine the circuits Output Offset, V_B, or Y-intercept value of the graph.

As an example of a linear amplification characteristic using the defined Y=MX+B quantities, consider the previous plots and examine the following.

    A) The amplifier circuit's **input** span (conceivably from a sensor's output) is plotted on the **x-axis**,

    B) The amplifier circuit's **output** span is plotted on the **y-axis**,

C) The slope of the corresponding input/output voltage spans represents the amplifier circuits operational gain. The slope (voltage gain, Av) is computed as;

$$Slope = rise/run = \Delta y/x = Output\ Change/Input\ Change$$

or,
$$A_v = \Delta V_{out}/\Delta V_{in}$$

D) The intersection of the input/output characteristic plot and the y-axis represents the circuits output offset value, "B". The output offset, or "B," value is often referred to as V_B and is computed by rearranging Y = MX + B algebraically to solve for "B." Therefore,

Since, $$Y = M*X + B$$

Rearranging, $$B = Y - MX$$

or, $$Offset = Output - (Gain*Input)$$

And "B" or V_B equals, $$V_B = V_{out} - (A_v * V_{in})$$

As a reminder, in the general form,
$$Y = MX + B$$
And in circuit form,
$$V_{out} = A_v V_{in} + V_B$$

where V_B is the "output offset" quantity computed from the input/output plot indicating the amount and direction of the *output voltage spans' displacement* along the y-axis.

To achieve the non-zero characteristic plot shift of V_B, A DC Vbias will be the biasing voltage (or offset bias voltage) applied to the amplifier circuit to achieve the output spans' vertical displacement up or down the Y-axis. Although the two quantities of V_B and Vbias sound and appear similar, they are almost always numerically different and should not be confused.

Note:

B, or V_B = Y-axis intercept quantity computed from the input/output plot, the output-bias quantity.

V_{bias} = DC Voltage applied to the circuit to create the above Y-axis intercept value shift of B.

Y=MX+B Amplifier Example

The following plot shows an input voltage span of 0 to 4 Volts and a corresponding output voltage span of 1 to 9 Volts. From the previous equations this results in a circuit gain,

$$A_v = \Delta V_{out}/\Delta V_{in}$$

$$A_v = 8V/4V$$

Therefore,

$$A_v = +2$$

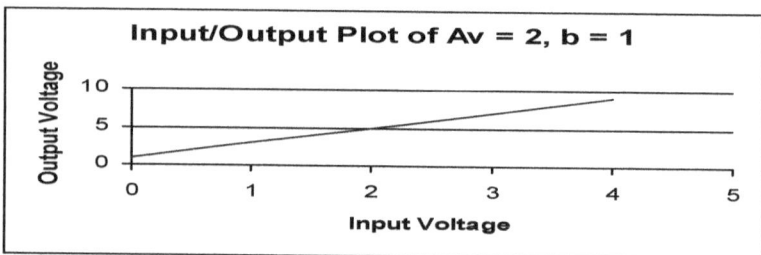

Figure 4 - Input / Output plot of the amplifier characteristic in the previous example. Note the slope and offset (Y-axis intercept) values.

Where the positive gain polarity represents a non-inverting or direct-acting amplifier function. Direct-action means the amplifiers output will increase as the input increases, in a direct manner. An inverting or reverse-acting characteristic would exhibit a <u>negative</u> gain polarity and input/output slope, (V_{in} increases, V_{out} decreases).

The output bias or y-axis intercept point of the graph is;
$$B = Y - MX$$

Or,
$$V_B = V_{out} - A_v V_{in}$$

Substituting corresponding V_{out} and V_{in} values from the given input and output spans,

$$V_B = +1 \text{ Volt}$$

Again, V_B represents the distance and direction of the output voltage span from 0 Volts along the y-axis. The required V_{bias} to create this output-bias, V_B remains to be determined and is contingent upon the specific circuit values used to design the amplifier.

Sensor Example

Being proportional devices, sensors can also be represented using straight-line proportional, Y=MX+B characteristics.

Assume it is desired to know the gain and offset of a pressure sensor. A graph similar to the amplifier input/output plot might show an input pressure span of 0 to 30 psi and an arbitrarily chosen output voltage span of -30 to 90 millivolts (-.030 to +.090 Volts).

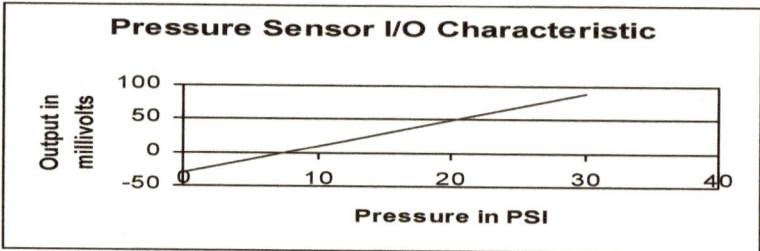

Figure 5 - Pressure sensor and I/O characteristic plot.

To determine the gain and offset associated with our pressure sensor, the mathematical process used in the previous example would also work here. However, since the units of input are now pressure, the units associated with the computations would represent psi for the input and Volts for the output. In addition, since the input and output are no longer both voltage, the symbol of Av is not used to designate gain. In cases where the input and output are different energy forms (such as mechanical and electrical in this case), the symbol for gain becomes, K, or Kx, for transducer gain.

Using the technique introduced in the previous example, the sensor gain becomes,

$$Kx = \Delta Vout/\Delta Pin$$
$$Kx = (-.03V \text{ to } +.09V)/(0psi \text{ to } 30psi)$$
$$Kx = .12V/30psi$$

Therefore, $Kx = +.04 \text{ V/psi}$

Or, $Kx = +40 \text{ millivolts per psi}$

Where the positive polarity represents a non-inverting, or direct-acting sensor operation. Again, direct-action means the sensor's output will increase as the input increases, the sensor input and output are operating in a direct manner. An inverting or reverse-acting characteristic would exhibit a <u>negative</u> gain polarity and input/output slope, (input increases, output decreases). The sensor gain represents the quantity of output *change*, which results from an input *change* of 1 psi. It is important to note that gain is a dynamic quantity that relates how much input or output values change under given conditions. Knowing gain without knowing the initial starting value of output offset, it is not possible to determine an output value corresponding to a specific input value, only the amount of output change can be determined.

The output offset or y-axis intercept point of the sensor's operational span is;

$$B = Y - MX$$

Or, in electronic terms,

$$V_B = V_{out} - (Kx * P_{in})$$

Substituting corresponding V_{out} and P_{in} values from the given input and output spans yields,

$$V_B = -.03 \text{ Volt}$$

Or, negative 30 millivolts is the output offset of the sensor.

Again, V_B merely represents the distance and direction of the output voltage span from 0 Volts (at 0 psi) along the y-axis. In this case the sensor will provide a non-zero quantity of minus thirty millivolts when 0 psi is applied to the sensor. Many sensors employ non-zero output offsets to assist in rapidly assessing problems.

Figure 6 - Low level output (millivolts) pressure sensors. Each of these sensors generates outputs of millivolts per psi of input pressure. The previous example was modelled after the Sensym SX30DN, which outputs –30 to +90 millivolts over zero to thirty psi. The SX30DN is at the lower right.

A sensor whose output offset is zero would exhibit no output voltage at the minimum applied input. Since the loss of electrical supply power to a sensor would result in the same output condition of zero volts, some sensor manufacturers intentionally shift the output span positively or negatively to differentiate between a zero input from the measurement, and a loss of power supply condition.

Figure 7 – This disassembled Viatran 3000 psi pressure sensor outputs a millivolt pressure span, and is a common sight in experimental pressure testing applications.

The shift away from zero volts in the output of the previous example is not intentional however. The sensor used in the example was modeled after the Sensym SX30DN, a 30 psi pressure sensor. The SX30DN utilizes a fully active semiconductor strain gauge bridge internally to sense and convert pressure into voltage. Since semiconductor strain gauges are difficult to match (like all discrete semiconductor components), the bridge is not completely nulled (or balanced, an output condition of zero volts) at 0 psi of input pressure. The result is a small amount of output offset voltage. Additional bridge and pressure sensor material is located in the following chapter.

The Purpose and Process of Calibration

Calibration as defined by the Instrument Society of America (ISA) is the procedure of checking, quantifying and correcting the graduations on a measurement instrument. Since amplifiers are commonly used for the purpose of sensor signal conditioning, the sensor, amplifier and associated output load become a "measurement instrument." In order to receive accurate data from the instrument, calibration adjustments to minimize the inherent error are commonly provided. These error-minimizing provisions are the gain (usually referred to as "span") and offset bias (often labeled as "zero") adjustments. An additional adjustment of linearity will occasionally be present to assist in assuring a proportional, linear and predictable relation between the variable at the input and the output span.

Span (gain) and zero (bias) are terms that relate to the graphing of straight-line, linear, Y=MX+B input/output relations. Simply stated, the *zero* (offset) adjustment determines the Y-axis intercept location, and the *span* adjustment will establish the distance between the output end points and slope of the input/output plot.

Figure 8 – A "rack" of signal conditioning amplifiers. Different types of sensors require different forms of amplifiers. The doors on three of the units are opened to expose the calibration adjustments of "span" and "zero."

Using the pressure sensor of the previous example, the output offset was computed as -30 millivolts and the gain was determined as 4 millivolts per psi. If provisions for offset variation were included with the sensor, the initial starting value would be variable within a fixed range. A variable offset could possibly allow for an output offset of 0 millivolts, or +30 millivolts at 0 psi if desired. Having variable gain available would modify the amount of output change within a fixed range around 4 millivolts per psi. Ultimately, variable gain determines the amount of output span observed as the input varies between the minimum and maximum values, or the distance between output end points. Variable gain could reduce the output span to possibly 100 millivolts, increase it to 150 millivolts, or allow setting it precisely to 120 millivolts.

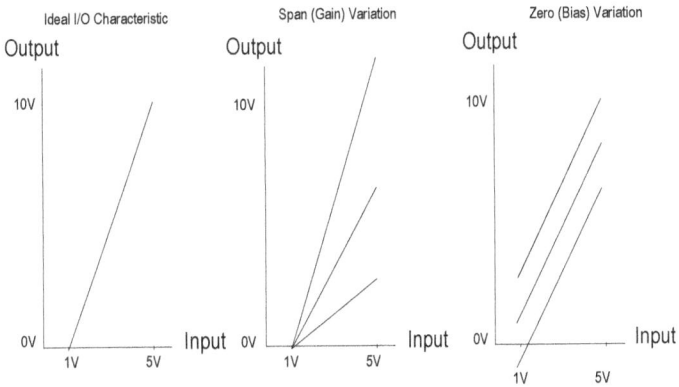

Figure 9 - Effects of independent gain (center) and offset (right) variation.

Amplifier Calibration

The effects of individual span and zero adjustments are demonstrated in figure 8. The example demonstrates the graphical changes created by varying gain and offset. As can be seen, the diagram at the left represents the ideal input/output characteristics of a direct acting amplifier. This characteristic plot could be the end result of the calibration attempt. The center figure illustrates the effect of varying the span (gain). A lower span is associated with a smaller distance between output end points and lower slope, whereas a greater span results in a larger distance between end points and a steeper slope.

The effects of bias variations, or zero adjustments, become obvious when the input span is assumed to be ideal and varied between 1 and 5 Volts. As is demonstrated in the graph at the right, a variable zero adjustment results in different values of the Y-axis intercept quantity, best observed if the input/output plots are projected below the 1 Volt minimum input value. A lower bias or "zero" setting is understood to suppress the output voltage span since some of the output values will be negative. This is called an "elevated-zero" span since zero is elevated towards the center of the operating span. A higher, more positive, Y-axis intercept voltage is understood to elevate the output span into a range where all output values are positive. This type of output is referred to as "suppressed-zero" span since zero is suppressed below the operating span. Output spans starting at zero volts and moving positive are often referred to as "true-zero" spans.

Figure 10 – A self-contained 120VAC-powered amplifier module for a type "J" (iron/constantan) thermocouple temperature sensor. The calibration adjustments are obvious at the top of the unit. The output signal is a 4 to 20 milliamp current loop. Current loop outputs are used when the signal must travel over significant distance.

To set the span and zero adjustments the ideal input measurement variable or voltage span must be applied and the individual adjustments set at the appropriate input extremes. An accurate measurement instrument is employed to measure the input and output quantities. Often when working with sensors, calibration standards are employed to assure accurate measurement of the input measurement variable.

When performing the calibration, it is highly beneficial to retain the desired input/output plot in mind and imperative that the concepts of Y-axis intercept and input/output slope be understood when performing the calibration procedure.

If a sensor is connected to the input of the amplifier undergoing calibration, the measurement variable to be measured should be applied preferably over the entire range of operation. The measurement variable will need to be measured using an accurately known standard or the results of the calibration process will be questionable. If a sensor is not attached to the signal conditioner to be calibrated, or if varying the measurement variable throughout its operating span is not feasible, the sensors' output (amplifiers' input) span must be simulated as closely as possible.

Figure 11 – A multi-purpose signal-conditioning amplifier. This unit will accept inputs from resistive, voltage and frequency generating sensors. The larger lock-down adjustments on the front panel are for calibration purposes.

Measurement System Calibration

Assume a pressure sensor, signal conditioning amplifier and digital panel meter (DPM) similar to the following diagram is to be calibrated. The amplifier will be the focus of the calibration effort since it contains variable gain and bias potentiometers. The ideal end result of the calibration process will be a 0.00 displayed when the input pressure is 0 psi and 10.00 will be exhibited when the input pressure is 10 psi. A pressure gauge of known and documented accuracy or mercury manometer will be connected as a measurement standard (accuracy reference) to the sensors' input to precisely measure the input pressure. After the system is assembled and sufficient warm-up time has passed, apply 0 psi to the sensor. Under this condition the sensor will output 1 Volt to the signal-conditioning amplifier. A biasing voltage applied to the amplifier is responsible for canceling the 1 Volt input and sending 0 Volts to the DPM. The panel meter will display 0 Volts as 0.00 psi.

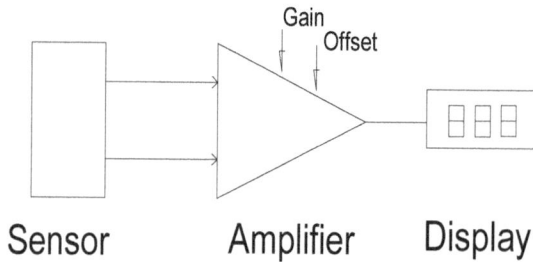

Gain
Offset

Sensor   Amplifier   Display

Figure 12- A simple measurement system. In this example the amplifier contains the gain and offset adjustments for system calibration although, calibration adjustments are commonly found at all three components.

Assuming the amplifier offset and gain potentiometers are both at random positions, the zero (offset) potentiometer should be adjusted for a display as close to the desired 0.00 as possible. This adjustment will probably require resetting since the present output is determined by the value of the gain potentiometer which is about to be adjusted.

As mentioned earlier, the input/output plot of pressure versus displayed value at the DPM should be kept in mind while performing the calibration. The previous bias adjustment was the first step in establishing the Y-axis intercept point. Varying the bias adjustment should be understood to shift the input/output plot vertically without effecting the slope. If the maximum pressure of 10 psi is applied and verified by the measurement standard instrument, the span (gain) potentiometer should be adjusted to display 10.0 psi on the DPM. Varying the gain has unfortunately altered the output-bias since the effect of the bias is determined by the amplifier gain, A_{Vbias}. The bias adjustment will usually require readjusting.

Reapplying 0 psi to the sensor, the bias potentiometer should again be set for a display of 0.00 psi. Raising the applied pressure to 10 psi, check and adjust for the required 10.0 psi display. After the input extremes have been calibrated several points at various locations in between should be checked. Since provisions for linearity are not included with our signal conditioner, any error associated with the mid-values cannot be corrected.

The calibration process with input biased operational amplifiers unfortunately requires setting the bias and gain value at least twice initially since the gain and bias functions are interactive. The value of one is dependent upon the other.

Figure 13 – A "Rack" of signal conditioning amplifiers, 9 of the 16 amplifiers is obvious. The calibration adjustments of span and zero are obvious on two of the units. Each module can output two signals (one voltage and one current) simultaneously, and requires two sets of output calibration adjustments as shown. The amplifier power supply is at the right of the photo.

27

Standards

In order to maintain consistency in measurement results, so 10 volts in Moline equals 10 volts in Beaver Falls, a system of measurement standards has been developed. Standards take many forms, everything from measurement procedures, to hardware items, to observing the response of materials when subjected to a physical quantity such as pressure or temperature. The National Institute for Standards and Technology (NIST) and other institutes around the world are responsible for developing, documenting, inventorying and distributing primary, secondary, tertiary and "working" measurement standards, and measurement procedures.

Standards in industrial measurement and experimental testing are used for the purpose of calibration. Calibration is the process of evaluating, quantifying and/or correcting the error that is inherent in every measurement instrument. Calibration procedures are relatively simple, the underlying idea or concept behind calibration is comparison of an accurately known standard value to an unknown or questionable one. By observing any difference between the standard and the "unknown," the error can be evaluated and through adjustment, minimized or if no adjustment is available the error can be documented and compensated for in future measurements. The importance of this comparison concept cannot be overemphasized. It is the foundation of the measurements and controls field and carries over into the daily activities of virtually every scientific "student" regardless of discipline.

Secondary and working measurement standards are commercially available and maintained by manufacturers for their own lab use throughout the world. Primary or prime standards are in NBS and other standards labs around the world and are used to calibrate and verify secondary and working standards. All standards should be traceable to the NIST or some other recognised standards institute to be of any value. As a simplified example, a digital DC voltmeter that is of questionable accuracy should be calibrated by a DC voltmeter calibrator that was calibrated by standard cells directly traceable to NIST. In this way the digital voltmeters' calibration is understood to be traceable to the NIST. If all voltmeters are calibrated to NIST standards, then 10 volts in Seattle will equal 10 volts in Miami and anywhere else.

The following figures exhibit tertiary and working standards, two encased wire-wound resistors and a 1.019v Weston standard unsaturated cadmium cell. The resistors are designed to minimize the effects of temperature, humidity and age. The resistance sits in an oil-filled, insulated wall container, and is used at a temperature of 77° F (25 C), at a power dissipation of .01 Watt. The Weston cell maintains a very constant voltage over long periods with only minor temperature effects. This particular cell was mounted inside a potentiometric bridge (a sensitive millivoltmeter) and was used to standardize, or calibrate the bridge before use.

Figure 14 - 100 and 1K Ohm measurement standard resistors.

Figure 15 - Internal construction of a standard resistor.

Figure 16 – A close-up photo of a standard cell used to standardise (calibrate) a millivolt potentiometric bridge for use with thermocouple temperature measurements. Notice the voltage rating in small print on the left side label of 1.0193 Volts.

Chapter 1 - Homework problems - Gain and Offset

Determine the gain and offset, including units for the following problems.

#1 - An amplifier is to convert an input signal span of 0 to 4 Volts into 1 to 9 Volts. Determine the amplifier gain (Av) and offset (b). Determine and include the units of gain and offset.

#2 - An amplifier is to convert an input signal span of 1 to 5 Volts into 1 to 9 Volts. Determine the amplifier gain (Av) and offset (b), include the units of gain and offset.

#3 - An amplifier is to convert an input signal span of 1 to 5 Volts into 0 to 10 Volts. Determine the amplifier gain (Av) and offset (b), include the units of gain and offset.

#4 - An amplifier is to convert an input signal span of .02 to .08 Volts into 0 to 10 Volts. Determine the amplifier gain (Av) and offset (b). Determine and include the units of gain and offset.

#5 - An amplifier is to convert an input signal span of -.05 to +.10 Volts into 1 to 9 Volts. Determine the amplifier gain (Av) and offset (b), include units.

#6 - An amplifier is to convert an input signal span of -.02 to .09 Volts into 0 to 3 Volts. Determine, including units the amplifier gain (Av) and offset (b).

#7 - A sensor outputs -.02 to +.09 Volts over an input span of 0 to 30 psi. Determine the sensor's gain (Kx) and offset (b). Determine and include the units of gain and offset.

#8 - A sensor outputs .00 to +.02 Volts over an input span of 0 to 20 pounds. Determine the sensor's gain (Kx) and offset (b). Determine and include the units of gain and offset.

#9 - A sensor outputs 10 to 80 millivolts over an input span of -50 to +150 degrees centigrade. Determine the sensor's gain (Kx) and offset (b), including the sensor's units of gain and offset.

#10 - A sensor outputs 2.73 to 3.73 Volts over an input span of 273 to 373 degrees Kelvin. Determine the sensor's gain (Kx) and offset (b). Determine and include the units of gain and offset.

#11 - Given the following Voltage versus Time graph assume the upper waveform is an output signal and the lower is the input signal of a signal-conditioning amplifier. Find the gain and offset values of the amplifier.

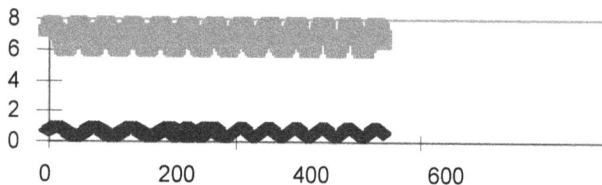

Figure 17 - Graph for problem #11.

#12 - Estimate the amplifier gain and offset
given the characteristic plot and data points following. (Consider using the minimum and maximum values associated with the data columns following the graph .)

Amplifier In/Out Characteristic

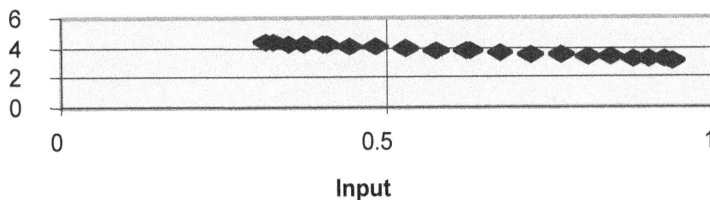

Input

Problem #12 - Sample Data Points

0.3125	4.375
0.3162	4.3677
0.3271	4.3457
0.3461	4.3079
0.3711	4.2578
0.4028	4.1943
0.4401	4.1199
0.4822	4.0356
0.5273	3.9453
0.575	3.8501
0.6238	3.7524
0.6738	3.6523
0.7214	3.5571
0.7666	3.4668
0.8081	3.3838
0.8453	3.3093
0.8777	3.2446
0.9033	3.1934
0.9216	3.1567
0.9332	3.1335
0.9375	3.125
0.9332	3.1335
0.9222	3.1555
0.9033	3.1934
0.8777	3.2446
0.8459	3.3081
0.8087	3.3826
0.7672	3.4656
0.7214	3.5571
0.6738	3.6523

0.625	3.75
0.5762	3.8477
0.5286	3.9429
0.4834	4.0332
0.4413	4.1174
0.4041	4.1919
0.3723	4.2554
0.3467	4.3066
0.3278	4.3445
0.3162	4.3677
0.3125	4.375

Chapter 2 – Resistance Bridges

The following chapter will introduce one of the most commonly used circuit configurations for sensor-based experimental measurements. The bridge circuit and its numerous variants are investigated from historical, operational and application perspectives. Chapter topics are included following.

Objectives:
Upon completion of this chapter, you should be able to:
- Explain the purpose and operation of common resistive bridge circuits
- Provide typical application examples of resistive bridges
- Determine the input and output resistances of bridge circuits
- Explain the operation of half and full bridges
- Address gain and offset concerns with bridges
- Incorporate single and dual active resistive sensors into full bridges
- Utilize bridge circuits to convert sensor resistive changes into voltage changes

Introduction

Sensors are available in a variety of forms and complexities. From simple potentiometric types to chemical analyzers, most employ operational concepts that require an understanding of basic analog circuit characteristics. The current proliferation of digital devices such as computers and analog I/O has not significantly altered the operation of most sensors. What is most likely, any "new" digital devices are an additional functional stage added upon the basic sensor that has been used reliably for decades.

Probably the most universally applied sensor characteristic is centered on resistance. Resistive sensors are employed to measure force, pressure, stress, strain, temperature, rotation, displacement and a number of other physical variables. Resistive sensors require *signal conditioning* to convert an induced change in resistance, to a change in voltage for amplification and further conditioning, such as analog to digital (A/D) conversion. When used in sensors, bridge circuits perform the resistance to voltage conversion function. Before proceeding, it is suggested that one have a handle upon DC resistive circuit characteristics as covered in an introductory electronics course, with an emphasis on series resistance, voltage divider circuits.

The Voltage Divider

Any series assembly of electronic components forms a voltage divider. To qualify as a voltage divider, a circuit needs to proportionally reduce the applied voltage. This can only be accomplished by a series circuit since parallel circuits are associated with a constant voltage and a divided total current. Mathematically, the voltage divider equation is derived from Ohm's law, and demonstrates that the larger the resistances, the larger the voltage drops. As an example, consider the following circuit.

12 Volts

R1 = 500

R2 = 300

R3 = 200

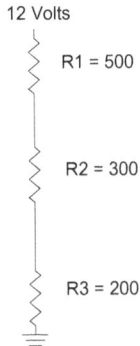

Figure 1- A simple series circuit or "Voltage Divider," distributes the voltage across the resistances in proportion to the individual resistance values.

The individual voltage drops can be determined by totaling the series resistance (Rt), finding the total current (It), and using Ohm's Law to solve for the individual voltage drops. Or,

$$Rt = R1 + R2 + R3$$
$$Rt = 500 + 300 + 200$$
$$Rt = 1000$$

$$It = Vs / Rt$$
$$It = 12V / 1000\Omega$$
$$It = .012 \text{ A or } 12 \text{ mA}$$

$$Vr1 = It * R1 = 12 \text{ mA} * .5K$$
$$Vr1 = 6 \text{ Volts}$$

$$Vr2 = It * R2 = 12 \text{ mA}* .3K$$
$$Vr2 = 3.6 \text{ Volts}$$

$$Vr3 = It * R3 = 12 \text{ mA} * .2K$$

$$Vr3 = 2.4 \text{ Volts}$$

Reviewing the results, note that R1 equals the sum of R2 and R3, and that the voltage drops across R2 and R3, equals the voltage across R1. In addition, R1 is equal to half of the total resistance and half of the applied voltage appears across R1. R2 is 30% of the total resistance, and 30% of the applied voltage appears across R2. Same for R3, 20% of the total resistance and 20% of the applied voltage appears across R3. This then is the concept of the voltage divider; each resistor will drop a proportion of the applied voltage. The proportion is determined by the ratio of the individual resistance to the total resistance. This proportion multiplied against the applied voltage yields the individual resistor voltage drop. Or,

$$R_1 / R_T = V_{R1} / Vs$$

This can be proven using Ohm's Law,

$$V_{R1} = I_T * R_1$$
$$I_T = Vs / R_T$$

$\therefore$
$$V_{R1} = (Vs / R_T) * R_1$$

Or,
$$V_{R1} / Vs = R_1 / R_T$$

And,
$$V_{R1} = Vs * R_1 / R_T$$

The equations demonstrate the proportional relationship between individual resistance and total resistance is equal to the ratio of individual voltage drop and total applied voltage. The final equation is most useful for rapid computation of individual voltage drops in series circuits.

The following chapter material on Bridges is developed around series "Voltage divider" circuits. Bridges are composed of variations upon the basic series, voltage divider circuit. The better-connected one is with the previous material, the better the following will be digested.

The Wheatstone Bridge

Contrary to popular opinion, Sir Charles Wheatstone did not create the Wheatstone Resistance Bridge in the early 1800's, but he did find a number of new applications for the circuit. The common Wheatstone Resistance Bridge appears sophisticated with diagonal resistors, potentiometric and compensation components included. However with a little persistence and applied series circuit concepts, the operation and purpose of each component in any bridge can be determined.

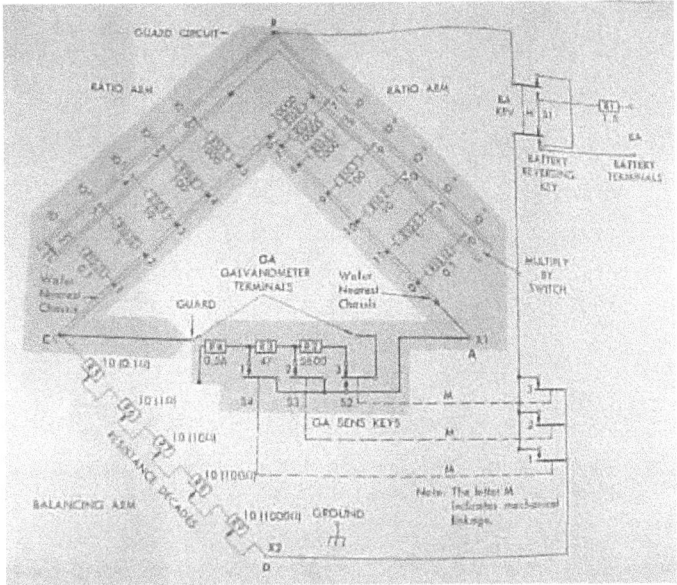

Figure 2 – A really bad copy of the schematic diagram of the Leeds and Northrup 4735 Wheatstone Resistance Bridge.

Bridge circuits of all types are employed in sensors, transmitters, controllers, communications, power supply and signal conditioning circuits, and are also used as secondary and working standards throughout measurement and controls fields. Bridges and the underlying concepts of comparison are found in mechanical form as well in ratio totalizers and force bridges.

A *Standard* component is used for purposes of comparison. Manufacturing and assembly operations everywhere keep standard components to determine if measurement instruments are operating within accepted accuracy tolerances, or to determine if the parts being manufactured are within acceptable tolerances. Within the Wheatstone Resistance Bridge, a series of standard resistances are used for purposes of measurement by comparison, usually to check the accuracy or resistance value of another instrument or component.

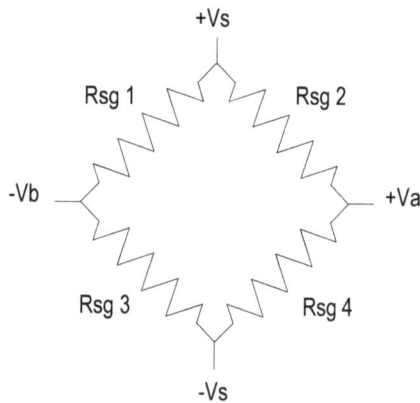

Figure 3 - Schematic representation of a 4-element strain gauge resistance Bridge. The bridge circuit is represnted by the four diagonal resistors in a rhombus configuration.

In operation, the unknown Rx, and a DC supply voltage are connected to the bridge and the sensitive galvanometer (a micro-amp ammeter) movement is observed. The standard being variable is changed until a zero, or Null indication is observed on the meter. Under this condition the bridge is understood to be at Null-balance, a condition of zero voltage differential across the bridge. Once a Null condition is achieved, the value of the unknown resistance can be read from the scale indicator associated with the standard resistance. The resistance scale indicator, and the variable standard resistances within the Wheatstone bridge are actually individual resistors connected to rotary switches where the value of each resistance is labeled on the switch.

The following photo reveals the inner workings of a Wheatstone bridge.

Figure 4 – The resistors used within a Wheatstone bridge are wire-wound to precise values. Each resistance is switched into the circuit through the use of a rotary switch. This photo is the backside of the Leeds and Northrup 4735 Wheatstone Bridge.

Figure 5 – The Leeds & Northrup 4735 Wheatstone Bridge for measuring resistance. The standard values are variable in increments and are shown as the five large rotary switches across the front mid-section. The large rotary switch at the top center is the multiplier. The resistance value under test is determined by observing the entered value across the center and multiplying by the value at the top center after a null is achieved. The terminals at the lower left connect the unknown Rx for resistance measurement, and allow the bridge to be used as a precision resistance when required as

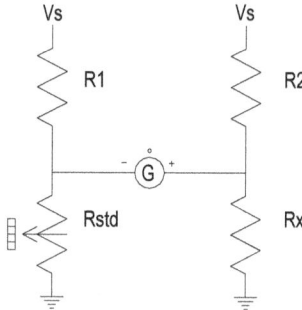

Figure 6 - The basic Wheatstone bridge (shown with the resistors drawn vertically instead of diagonally) compared an unknown resistance with an accurately known standard. The standard resistance was varied until the galvanometer at center indicated a null-balance condition. The value of the unknown was then read from the scale.

In terms of operation, if R1 and R2 are equal, the voltage across the galvanometer will approach zero volts as the standard resistance approaches the value of the unknown. R1 forms a series, voltage divider circuit with the standard resistance. As the standard is varied, the voltage across the standard resistor to the left of the galvanometer will vary in proportion. Once connected, the unknown Rx and R2 also form a series voltage divider creating a voltage on the right side of the galvanometer. Any variation of the standard resistance causes a variation of the voltage across the galvanometer and a variation in the current through the galvanometer. As the unknown and standard values become equal, the voltage at each side of the meter become equal and the differential voltage across the meter becomes zero, resulting in zero current and a Null meter indication. The relationship between the circuit components can be best observed at Null-balance. Given the voltage at each side of the bridge is equal, or Va = Vb;

Since,

$$Va = Vb,$$

$$V_{Rstd} = V_{Rx}$$

Substituting voltage divider equations for Rstd and Rx,

$$Rstd/(R1+Rstd)= Rx/(R2+Rx)$$

Rearranging,

$$RstdR2+RstdRx = RxR1+RxRstd$$

Canceling common terms of RstdRx yields,

$$RstdR2 = RxR1$$

Or, at Null-balance,

$$Rstd/Rx = R1/R2$$

In addition,

$$Rstd/R1 = Rx/R2$$

At Null-balance, the ratio of the resistors across and vertically is equal.

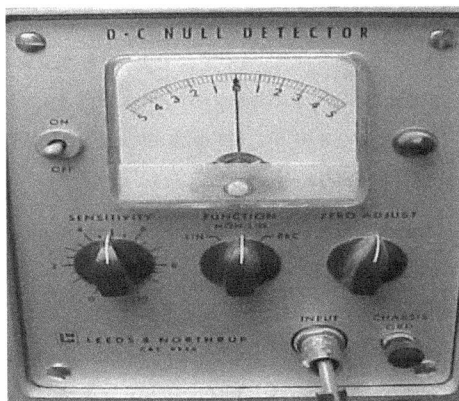

Figure 7 – A DC Null Detector is comparable to a galvanometer in function, but is actually a high impedance, variable gain DC amplifier. This unit connects across the Wheatstone bridge to indicate a condition of bridge Null.

Obviously, under the condition of R1 = R2, the range of the Wheatstone bridge becomes rather narrow and limited to the range of variation associated with the standard. However, the derived Null-balance ratios imply R1 or R2 or both can be made variable in incremental steps, to extend the range of the instrument to measure unknown resistance. Typically the ratio of R1 and R2 are varied in incremental steps from .001 to 1000, resulting in a measurement range of unknown resistor values from milli-ohms to Mega-ohms, with a single limited range variable standard resistance. Mathematically this can be verified as follows

At Null-balance,

$$Rstd/Rx = R1/R2$$

And,

$$Rx = Rstd (R2/R1)$$

Or,

$$Rstd = Rx (R1/R2)$$

The extended range of measurement allows Wheatstone bridges to find use in a variety of applications. From accurate measurement of an unknown or questionable resistance, to calibration of measurement apparatus and process instruments requiring specific values of resistance (the Wheatstone can also be used as a precision resistor substitution), to locating underground short circuits in power cables. Whenever an accurate and precise resistance measurement is required, a Wheatstone resistance bridge should be considered.

RTD's and Stain Gauges

Before beginning a discussion of bridge circuit applications, a word about two common resistive sensors is in order. The following chapter material will reference strain gauges and RTD's frequently. RTD is an abbreviation for Resistance Temperature Detector or Resistive Thermal Device. An RTD is composed of a pure metal and housed in variety of styles including probes, washers, magnetic mounts and patches for gluing to a device under test. Platinum, copper and nickel types are available with Platinum being the most widely used. The RTD is understood to be linear although not perfectly, and exhibits a positive resistance temperature coefficient, meaning the resistance increases as the temperature increases.

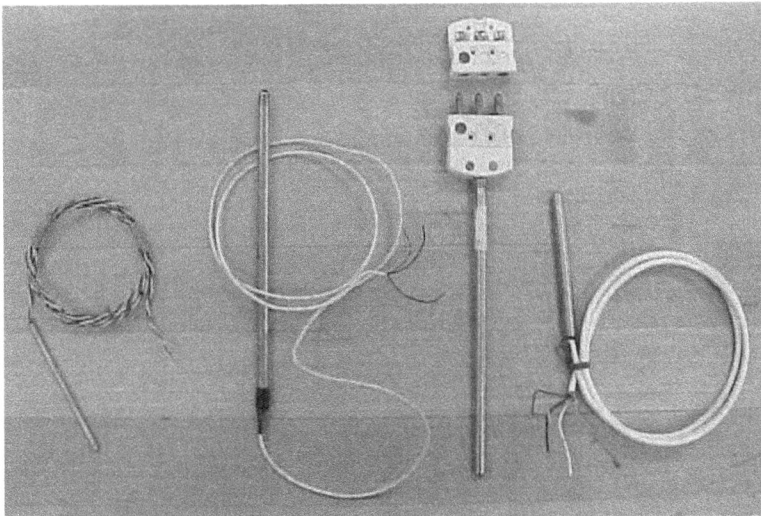

Figure 8 – 2, 3 and 4-wire RTD temperature probes. All RTD's are temperature sensitive resistances and require only 2 wires. The 3 wire RTD (2 units in the center of the photo) allows for compensation of leadwire resistance. The 4 wire RTD (right) uses a current generator to pass a small and constant current through the temperature sensitive RTD element on 2-wires, and provide a temperature proportional output voltage on the other 2 wires. The 2 wire RTD (left) is used in a lab environment where leadwire length is minimal and the temperature remains relatively constant.

Typical Platinum RTD's will exhibit 100 Ohms of resistance at zero degrees Centigrade, and will vary by .385 Ohms per each degree Centigrade. Platinum RTD's in a 100 degree Centigrade application will appear as a 138.5 Ohm resistor. A bridge circuit is required to convert the changing resistance of the RTD into a proportionally changing voltage. RTD's will be used as examples of resistive sensors in the following bridge circuit material.

The Strain Gauge is another common form of resistance sensor, especially in experimental and product development applications. Probably the most widely employed sensor, strain gauges are found within most pressure sensors and are often individually glued to mechanical and material samples to measure the stress (applied pressure) and/or strain (the deformation) that result from an applied force.

A strain gauge is a convoluted length of metal-alloy wire configured to be sensitive to strains along a specific axis of pressure and motion. The underlying principle of operation results from the equation for resistance of a length of wire.

$$Resistance = (\Re ho * Length) / Area$$

Where Rho ($\Re$) represents the resistivity coefficient of the wires' material, length represents the wires' length, and area is the wires' cross-sectional area.

Strain gauges are applied to a sample under test using epoxy glue or other adhesive. As an applied stress causes a material sample to yield, a strain gauge properly glued to the sample will also deform. Should the resulting strain occur along the gauges' sensitive axis, a proportional change in resistance will be exhibited. Using the strain gauge example of figure 8, tension force(s) applied in the direction indicated by the arrows will cause the gauges' length to increase, the area to decrease, and the gauge's resistance to increase. According to the previous equation, as the length increases, the resistance increases. Applying force(s) in a manner opposing the direction of the arrows will result in a resistance decrease. Common strain gauges exhibit 120, 350 or 1000 Ohms of resistance under zero applied strain. Typical values of resistance change are on the order of fractional Ohms, with milli-Ohms being common and should never exceed about .1 Ohms of resistance change. When configured into bridges to convert the resistance change into a corresponding voltage change, strain gauge bridges can be expected to output signals within the microvolt (10^{-6}) to millivolt (10^{-3}) range.

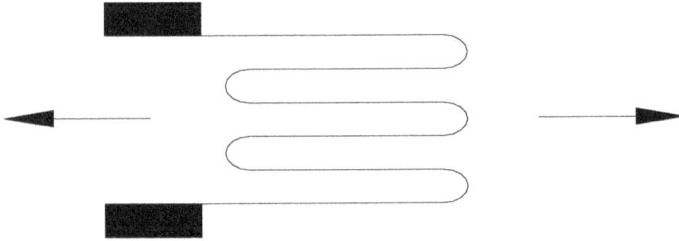

Figure 9– A strain gauge. The arrows indicate the sensitive axis of motion. Tension forces applied in the manner indicated by the arrows will result in a proportional increase of resistance by the strain gauge.

Recent developments in semiconductor technologies have seen a proliferation of semiconductor strain gauges. Semiconductor strain gauges exhibit resistance variations of 100 to 200 times greater than conventional metal alloy strain gauges but are also understood to be highly non-linear. Semiconductor strain gauges are finding use with pressure sensors where the amount of strain is minimal and the resulting change in resistance is considerable because of their high sensitivity.

It should also be mentioned that a basic principle of instrument and measurement science states that the application of any force, regardless of how large or small, will result in a deformation that can be used to determine the magnitude of the applied force. For those who are fortunate to work with strain gauges, time and again the strain gauges' sensitivity will reinforce the previous statement. And although the output signal from a strain gauge bridge will usually require additional conditioning, specifically in the form of amplification (and usually a lot of amplification), the strain gauges' intrinsic simplicity of operation and application makes it one of the most, if not the most universally applied sensor. Strain gauges are referenced frequently in the following text.

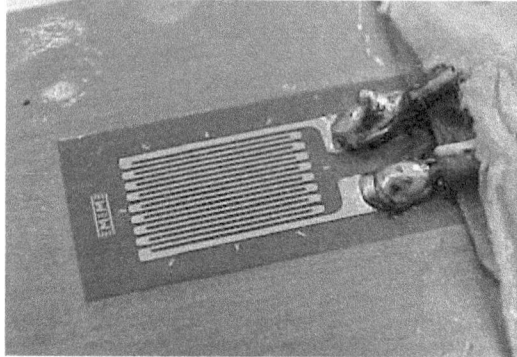

Full-Bridge Sensors

In process automation and especially experimental testing applications, resistance bridge circuits can be found embedded as the active element within a sensor, or in the "front-end" of a signal conditioner converting resistance changes, induced by temperature or other physical variation, into voltage for further conditioning. In these applications, the galvanometer of the Wheatstone resistance Measurement Bridge is eliminated, and the voltage difference across the bridge is considered the output signal.

In use, the bridge is initially assumed to be "null-balanced" when zero physical variation is applied, and gradually unbalances as the physical variable increases. In reality, the bridge is rarely completely nulled at minimum input and may even exhibit a considerable positive or negative output offset voltage initially. In this case, additional circuitry may be employed to achieve a null condition, or as close as reasonably possible since true zero never does really exist in micro-measurement applications (zero becomes a matter of decimal places). Should the initial condition be a negative offset voltage, as the input variable to the sensing element(s) within the bridge increases, the output rises to zero volts and continues into a positive output value. As the bridge becomes unbalanced with the increasing input, a *differential output voltage* is generated.

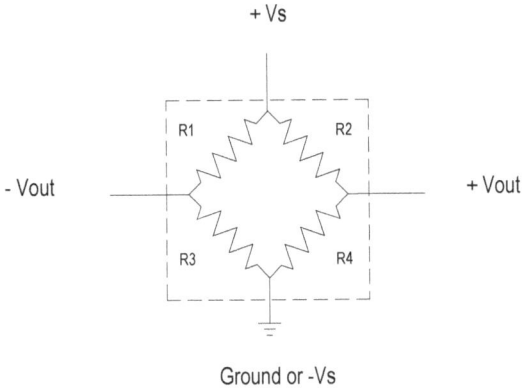

+ Vs

R1   R2

- Vout   + Vout

R3   R4

Ground or -Vs

Figure 11 - A typical resistance bridge as used in sensor applications. Note the two outputs and corresponding polarities. In this example all four resistor elements are included within a sensor housing indicated by the dashed line around the bridge. The diagram is representative of low-level pressure sensors offered by a number of manufacturers.

The differential output is the result of having the voltage at each output (with respect to ground) moving in different directions. This is caused by the manner with which the individual sensing resistors are connected, and the physical variable(s) applied internally. Upon inspection of any four-arm bridge-type sensor, one will notice that each output has an associated polarity. The polarity indicates the direction that particular output will move when the input is applied and increased. Note this does not mean the polarity of the voltage at that output location, but the direction the output will change when the input variable is applied. Negative polarity indicates the output will be negative-going with an increase of the input variable, and positive implies the output will be positive-going with an increase in the applied physical variable.

The output polarities also imply something about the operational conditions of the individual resistors that compose the bridge. Typically, the resistor across each bridge output will change its resistance in a manner indicated by the output polarity. If the polarity is negative, the resistor at the negative output (to ground or –Vs) will be decreasing in resistance value. At the positive output the resistor between the output and ground (or –Vs) will be increasing in resistance. The polarities associated with the resistor labels in the following diagram represent the direction the individual resistance changes when an input variable is applied. As is the case with most housed, integral bridge sensors, all four internal bridge components are made sensitive to the input variable. In which case resistive sensors in opposite legs will vary their resistance in a corresponding manner, and resistors in adjacent legs will vary in an opposing manner. The following diagram demonstrates the phase relationship among the four active, internal bridge components.

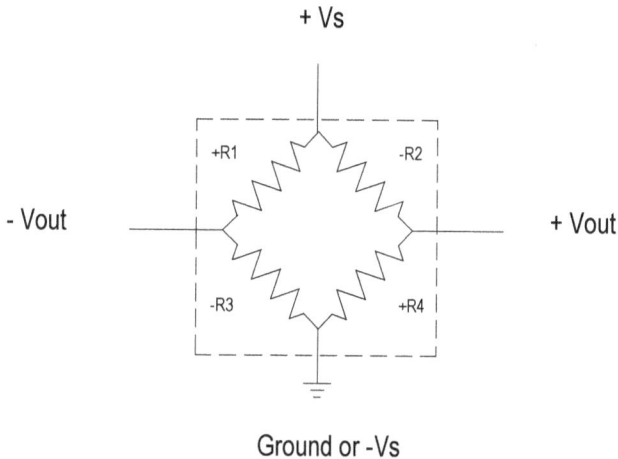

Figure 12 – A common bridge circuit representing the operation of many enclosed sensors. The polarities represent the direction of the resistance change with an increase in the applied physical variable.

With all circuits employing differential outputs, where neither output is connected directly to ground, interpreting the output signal requires measuring (or amplifying) the *difference between the two voltages at the positive and negative output leads*, independent of a reference to ground. Usually a ground connection is available but is not required to determine the output signal. Only the difference between the two outputs is required to determine Vout. Likewise, **when asked to measure the output from a bridge, the voltage measurement should be made across the (positive and negative) outputs, and not from either output to ground**. The bridge output voltage is typically in the range of microvolts to millivolts, and almost always requires further amplification for the output voltage to be of any significant use.

Figure 13 – An older generation of analog bridge amplifier. This module is the "Front-end," or input signal amplifier to a data acquisition system used to log sensor output data. Although still commonly used, the current generation of data acquisition signal conditioning amplifiers is software driven. The Balance potentiometer in the photo is used to assist in "Null-balancing" resistive bridge sensor circuits. The "Cal" potentiometer at the bottom of the board determines the sensor circuits' output to the data logger at a pre-determined value of applied physical variable. Setting these two adjustments is understood to "Calibrate" the data acquisition system.

Most experimental and product testing measurements employ pressure, temperature, stress/strain, vibration, acceleration, displacement, velocity and/or other transducers whose internal electronic components are one, two or four physically sensitive resistors connected in one of a variety of bridge circuit configurations. The resistive sensing elements of the bridge generally assume the form of individual strain gauges or a combination of individual resistive sensor(s), and precision fixed value, low tolerance, and low temperature coefficient resistors. Inside most pressure sensors, strain gauges are commonly mounted to semi-rigid diaphragms and encapsulated into a nylon or plastic housing, (such as the Sensym SX series of pressure sensors in the Appendix).

Figure 14- Low-level (millivolt output) pressure sensors such as the ones shown here utilize semiconductor strain gauges and flexible diaphragms, generating an output in millivolts per psi of pressure. Semiconductor strain gauges develop large resistance changes under low applied forces but can be highly non-linear. Note all contain four connection leadwires, an indication of a full-bridge internally.

Strain gauges are also commonly applied to ductile members of mechanical structures and electronically connected into a bridge circuit configuration (similar to the Entran and Omega load cells found in the Appendix). And, as is demonstrated in the next section, it is also common to find a single strain gauge or RTD within a four-arm bridge as well.

Single Active Element Bridges and Half-Bridges

Resistive bridges may contain multiple (2, 4 or more) resistive sensing elements depending upon the desired sensitivity, expense, application demands and the nature of the sensing elements. Bridges containing a single sensor are referred to as single active (sensor) element bridges or quarter active bridges, or simply quarter bridges. Two and four active sensor bridges are called half and full (or four active element) bridges respectively. The enclosed sensors discussed in the previous section were assumed to be full bridge circuits since all four of the resistive components were assumed to be sensing and changing with an applied input variable.

Figure 15 -A 3000 psig Viatran pressure sensor utilises two strain gauges (barely visible in the upper right assembly) mounted on a metal diaphragm. The bridge completion resistors are visible on the adjoining circuit board.

Quarter bridges are rarely employed without additional bridge completion components since having only a single component in the circuit would result in an output voltage equal to the applied source voltage. When quarter bridges are used without additional bridge completion resistors, a *constant current generator* IC (integrated circuit) is placed in series with the single sensing element. The current generator develops a constant value (1, 5 or 10 milliamps is most common) of current, independent of imposed disturbance variables such as ambient temperature. The output voltage signal is taken directly across the sensor, as indicated in the following diagram.

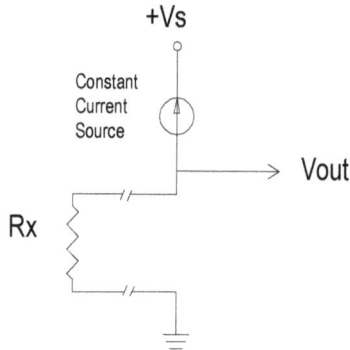

+Vs

Constant
Current
Source

Rx

Vout

Figure 16 – A quarter bridge employs a single sensor Rx and a current generator. The output signal is taken across the sensor. Circuits of this type are common with RTD temperature sensors and single strain gauges. Temperature sensors for this purpose are available containing 4 leads, two for the current supply and two for the output to minimize the effects of leadwire runs.

In a single active element half bridge (figure 15), the sensing element (shown as resistor Rx in the previous figure), is indicated as being removed from the bridge to place emphasis on the remote location of the sensor from the circuit and to draw attention to the lead wire resistance. Bridges of this type often employ a single resistive sensor, such as a strain gauge or temperature sensor at the point of measurement and a single fixed resistance at a remote location (where the data acquisition system resides) to complete the bridge. It might be appropriate to again mention the original idea behind the bridge in sensor applications was to convert a resistance change into a voltage change. Therefore, bridge completion as referred to previous implies the insertion of an additional component(s) for the purpose of converting the resistance change into a voltage change.

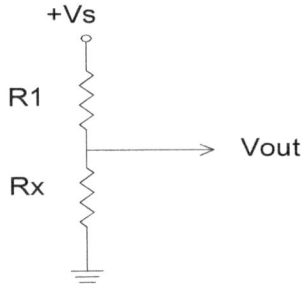

Figure 17–Schematic representation of a single active element half-bridge, or voltage divider circuit. The lower resistor labelled Rx is the single sensing element. The upper resistor is a precision, low temperature coefficient fixed value resistor. The circuit converts a change in resistance to a proportional change in voltage but exhibits an output offset of half the supply at the minimum input condition.

Half-bridge circuits are common in experimental applications where a test fixture is assembled utilizing a single sensor to measure a specific characteristic associated with the device under test. The fixed resistor (R1 in the following figure) utilized in the half-bridge is selected to be a value equal to the sensors' resistance value at zero applied input.

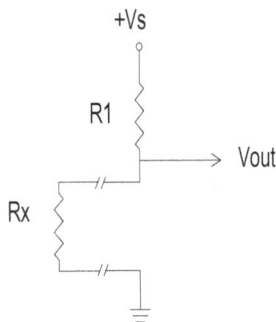

Figure 18 - A single active element employed in a half bridge. The sensor, Rx, is indicated as being remote to emphasise the effects of the leadwire resistance.

As an example, a single 120-Ohm strain gauge will employ a 120 Ohm fixed resistor as the bridge completion component, and a single 100-Ohm RTD would require a 100 Ohm fixed resistor <u>if the initial temperature were 0 degrees</u> <u>Centigrade</u>. If the initial temperature of the RTD were higher than 0° C, a larger value of fixed resistor equal to the initial RTD resistance would be required. Often a variable resistor is used in the location of R1 to set the initial bridge output condition in these cases.

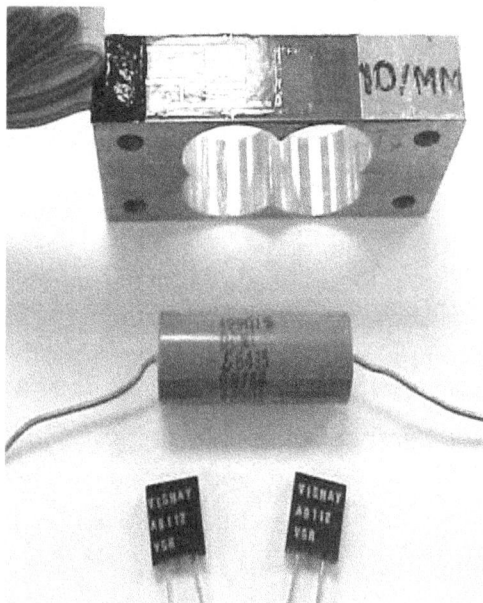

Figure 19 – An inexpensive two active element load cell (top) with two forms of bridge completion resistors (center and bottom). The resistor in the center is 499 Ohms at .01% tolerance, the two resistors at the bottom are low temperature coefficient, 01% tolerance, 120 Ohm bridge completion resistors intended for use with strain gauges.

From the previous diagram (figure 17), using half-bridges will result in a minimum output value of one-half the supply. Since the value of both components is equal initially, the initial output will equal one-half the bridge supply voltage. If the applied supply voltage is 5 volts, the initial minimum output from the bridge is 2.5 volts. The large amount of initial voltage, or offset, (2.5V) can impose a severe limitation on the amplifier gain following the bridge. For this reason the initial bridge output is usually subtracted, or nulled, before amplification.

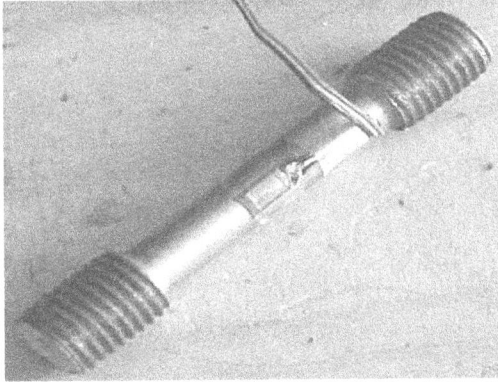

Figure 20 – A standard size and shape of metal sample prepared with a single strain gauge for a tensile test.

To understand the reason behind the gain limitation associated with half-bridges, consider a single active strain gauge in a half-bridge with a 5 Volt supply. With the strain gauge and fixed resistor equal in value initially, say 120 Ohms, the output would vary from 2.5 Volts as the physical variable begins to change. The 2.5 Volt "pedestal" that the output is riding upon (starting from) must be nulled-out or the gain of subsequent amplifier stages must be limited to avoid saturation, or generating an output that is approximately equal to the amplifiers' maximum output, usually understood to be the amplifiers' supply voltage. More specifically, if the bridge circuit outputs 2.5 Volts at the minimum input condition and 2.6 Volts at the maximum input condition, the signal span is .1 Volts with a DC component of 2.5 Volts. If the amplifiers' supply/saturation voltage equals 12 V, the DC amplifier gain would be limited to 4.615 before the amplifier output reached 12 Volts. With the 2.5 Volt DC component "nulled-out", the half-bridge output signal span becomes 0.0 Volts to 0.1 Volts, and the gain of the signal conditioning amplifier can be increased to about 120, resulting in substantially improved *resolution* of the physical variable. The following example demonstrates the limitation imposed on amplifier gain by using the half-bridge.

Max DC Gain before Nulling the 2.5 Volts, assuming the maximum amplifier output is equal to the amplifiers' supply of 12 Volts:

$$Av = \Delta Vout / \Delta Vin$$
$$Av = 12\ V / 2.6\ V$$
$$Av = 4.615$$

Max DC Gain after Nulling the 2.5 Volts:
$$Av = \Delta Vout / \Delta Vin$$
$$Av = 12\ V / .1\ V$$
$$Av = 120$$

The higher gain is desirable since it provides better observation of small changes in the measured variable, improving the measurement circuits' *resolution*. To remove, or "Null-out" the 2.5-Volt DC voltage the .1Volt signal is riding upon, an additional two-resistor voltage divider, or *Offset-Null* circuit is assembled to output a constant 2.5 volts. When the signal and the fixed 2.5 volts are applied to a *differential* amplifier, the amplifier "sees" and amplifies only the difference between the two inputs. One amplifier input varies from 2.5 to 2.6 Volts while the other remains at 2.5 constantly. Differential amplifiers (also called "subtracting" amps) amplify only the difference between the amplifiers' two inputs, which will appear as 2.5V and 2.5V at the minimum input variable, and 2.6V and 2.5V at the maximum input value. The result, as seen by the amplifier is 0.0 to 0.1 Volts, as the physical input variable to the bridge changes from minimum to maximum.

Interestingly, when the two circuits (half-bridge with sensor and Offset-Null voltage divider with constant 2.5Volts) are placed together and side-by-side, a single active element full-bridge is formed. The active (left) side of the bridge is developing a voltage in proportion to the physical variable applied to the sensor. While the inactive (right) side of the bridge is responsible for Nulling, or removing the DC component that the signal portion of the output, (the voltage that changes with the input variable), is "riding" upon.

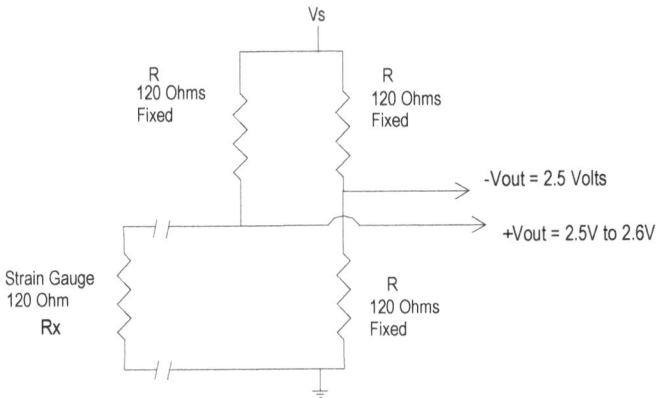

Figure 21 – A single active element, full-bridge. The series circuit to the right of Vs (-Vout) is used to generate a voltage equal to the output offset of the sensor voltage. When the outputs are subtracted (differentially amplified), only the amplified, sensor induced signal will remain.

In summary, half-bridges are commonly used when a single resistive sensor is necessary to measure a physical variable. To eliminate the initial DC offset from the half-bridge output, an additional series voltage divider circuit composed of two fixed resistors is used in conjunction with the half-bridges' completion and sensing resistors. The resulting circuit takes the form of a single active element, full bridge. Again, since the outputs are differential (require subtracting), the voltage at –Vout will be subtracted from the voltage across the sensing strain gauge (+Vout), resulting in an output with an initial offset of close to zero volts rather than one-half of the bridge supply voltage, Vs.

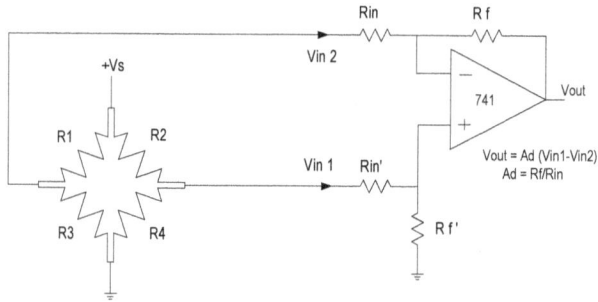

Figure 22 – Bridge and simple Op Amp Differential Amplifier combination. A single, two or four active Element Bridge could be used as the sensor.

Dual Active element Bridges

Occasionally two active elements are utilized in bridge circuits. Two sensing elements provide the benefit of greater sensitivity to the physical variable being measured, less susceptibility to adverse effects from the device under test and the ambient, and require less gain from the subsequent amplifier. Two active element bridges are quite common in strain gauge applications, where many forms of strain gauges are available off-the-shelf with two gauges internally. Torque measurement (twisting action) strain gauges for example require two sensors, and are usually assembled on the same gauge bond to measure torque in opposite (positive and negative) directions.

Figure 23 – A torque strain gauge is actually two resistive gauges on a single bond.

The specific circuit configuration of a two active element bridge is contingent upon the sensors and the application. If one considers the operation of the bridge and recalls that the output signal will be the difference between the output leads, the location of the sensors within the bridge becomes more apparent. The bridge configuration is also determined by the nature of the measurement and the information desired. Two temperature sensors measuring the same physical variable would need to be connected in a manner to avoid canceling each signal. Whereas two strain gauges mounted on each side of a cantilevered (bending) beam will measure the same applied force in opposing directions. Referring to the basic four active element bridge in the following figure, a differential output voltage will be developed when the bridge resistors vary in a manner to cause the outputs to move in opposing directions. Using the indicated output polarities and assuming a two active element bridge, this will occur when R1 and R4 increase, or when R2 and R3 decrease. Assuming a fully active, four-element bridge, the outputs will move in the indicated polarity directions when R1 and R4 increase and when R2 and R3 decrease.

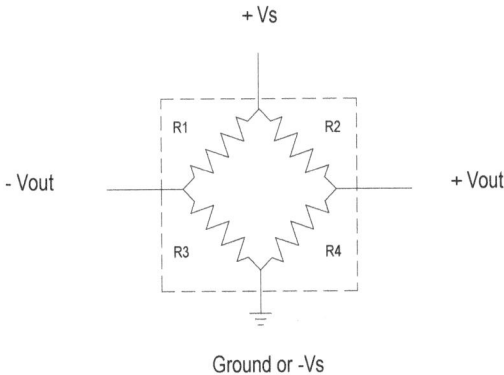

Figure 24 - A full-bridge with differential outputs indicated.

Given the previous temperature measurement example using two RTD's, to avoid canceling the signal from each, the RTD's should be located in opposite legs and not adjacent legs of the bridge. Assuming both will be subjected to the same temperature environment, if both were connected as R3 and R4 in the previous figure, the resulting output voltages would both increase with an increase in temperature. Since the output signal will be the difference between the two output leads, if both voltages are increasing or decreasing together, the net output difference would be zero, and the output signals would cancel. Connecting one RTD in the location of R4 and the other in the location of R1 would cause the –Vout to move negative and the +Vout to move positive with an increase in temperature, and subsequent increases in RTD resistance. The end result is twice the output voltage signal over using a single RTD.

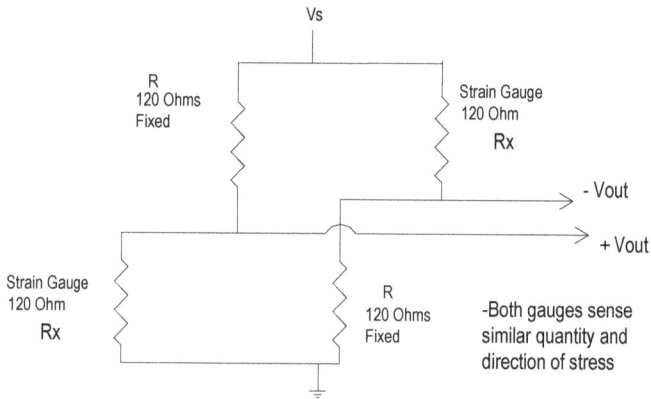

Figure 25- A two active element strain gauge (could also be 2 RTD's) bridge with the gauges placed in opposite legs. Both gauges will experience the same applied force (stress) and direction of force in this circuit configuration.

Strain gauges are similar, in that the application must be considered before connecting two (or four) gauges into a bridge. Given a metal column with two strain gauges attached and expected tension or compression loads (force), the gauges would each operate in an in-phase (same direction and quantity of applied force) manner by exhibiting a resistance increase with an applied tension (stretching) force. Compression forces would result in each gauge exhibiting a decrease in resistance. In this application, the gauges would be connected in opposite legs similar to the RTD's, so that each resistance change would cause the bridge outputs to move in the appropriate direction.

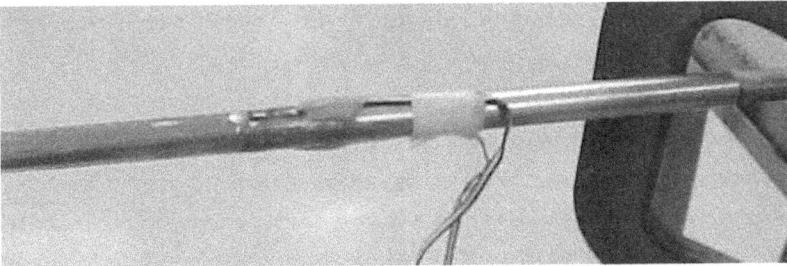

Figure 26 – A bending beam with two strain gauges, one mounted on each side, requires the gauges to be mounted in adjacent legs of a bridge.

However, a bending beam with gauges on each side would require both gauges to be connected in adjacent legs (R3 and R4 in the figure 22 bridge diagram) to establish the correct output polarities. In a bending application, the upper gauge will be in tension (increasing resistance) and the lower in compression (decreasing resistance) when a force is applied. The gauges must be connected in a manner to be consistent with the indicated output polarities.

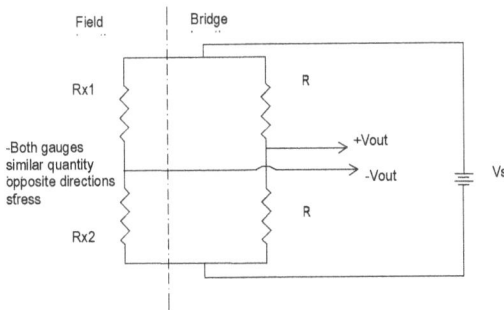

Figure 27 - A two active element (indicated as being in a "field" location) strain gauge bridge with the gauges placed in adjacent legs. To cause the output connected between the strain gauges to move in the direction indicated, Rx2 would be in compression (decreasing resistance) and Rx1 in tension (increasing resistance). The fixed resistors in the right half of the bridge (indicated as being at the location of the data acquisition system) are used for Offset-Null purposes.

Offset Null Considerations

As was demonstrated with single active element full bridges, the static side of the bridge can be seen as performing a biasing (offset nulling) function since this portion of the circuit contains no physically sensitive resistors or sensors. The Offset-Null, or *Biasing* side of the bridge occasionally includes a potentiometer, providing a variable offset voltage from a series divider. The potentiometer wiper provides a variable biasing voltage that is used to determine the bridges' minimum output voltage under the minimum input variable condition. As mentioned previous, the voltage provided by the circuit under a condition of zero applied physical variable is often referred to as the Null, or Null Balance output voltage. A potentiometer performing in this capacity is usually named a "zero" adjustment, and is often labeled "Bridge Zero," "Offset Null," "Balance," or "Bridge Null" on schematic diagrams and front panel adjustments on data acquisition systems. When the pot is adjusted for a differential output of zero volts under a minimum input variable condition, the bridge is understood to be "Null-Balanced."

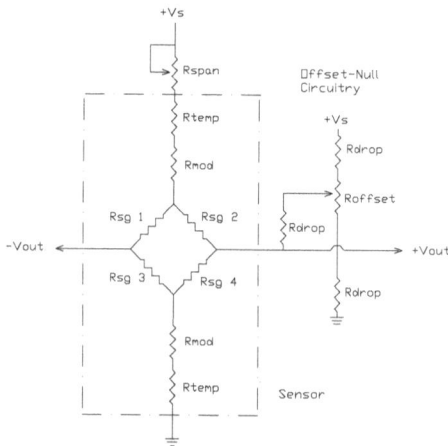

Figure 28- Schematic representation of a 4-element resistance bridge employed within a load cell (dashed line). External components are used to set the gain and Offset-Null (dotted line).

To understand the operation of the offset nulling circuit, the bridge must first be seen as a Thevenin equivalent circuit. Since that can be painful to a second or third semester ET student, picture all of the bridge components connected to the Vout terminal nearest the bottom of the Rdrop resistor, as being a single resistance with a small amount of undesired voltage present. The circuit connected between Vs and ground on the right, and composed of 3 dropping resistors and an offset resistor is designed to provide a small amount of variable millivoltage to the output lead to cancel any residual and undesired offset millivoltage. This will be performed so that the bridge can start its' measurement from an initial condition of zero in the output. The three series resistors from the +Vs location form a voltage divider circuit that reduces the supply voltage to value slightly above and slightly below the voltage that should exist at the output, one-half of the supply. All bridges are designed to output one-half of the supply at each output, similarly the external offset nulling circuit is designed to vary the voltage at the output location slightly above and slightly below one-half of the supply.

As an example, assume the supply is five volts. The voltage at each bridge output (-Vout and +Vout) would be approximately 2.5 volts, or half of the supply voltage. The voltage at each side of the potentiometer is the voltage available at each extreme of the wiper position. The voltage extremes are usually designed to be about half a volt above and half a volt below 2.5 Volts. Using a 5 volt supply, the wiper will exhibit three volts above, and two volts (measured in reference to ground, plus and minus .5 volts measured differentially with reference to the other side of the bridge) in the lower position of the potentiometer. The two vertical resistors on each side of the pot will each drop 2 volts, leaving a 1 volt drop across the pot. This condition is shown in the figure following.

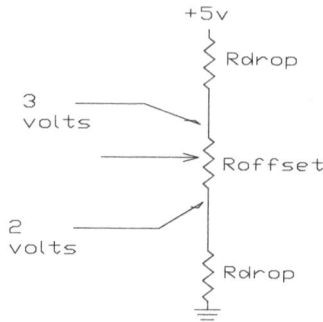

Figure 29 – An offset null circuit applied to a bridge to assist in zeroing the output. Each Rdrop is dropping 2 volts, the potentiometer has a 1 volt drop. The wiper exhibits 2 to 3 volts (with respect to ground) when varied, or -.5 to +.5 of differential output voltage equivalent.

If the circuit were to be connected as shown to the bridge output, a +. 5 volt to -. 5 volt offset bias voltage range (measured differentially, 2v to 3v measured with reference to ground) would be observed at the output lead when the potentiometer is adjusted. This is the result of having 2.5 volts at the bridge output, and varying the potentiometer above and below this value by .5 volts. However this voltage span would be considered far too excessive since only a few millivolts each side of 2.5 volts is required to zero most bridge circuits. To reduce the existing +.5 to -.5 volt span to a few millivolts, the voltage variation available at the wiper of the potentiometer needs to be reduced further. This can be accomplished through the use of an additional dropping resistor and the Thevenin equivalent resistance of the bridge output.

The Thevenization process takes into consideration all resistance values seen looking into the output terminal, while visualizing all possible paths to ground. For this circuit, the total resistance would be the parallel paths of Rsg4 + Rmod +Rtemp to ground, and Rsg2 + Rmod +Rtemp through the 5 volt supply to ground (voltage sources exhibit a low internal resistance and are seen as an additional path to ground).

Or,

$$(R_{sg4} + R_{mod} + R_{temp}) \parallel (R_{sg2} + R_{mod} + R_{temp})$$

Since the values of the previous circuit are unknown, assume the total resistance seen looking into the +Vout terminal is 1K ohms. This becomes the output resistance value required for the dropping network to reduce the +.5 (3 V to ground) to -.5 (2 V to ground) volts, to plus and minus a few millivolts. At this point only a single additional component (Rdrop) is required to further reduce the potentiometer voltage span to a millivoltage. Once the equivalent bridge resistance is known or approximated (again, assume 1K ohm), determining the remaining dropping resistor is merely a matter of knowing how much to reduce the voltage at the wiper of the potentiometer. Let's say we want to vary the voltage plus and minus 5 millivolts instead of the current plus and .5 volts (500 millivolts). The voltage needs to be reduced to $1/100^{th}$ of its current value. Knowing the equivalent bridge resistance is 1K, the remaining dropping resistor needs to be 100 times larger, or 100K ohms.

In summary, since the voltage variation is bipolar it can be applied to either bridge output that is convenient for use. Driving the output responsible for the undesired offset to exactly 2.5000 volts is not as important as driving the output differential to zero volts.

The circuit appears in the following diagram.

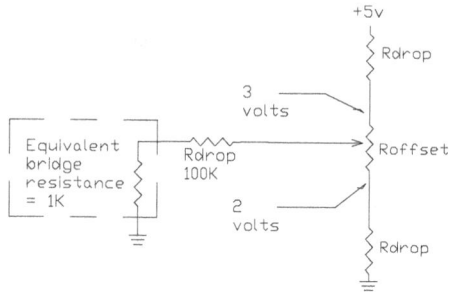

Figure 30 – The complete external bridge Nulling circuit. **The divider on the right reduces the 5 volt supply to a few fractions of a volt, and the 100K Rdrop in conjunction with the equivalent bridge resistance reduces the fractional voltage span from the wiper to a few plus and minus millivolts. Virtually any value of resistance can be used to establish the 2V and 3V voltages, as long as the current draw from the supply is not excessive. A 10K potentiometer and two 20K dropping resistors would suffice.**

Analyzing an external offset Nulling circuit appears overwhelming upon first exposure, but is nothing more than first semester circuit analysis. As we shall see, *designing* a Nulling circuit is even easier.

External Offset Nulling Circuit Design

The following is not intended to create circuit designers but merely introduce the student to simple circuit design concepts. Although circuit design may not be the principle interest of the student, circuit modification and limited design is a real possibility when employed in fields related to experimental testing. For this reason the following section and the remainder of the text will assume a conceptual perspective. Once familiar the student will most likely find the design perspective to be less demanding than the analytical.

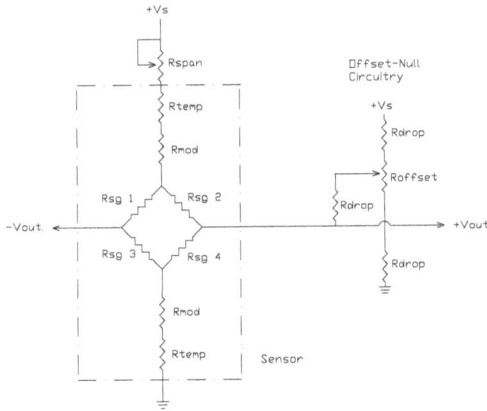

Figure 31 – A common load cell with external span (gain) and offset (Offset-Null) circuitry.

As an application example, consider the previous schematic diagram of figure 29. The diagram is taken from a common load cell. Within the dashed lines additional components in series with supply voltage are included to assist in determining and compensating for the load cells' temperature induced gain variation. The additional components are common in more precise load cells and are discussed in detail in the following section addressing bridge gain considerations.

A load cell is a four active element full-bridge, with four strain gauges applied to a metal (iron or aluminum typically) frame. The frame has been precisely machined to distribute an applied load (force) in a predictable manner throughout the mass of the frame. Located at stress points throughout the frame are four strain gauges to be included in the bridge circuit. As the load is applied, the stress is distributed equally throughout the frame but is applied to each pair of strain gauges in an opposing manner. The result is two gauges in tension and two in compression during an applied load.

Figure 32 – An internal view of a "pancake" compression load cell. The strain gauge leadwires are apparent in the upper right and lower left sections although the strain gauges themselves are barely visible. The load is applied to the ring in the center, causing the center section to flatten and spread outward. The 4 strain gauges mounted within the eight rounded areas experience tension or compression depending upon their location within the machined areas.

Other components are usually included outside of the bridge itself, but inside of the load cell unit to aid in predictable operation over a range of application and manufacturing assembly extremes. Some of the more common components have been included in the previous circuit diagram. In the diagram the components take the form of resistors and are labeled R_{temp} and R_{mod}. The R_{temp} components are resistive temperature sensors used to keep the load cell gain (usually equating to millivolts per pound) constant over a narrow range of temperature. As the temperature increases at the load cell, the bridge gain has a tendency to increase. This would result in an output increase that might be the result of a temperature variation and not a load increase if the bridge were not temperature compensated using the sensors. The R_{mod} components are also resistors installed at the factory to assure that all units of the same model type exhibit the same gain.

In examining the operation of the biasing, or offset null circuitry, one must first realize that not all bridges are at a condition of null-balance initially. In the case of load cells, the more expensive units come very close to exhibiting a null initial measurement condition. Whereas less expensive units almost always require some form of external nulling circuitry. In either case, most resistive bridge sensing circuits require an external and variable offset-bias circuit to provide long-term calibration provisions and to aid in counteracting the effects of age, drift, accumulating residual loads, metal fatigue and so forth.

As mentioned, the idea behind the external nulling circuitry is to slightly influence one of the bridges' outputs with a small quantity of bipolar millivoltage. Usually the amount of influence is considerably more than a bridge might require during the course of a normal lifetime, but not such a large variation that the voltage becomes difficult to set. As a rule of thumb, plan on designing a plus and minus variation equal to the value of the voltage supply in millivolts. As an example, a bridge using a 5 volt supply would be designed to apply a 5 millivolt span (plus and minus 2.5 millivolts) to either of the bridge output leadwires. A more precise value of offset-bias variation can be determined from the sensors' specifications which will give an indication of the expected deviation from null-balance off the shelf, and the expected drift (variation) over time and changing environmental conditions.

To design such a circuit requires two relatively simple items, an awareness of the bridge resistance seen looking into either output and an approximate idea of the desired variation in millivolts at the bridge output. To determine a ballpark figure for the offset adjustment span, review the specifications provided by the sensor manufacturer. To determine the approximate (absolute precision is rarely required when designing for variation) bridge resistance, Thevenize the circuit from the specific output through the bridge to ground.

If the supply voltage quantity is used to determine the variation in millivolts, place a potentiometer between the supply and ground. About any value of potentiometer will be sufficient as long as the current draw on the supply is minimal, 10K to 1 Megohm is common. Leave the wiper hanging open as this is where the final component will be connected, between the bridge output lead and potentiometer. Currently a potentiometer is assembled with a variable voltage ranging from ground to the supply value available at the wiper, measured in reference to ground. However, this same voltage becomes positive one-half of Vs to negative one-half of Vs when measured in reference to the other bridge output, which is how the circuits' output will be used. Once the Thevenized equivalent bridge resistance is known, the remaining dropping resistor to be connected between the potentiometer and the bridge output can be determined. The ratio of the dropping resistance to the bridge resistance will equal the ratio of the current wiper voltage (± one-half of Vs) to the desired variation at the bridge output. Or,

$$(Rdrop \div Bridge\ Requiv) = (\pm\ one\text{-}half\ of\ Vs \div \pm millivoltage\ variation)$$

Or, the dropping resistance will equal;

$$Rdrop = Requiv * (\pm Vs/2 \div \pm millivoltage)$$

As an example, assuming a bridge equivalent of 1K, a 5 volt supply and a desired variation of plus and 10 millivolts, the Rdrop would equal;

$$Rdrop = 1K\ (\pm\ 2.5V/\pm.01V)$$
$$Rdrop = 250K\ ohms$$

Two additional dropping resistors above and below the potentiometer in series with the supply (as in the previous load cell diagrams' offset-nulling circuit) are normally added to assist in reducing the voltage available at the potentiometer. Although not necessary, they may be added to enhance the fine-tuning of the bridge null if desired. The circuit, including a 100K potentiometer would appear as follows.

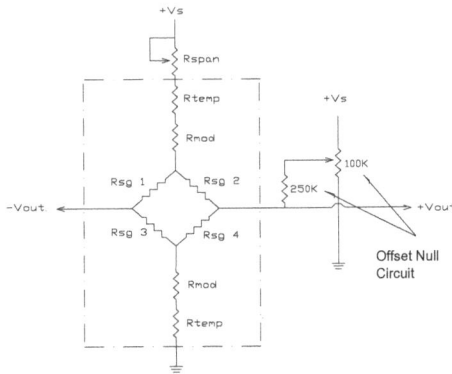

Figure 33 – Load cell bridge circuit with external offset null circuit.

Again, the intent of the preceding is not to create a designer but to introduce the student to the concept of circuit design. Although at this point the previous example probably appears tedious and unappealing, the offset-null circuit could be designed mentally without requiring a pencil to paper computation. As will be seen in subsequent chapters most linear circuit design will follow this example of requiring minimal computation if one is comfortable with the basics of circuit concepts.

Bridge Gain Considerations

Since any bridge is merely an application of series resistance voltage divider circuits, it can be proven that a bridges' differential output voltage is in direct proportion to the amount of supply voltage provided to the bridge. The following demonstrates the relation between bridge supply voltage and output.

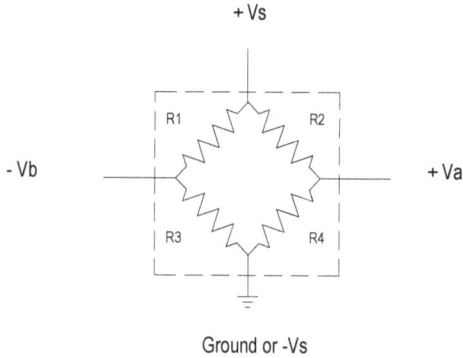

+ Vs

- Vb

+ Va

R1   R2

R3   R4

Ground or -Vs

Figure 34 - Circuit example to be used for Bridge Gain considerations.

$$Vout = Va - Vb$$
$$Vb = Vs[R3/(R1+R3)]$$
$$Va = Vs[R4/(R2+R4)]$$
$$Vout = Vs\{[R4/(R2+R4)] - [R3/(R1+R3)]\}$$

As can be seen from the equation, the bridge output varies in a direct proportion to the applied supply voltage, Vs. In this way varying the supply voltage to the bridge varies the output voltage span and as such, is understood to control or determine the bridge "Gain." Gain is defined as being the distance between output voltage-span end points, and is occasionally referred to and labeled on schematics as *Span*. This concept is most often exploited by inserting a potentiometer, configured as a rheostat between the supply voltage and the bridge. The bridge gain, or output voltage span is determined by setting the potentiometer under pre-determined conditions at the sensor input.

Another characteristic commonly associated with bridge type sensors is *Sensitivity*. Sensitivity is an indication of the bridges' response to a change of the input variable being measured, and to variations in the bridges' supply voltage. Expressed in terms of output change per supply voltage per input variable being measured, the

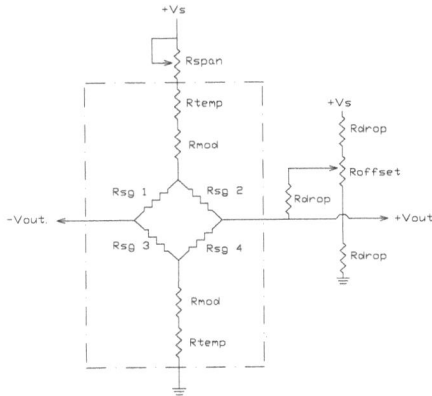

Figure 35- A load cell diagram. The diagram indicates temperature compensation and modulus resistors within the load cell, and bridge gain (Rspan) and null offset (Roffset) calibration adjustment external.

Sensitivity characteristic can be converted to Gain if the amount of supply voltage to be used is known.

$$\text{Sensor Sensitivity} = \Delta Vout / Vs / \Delta \text{ Input variable}$$

Since,

$$Kx \text{ (gain)} = \Delta Vout / \Delta \text{ Input variable}$$

$$Kx \text{ (gain)} = (\text{Sensitivity} * Vs) = \Delta Vout / \Delta \text{ Input variable}$$

Therefore, multiplying the sensors' sensitivity by the required voltage supply results in the sensors' gain, Kx. As an example, assume a low-level (output voltage in millivolts) pressure sensor exhibits a sensitivity of .4 mV/V/psi, and is to be used in conjunction with a 5 volt power supply. From the previous equation, the sensors' gain can be determined.

$$\text{Pressure Sensor Sensitivity} = \Delta Vout / Vs / \Delta \text{ Pin}$$

$$\text{Pressure Sensor Gain, Kx} = \text{Sensitivity} * Vs$$

$$Kx = .4mV/V/psi * 5V$$

$$Kx = 2mV/psi$$

The pressure sensor will output 2 millivolts for every psi of input pressure *change*, when a 5-volt supply is used, (the characteristics of sensitivity and gain are both dynamic, being associated with the slope of the sensors' input/output plot). A 10-Volt supply would double the pressure sensors' output voltage span and gain. Sensors that exhibit this type of proportional relationship are understood to be *Ratiometric*. Ratiometricity is occasionally referenced in sensor specifications, and represents the directly proportional relationship between a sensor output span (or gain) and the applied supply voltage. All bridges are Ratiometric, since bridge output spans are determined by the amount of supply applied. It should also be understood that ratiometricity also applies to the offset-null specification, doubling the supply voltage will also double the amount of initial output offset inherent within the sensor.

The addition of the Gain (often called "Span") and Offset (or "Zero") features allow the bridge to be adjusted to operate within a limited range of values. As such, the bridge and included sensor can be seen as a single functional component assembled for the purpose of converting some physical quantity into a proportional voltage, whose characteristics can be "fine-tuned" through the use of the gain and offset adjustments. Providing the bridge with a gain variation in conjunction with an offset adjustment contributes the necessary components for linear, Y=MX+B characteristic performance. As mentioned in the previous chapter, the gain adjustment provides the bridge with Slope (M) control and the offset provides Y-axis intercept (B) manipulation. Although under most conditions the Y-axis intercept will be set to zero volts for a Null-balance initial bridge output. Unfortunately, only the four active element bridge is the only truly linear bridge circuit. The single and double active element bridges exhibit only a slight non-linearity however and are considered by many to be linear.

Bridge Compensations

To avoid considerable error when using resistive sensors, the leadwires, connections and any other resistance contributing circuitry must be considered and if possible, eliminated as a possible source of error. Ancillary circuit resistance and it's susceptibility to temperature, age and any other resistance varying influences will contribute to the output of a bridge circuit, and provide false information about the device under test. Bridge circuits employed as resistive sensor signal conditioners are capable of exceptional sensitivity and precision. To maintain a high degree of accuracy over a wide application range, the bridge must be compensated against errors that can influence the resulting data taken from the sensor(s).

Bridge circuits are compensated against temperature induced gain variations by connecting a temperature sensitive resistance such as an RTD (or thermistor) in series with the bridge circuit. As the ambient temperature around the bridge changes, the individual sensors within the legs of the bridge become more or less sensitive. With strain gauge bridges, this typically results in an increased overall bridge gain. Strain gauge manufacturers provide characteristic curves indicating the sensitivity to temperature variation with each set of gauges to assist with compensation. The correct temperature sensor located in series with the bridge senses the change in the ambient and allows more or less supply to the bridge, thereby compensating against thermally induced gain variations.

As an example of compensation circuit operation, assume the bridge becomes more sensitive with an increase in ambient temperature. The result would be an output voltage differential that increases with a change in ambient as well as the applied measured physical variable. Sensing a change in the ambient, the temperature compensation component will increase its resistance to allow less supply voltage to the bridge, effectively decreasing bridge sensitivity and compensating for the ambient temperature induced gain variation. Most load cell manufacturers include compensation resistances within the load cell housing and circuit, as represented by the R_{temp} resistors in the schematic diagrams of the previous section.

It should be noted that bridge offset null will also be sensitive to temperature variations but not as drastically. When the bridge is calibrated for a non-zero minimum output value, variations in the span (bridge supply voltage) will effect the minimum and maximum output values. This will require multiple back and forth adjustments between the offset-bias and span before the desired calibration is achieved. Non-independent adjustments of this type are referred to as being interactive. Non-interactive span and zero adjustments are available on most sensor signal conditioning modules and greatly simplify the calibration process.

If the distance from the sensor to the bridge is significant, the leadwire resistance between the bridge and the sensor must also be considered as a source of measurement error. If the sensor leadwire resistance were to remain constant, the resulting effect upon the output voltage could be eliminated with the offset null, or bridge zero adjustment. However every component in a measurement system is sensitive to temperature variation. To clarify, assume the physical input variable were to be held constant and the temperature elsewhere in the measurement system (especially at the sensor's lead wires) were to change. As the temperature of the leadwires change so does the wire's resistance, this changes the resistance in the sensor leg of the bridge and results in an "apparent" change of the measured physical variable at the outputs' display. The output would deviate as a result of the temperature variation applied to the circuit instead of the sensor. An output change created in this manner would be providing false information about the physical variable at the sensor. When measuring physical deformation (strain) with strain gauges, an error of this type is often referred to as an "apparent microstrain" since the indicating display would yield a false strain value. If the undesirable effects of temperature change can be accurately determined, additional signal conditioning circuitry can eliminate and compensate for temperature induced errors.

Careful selection of the bridge completion (in cases of single or two active elements) and data acquisition system components can minimize error and drift (subtle changes induced by time, temperature or other factors) effects. Low temperature coefficient components are commercially available from numerous component manufacturers and will improve the stability and accuracy of the output indication. But quality components can do only so much in minimizing temperature effects. Occasionally circuit construction is altered for compensation purposes. The following figure demonstrates a common method of sensor leadwire compensation.

The connection of the supply to the bridge is made at the sensor rather than at the bridge. This connection requires three leadwires, instead of two, to be connected from the bridge to the resistive sensor but will eliminate any temperature effects upon the leadwire resistance and bridge output value. By making the supply connection at the sensor the individual lead resistances are placed in adjacent legs of the bridge which causes the voltages at points A and B to vary equally, and in the same direction from temperature changes. Since the bridges' output is a function of the difference in potential, Va - Vb, if both Va and Vb increase or decrease the same amount as a result of temperature changes, the effects will be offsetting and will cancel. The end result will be no change in the indicated physical variable at the display from leadwire temperature effects.

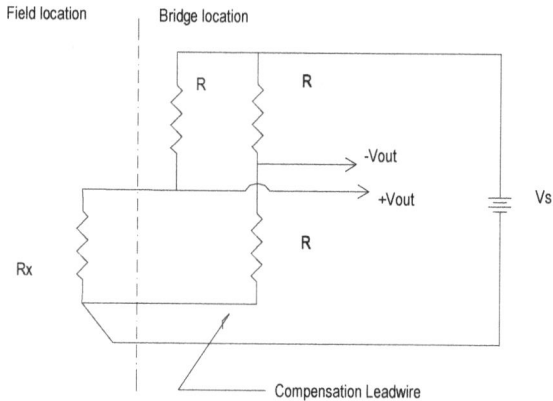

Figure 36 - A single active element, full bridge with sensor leadwire compensation. The additional
leadwire connection at the sensor places one of the leadwires in series with an adjacent bridge circuit
component, thereby cancelling the effects of leadwire induced error.

In this circuit the value of Rx changes with some physical variable causing the
voltage +Vout on the left side of the bridge to change in direct proportion to the
resistance of Rx, and in direct proportion to the physical variable as well. The
voltage at point B is maintained constant in the single element bridge, and the
output voltage is the result of the difference of potential developed across the
bridge. At null balance (minimal or zero physical input applied) all of the bridge
resistances are approximately equal and the differential output voltage is close to
zero.

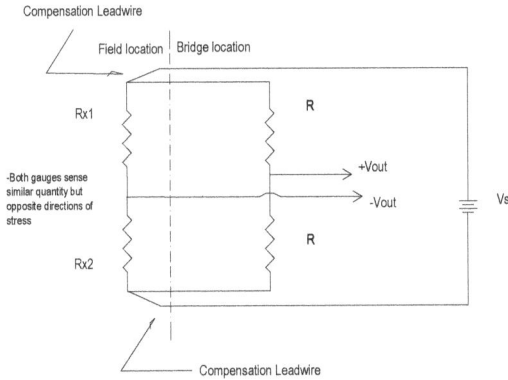

Figure 37 – A two active element bridge with both sensors connected for leadwire compensation. In this circuit, both sensors measure the same physical quantity but respond by changing in opposite directions. The sensors are connected in adjacent legs of the bridge to avoid cancelling their signals.

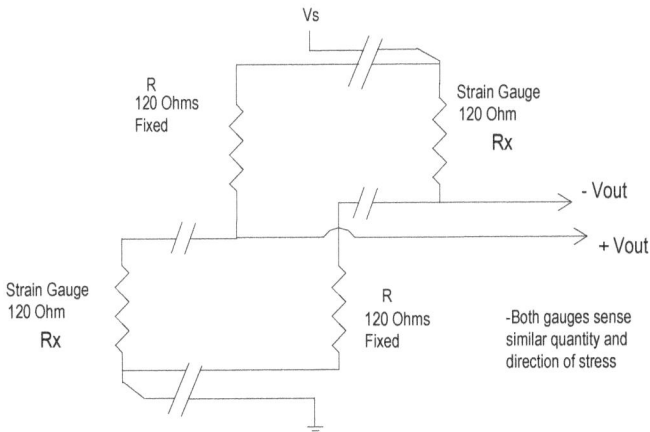

Figure 38 – Another two active element bridge with leadwire compensated sensors. In this example, the sensors are each measuring the same physical quantity and responding in a similar manner. The sensors are placed in opposite legs to have opposite effects upon their respective outputs.

"Dummy" gauges are another method of compensating strain gauges against thermal error. A dummy gauge is an additional strain gauge placed adjacent to but 90° away from a single active strain gauge. The two gauges are located in close proximity and undergo the same temperature of the sample being tested. In the bridge circuit both gauges are wired in adjacent legs. Since both vary in gain in the same thermal manner, any temperature-induced error is offsetting and canceled in the bridge output.

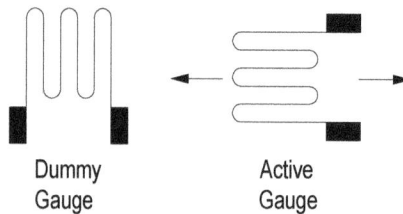

Dummy   Active
Gauge   Gauge

Figure 19 – Active and dummy strain gauge orientation. If applied as two gauges, both gauges would be applied in close proximity on the sample. Both gauges can also be purchased in this orientation as a *Poisson* gauge. Poisson connected gauges are used to correct for transverse strain errors but can also be used for thermal compensation.

In summary, bridge circuits are normally compensated against errors caused by temperature variations. Ambient temperature variations applied to the bridge completion circuitry can cause the output to change even when the measured physical variable is not. Leadwire compensation and bridge gain compensation techniques are commonly applied to counteract the undesired thermal effects. Compensation against bridge Null offset drift is rarely applied but is commonly countered through calibration procedures prior to use.

Sensor Loading and Bridge Circuit Reduction - Thevenization

When connecting a bridge to an amplifier one should consider more than output signal levels and voltage gains. To properly transfer the signal from one stage to another, such as a bridge to an amplifier, the internal resistances of each stage must also be considered.

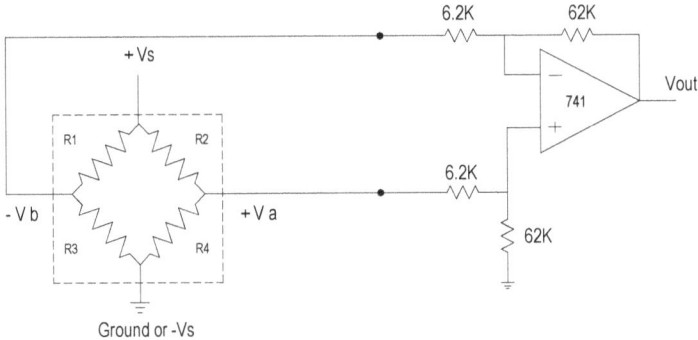

6.2K   62K

+Vs

R1   R2

741   Vout

-Vb   +Va

6.2K

R3   R4

62K

Ground or -Vs

Figure 40 - Cascading stages, such as connecting a sensor to an amplifier, always requires output resistance and input resistance matching considerations for maximum signal transfer from one stage to another.

Each stage of a measurement or other electronic system appears as a resistance or "load." The stage will exhibit resistance to the supply connected across it, and to the input circuit or output circuit connected to it. The all-encompassing name for the resistance a circuit exhibits to a supply or signal stage is Impedance. Although impedance is an AC quantity composed of not just resistance, but the vector sum of resistance and an AC effect called reactance. Since most sensors are DC (AC sensors, like the Linear Variable Differential Transformer, usually require an AC supply voltage), the most accurate term assigned to the resistance exhibited by a circuits' input and output is Resistance. Signal sources (such as sensors) and signal conditioners (such as amplifiers) exhibit internal resistance in two forms, input resistance (Rin) and output resistance (Rout).

To understand why the consideration of input and output resistance is important, the common series circuit or voltage divider needs to be considered. When two stages are connected in *Cascade* or series sequential (one follows the other), the input of the second stage is placed at the output of the first. An equivalent circuit can be assembled to represent the connection of the input and output resistances, including the equivalent source voltage available to the second stage. The equivalent circuit would take the appearance of a series circuit, similar to the following.

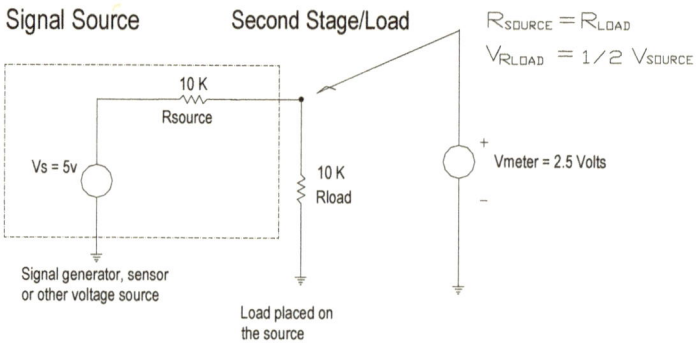

Figure 41 – A signal source connected to a load may be losing voltage if "impedance" (resistance) matching is not considered. With the source and load resistances equal, the output voltage will be one-half of the output voltage value when the load is disconnected. A change in voltage when a load is connected is an indication of "Loading effect" and should always be avoided.

In the diagram, a 5-volt source is attempting to pass an output voltage to an amplifier whose input appears as a 10K resistance. The source also contains a 10K value of output resistance. As can be observed in the diagram, a voltmeter placed across the amplifier input (10K load) will measure one-half of the available 5 volts, or 2.5 volts when the load is connected. However when the load is disconnected, the voltmeter would read 5 volts since the effective resistance placed on the circuit is now infinite. Connecting sensors to amplifiers, or amplifiers to amplifiers to amplifiers is no different than that depicted in the diagram. All cascaded stages can be reduced to a series circuit with an output resistance, input resistance and voltage source.

As is demonstrated in the accompanying figures, it is desirable to keep the input resistance to any stage as large as possible to avoid the reduction in signal voltage when the load is connected. The effect of having the signal voltage change substantially (all output voltages change somewhat since no load is infinite) is referred to as the "loading" effect. Loading is observed in hydraulics, pneumatics, mechanics and any other area of physics when a load demand is requesting more kinetic energy than what the source can realistically deliver. Anybody fortunate enough to be standing in a shower during a Super bowl commercial break has an appreciation of loading effect. The effect in electrical and electronic circuits is no different.

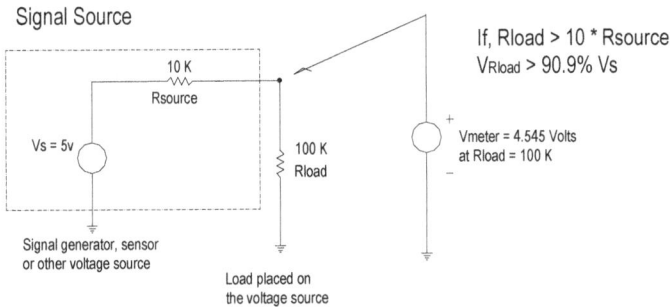

Signal Source

10 K
Rsource

Vs = 5v

100 K
Rload

Signal generator, sensor
or other voltage source

Load placed on
the voltage source

If, Rload > 10 * Rsource
VRload > 90.9% Vs

Vmeter = 4.545 Volts
at Rload = 100 K

Figure 42 – Increasing the amplifiers' input resistance to 100K (10:1 rule) increases the available voltage at the amplifier input by over 2 volts. Input resistance must be larger than or equal to 10 times the output resistance to avoid loading effects.

To avoid the loading effect, the load resistance must be at least ten times the value of the previous stages output resistance. From figure 32 it can be seen that increasing the load placed upon the first stage results in a larger value of voltage across the load, and considerably less change in voltage should the load be disconnected and re-connected to the source. From this condition the 10:1 rule of thumb applies, the load should be at least ten times greater than the previous stages' output resistance for *maximum signal transfer* from the source to the load. Signal generators will always indicate the internal, or source resistance on the face of the signal generator, usually near the output connector. This is done to facilitate impedance matching (since the signal generator is an AC output device, impedance used here is correct) the signal generator to the subsequent stage.

The amount of input resistance an amplifier, Panel Meter or other signal-conditioning device offers is available from the component specification sheet. It will be labeled as input resistance, Rin, or occasionally input impedance, Zin (even if impedance does not necessarily apply, as in a DC component). Output resistance is also specified on specification sheets. A DC powered (and AC powered) amplifier will exhibit both input and output resistance quantities. Input resistance is what the input signal will "see," and the output resistance will determine the maximum amount of signal transferred to the next stage.

Sensors are slightly different than amplifiers. Since the true input to a sensor is usually some mechanical energy form such as pressure, level, force, flow, temperature, position and so forth, sensors should always exhibit a large amount of "Input Resistance" to the variable being measured. In order to make an accurate measurement the sensor cannot be expected to alter the measured variable in the slightest manner. Hence the sensor-input resistance will always be large. However if a sensor requires a power supply connection, the amount of resistance offered to the power supply will be indicated as input resistance in the sensor specifications. The sensors' output resistance specification is the equivalent series circuit resistance as demonstrated in the previous circuit diagrams. Output resistance is often called internal resistance, source resistance or Thevenin resistance as well.

Sensors purchased with full bridges internally will indicate the output, or Thevenin, resistance directly in the sensor specifications to facilitate matching the sensor to an amplifier. However if the bridge is assembled manually, as in the case of RTD's or strain gauges, one may need to compute the value of the circuit output resistance. To determine the value of any circuits' internal or output resistance, Thevenins' Theorem is utilized.

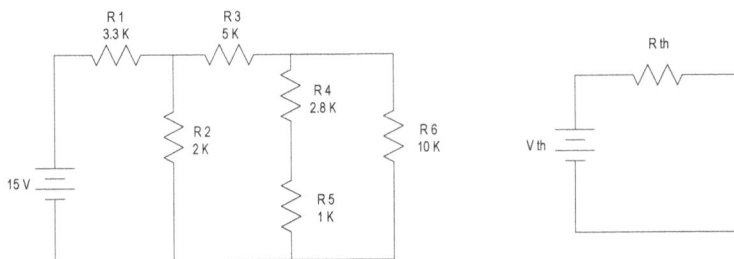

Figure 43 - A Thevenin circuit reduction converts a complex circuit such as on the left into the simplfied, 2-component circuit on the right.

Of all the circuit concepts covered in an introductory electronics course, none has such a universal appeal as Thevenin's Theorems. If well digested, the concepts represented by Thevenin's will provide a solid foundation from which any subject in electronics can be mastered. And although the mere mention of the name wreaks panic throughout the classroom, the process of circuit reduction using Thevenins' technique is not complicated if presented correctly. As an added benefit, the resulting procedure for bridge circuit Thevenization is identical regardless of the components within the bridge.

To begin, the process of Thevenization will yield two items, an internal equivalent resistance and a no-load (disconnected) output voltage value. To determine the no-load output voltage value, remove the load connection and measure or compute the value of voltage that will be available at no-load. This becomes the value associated with the voltage source in the previous diagrams.

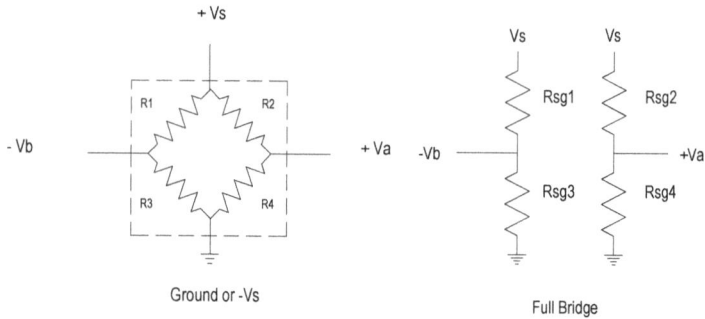

+ Vs

R1   R2

- Vb   + Va

R3   R4

Ground or -Vs

Vs   Vs

Rsg1   Rsg2

-Vb   +Va

Rsg3   Rsg4

Full Bridge

Figure 44 - The classic full-bridge in a rhombus configuration (left) and redrawn (right) to assist in performing the Thevenin analysis. The Thevenin voltage equals the difference in the two outputs, Vth = Va – Vb.

Next, to determine the internal resistance, or Thevenin resistance, remove any load and replace the circuit's source with an equivalent amount of resistance. If the source is a voltage supply, a small resistance or short is usually used since voltage supplies typically exhibit a small amount of internal resistance. A current source appears as a high internal resistance and would be substituted with an open circuit.

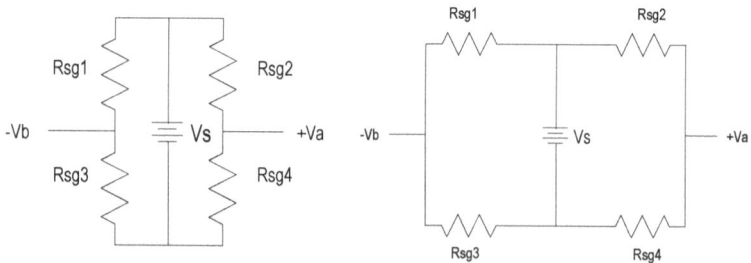

Rsg1   Rsg2

-Vb   Vs   +Va

Rsg3   Rsg4

Rsg1   Rsg2

-Vb   Vs   +Va

Rsg3   Rsg4

Figure 45 - To assist in Thevenizing, the circuit is redrawn again and the supply is labelled as Vs, DC voltage source (left). The individual bridge resistors are shown horizontally to assit in determining the resistance between the output connections (right). The DC source will be replaced with a short in the next diagram, to represent the resistance added by the voltage source.

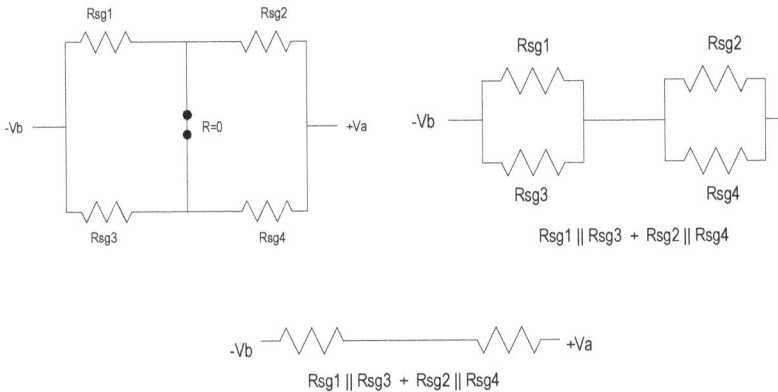

Figure 46 - The voltage source is replaced with an equivalent internal resistance of zero, and the circuit is redrawn to facilitate computation of the Thevenin, or output resistance. Given the indicated equations, if all resistances within the bridge are equal (such as 120 Ohm strain gauges), the Thevenin resistance will be 120 Ohms.

At this point only resistive components should be obvious in the bridge circuit, to determine the Thevenin resistance measure or compute the resistance available between the output terminals where the load is connected. Once this is accomplished the internal resistance and open-circuit, no-load voltage can be associated with an equivalent circuit diagram, and the effect of an output load can be determined. The resulting equivalent circuit follows.

Signal Source

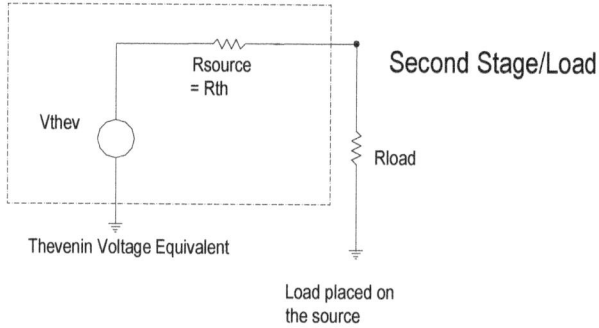

Figure 47 – Once the Thevenin voltage and resistance is determined, the effect of connecting a load across the bridge output can be easily determined.

However the circuit shown in the previous figure is not the only form of diagram representing the bridge circuit. In fact, the equivalent circuit introduced in Figure 36 is misleading when one considers the bridge is a differential output circuit, and the equivalent implies a single-ended output stage. Given the realization that the bridge is a differential signal source, the following diagram would prove to be more representative of the bridges' equivalent circuit.

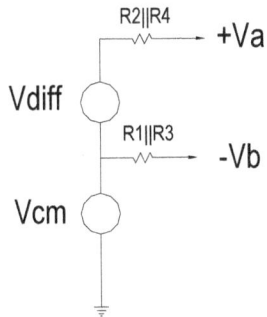

Figure 48 – The equivalent circuit symbol for a differential signal source indicates all voltages common to both outputs are labelled as Vcm or common-mode voltage, and the difference between the output leads is the output signal, Vdiff. Note the resistances contributing to the output resistance are also indicated. The resulting output voltage will be the difference between Va and Vb.

Chapter 2 Homework Problems

#1 – Given a Wheatstone Bridge with the following component values, determine the unknown resistance, Rx.

R1 = 10K

R2 = 1K

Rstd = 103.10

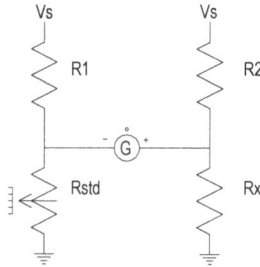

Figure 49 - circuit diagram for use with problems 1-3.

#2 – Given a Wheatstone Bridge with the following component values, determine the unknown resistance, Rx.

R1 = 100K

R2 = 1K

Rstd = 54610

#3 – Given a Wheatstone bridge with the following component values determine the unknown resistance, Rx.

R1 = 100K

R2 = .1K

Rstd = 25487

#4 – A Platinum (Pt.) RTD measures 119.4 Ohms. The gain of the sensor is .385 Ohms per C°, and the offset is 100 Ohms at 0°. Determine the RTD temperature.

#5 - A Platinum (Pt.) RTD is immersed in a solution of –20 Centigrade degrees. Determine the RTD resistance.

#6 – A single-axis strain gauge of 120 Ohms, GF = 2.05 is experiencing a strain of 500 micro-strain. Determine the change in resistance.

#7 - A single-axis strain gauge of 350 Ohms, GF = 2.05 is experiencing a strain of 500 micro-strain. Determine the change in resistance.

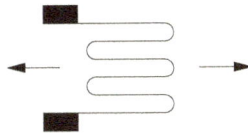

Figure 50 – Strain gauge diagram for use with problems #6 and #7.

#8 – In the following diagram, the single strain gauge of the previous problem is connected into a half-bridge by connecting an additional 350-Ohm resistance (R1) in series with the strain gauge (Rx). Determine the bridge output voltage at zero applied strain and at 500 microstrain.

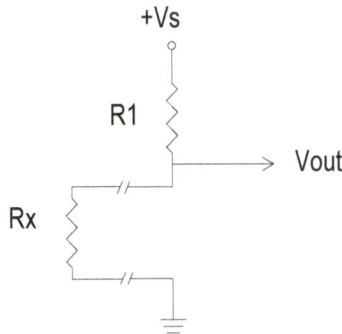

Figure 51 – Diagram for use with problem #8.

#9 – Copper exhibits an alpha resistance-temperature coefficient of .393%/C°. As an example, 100 feet of 20-gauge wire exhibits a resistance of 1.015 ohms at 20° C (room temp). If the temperature of the wire goes up 1.0°C the resistance will change by 0.00399 ohms (1.0 degrees * 0.00393 per degree * 1.015 ohms = 0.00399 ohms). How much microstrain applied to a 350-Ohm strain gauge with a GF = 2.10 would yield the same resistance change?

#10 – Does the previous problem support the need for using strain gauges composed of copper alloys to minimize temperature induced resistance changes?

Chapter 3 - Introduction to the Operational Amplifier

This chapter provides an introduction to operational amplifier characteristics such as gain, input/output resistance, packaging, pin connections, current draw and power supply requirements. The internal construction of the op amp will be investigated, differential signals and inputs are introduced, open and closed loop circuit connections are demonstrated and fundamental specifications are presented. Chapter topics are included following.

Objectives:
Upon completion of this chapter, you should be able to:
- Explain the purpose and operation of general purpose operational amplifiers
- Provide typical application examples of operational amplifiers
- Interpret essential specifications of operational amplifiers
- Assemble a simple op amp comparator
- Explain the purpose of negative feedback with operational amplifiers
- Explain the operation of inverting and non-inverting op amp circuits
- Assemble simple inverting and non-inverting op amp circuits
- Design simple inverting and non-inverting op amp circuits for specific gain values

Introduction

The operational amplifier (designated for its ability to perform mathematical operations) represents the culmination of analog integrated circuit development. An integrated amplifier module, the "op amp" features many characteristics that approach ideal amplifier specifications. High input impedance, low output impedance, high gain, differential amplification, linear operation and minimal error characteristics are associated with operational amplifiers. Each of these characteristics is defined in subsequent text.

Probably the greatest asset of an op amp is the ability to establish the operating function and specifications with a minimum of externally connected components. In the case of DC and low frequency AC amplification the amplifier's gain (output to input ratio) is determined by the ratio of two resistors. No characteristic curves as with transistors are necessary, only a fundamental understanding of negative feedback amplifier concepts and a couple of basic equations are required to design the vast majority of amplifiers.

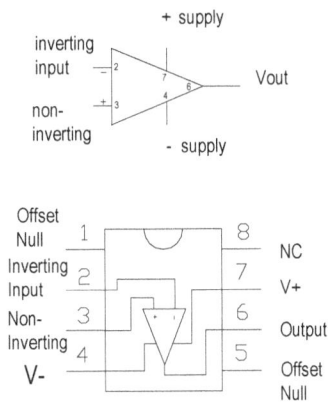

Figure 1- The "Open-Loop" operational amplifier (top), and the integrated circuit pin-out (below).

This chapter will introduce the operational amplifier and a few of its more useful specifications, circuit configurations and operational characteristics. As the text progresses the vast majority of op amp specifications will be investigated and in most cases tested. In addition, many useful applications of sensor signal conditioning, signal voltage scaling, AC amplification, data acquisition and measurement system calibration will also be provided. The focus of all op amp applications in this text will be as sensor amplifiers, the original purpose for the linear integrated circuit.

The Open-Loop Op Amp

Most general-purpose op amps are available in 8 or 14-pin integrated circuit packages. Not all of the pins need to be connected for operation. Two pins are allocated for differential inputs, one pin for the single-ended output and two pins for the bi-polar (positive and negative) power supply connections. With the power supplies connected and no external gain or function determining components the op amp will exhibit a high internal or "open-loop" voltage gain, Avol. The term "open-loop" is used because no external "feedback" components are utilized to connect the output back to the input, which would effectively reduce or control the gain. Typical open-loop voltage gain (Avol) quantities are around 200,000 and higher. Ironically the op amp's open-loop gain is so large, it is rarely used open-loop, or without external components. The op amp's operation in this configuration is very unpredictable. As will be demonstrated later, the larger an op amp's Avol the more precise and predictable its closed-loop (negative feedback) operation. Open-loop voltage gain and a typical open-loop application will be discussed later in this chapter. For now the op amp should be seen as a self-contained high gain amplifier component or module.

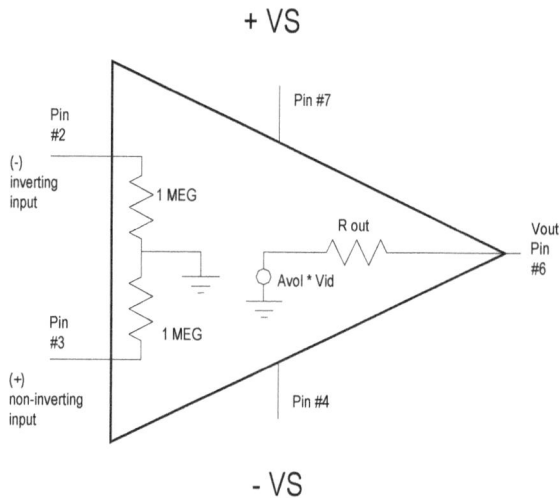

Figure 2 - equivalent internal circuitry of an operational amplifier.

Internal Construction

The op amp is composed of numerous discrete transistor circuits internally. Typical general purpose op amps contain transistor differential amplifiers in the first two stages, followed by a transistor level shifting or biasing amplifier, and a two-stage transistor pre-amp and milliwatt level power amp in the final stages. Additional transistors are included for operational and specification enhancement. Internal circuit equivalent diagrams are available from all manufacturers and are usually included within the op amp's specifications. Upon inspection, the differential amplifiers, level shifter and power amps are obvious if one is familiar with these discrete transistor circuits. Overall each of the above stages contributes to the large internal voltage gain, Avol.

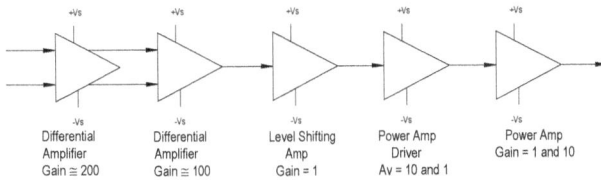

Figure 3 - General purpose operational amplifiers commonly contain five internal stages.

One need not be completely familiar with the construction and internal operation of an op amp to design, apply or analyze op amp circuits however. Becoming familiar with a few common amplifier circuits, their respective equations and negative feedback principles will yield results that can be applied to the vast majority of all op amp applications. However, the better one understands internal construction and operation the better the understanding of op amp limitations, specifications and applications.

Figure 4 – Linear integrated circuit operational amplifiers. Clockwise from top center, an exposed RCA 741 op amp "chip," a Harris 741 op amp, a Motorola 741 op amp, an Analog Devices precision 711 op amp, an RCA 741 op amp, a National Semiconductor dual op amp (2 op amps in a 14 pin DIP package), a National Semiconductor quad op amp (four 741 op amps in a single 14 pin DIP package). Most of the op amps shown display a four-digit date code (e.g., 8924, 9036, etc.), representing the year and week the IC was manufactured.

Power Supplies

Operational amplifiers and associated linear integrated circuits commonly require *bipolar* supplies. Bipolar supplies are connected as two voltage sources of opposite polarities. Plus and minus 9-V, 12-V, 15-V and 24V are among the commonly available bipolar power supplies.

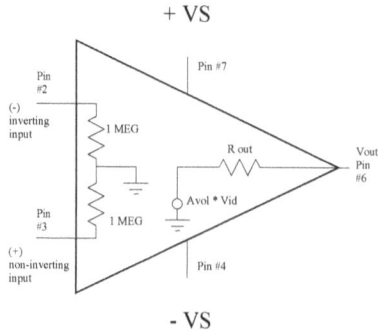

Figure 5 – Operational amplifiers and other linear IC's often require bipolar (dual polarity, plus and minus) power sources.

98

However, many power supplies are not available as bipolar supplies, like batteries, and must made into a bipolar supply. Switching supplies, also called DC to DC supplies, are available to perform an efficient conversion of DC voltage to a different DC voltage while minimizing the power lost in the conversion. Switching supplies are commonly used where the amount of power consumed by the load or lost in the conversion is considerable. Switching supplies are common in computer applications where several supply voltages are required from a single source (AC or battery) and the supply will be powered for extended periods.

The procedure for conversion of a unipolar supply into a bipolar supply is relatively simple, and results in two supplies of opposite polarity (plus and minus), each equal to about half the value of the originating power source. The conversion also creates a common ground, or zero-volt location between the two bipolar voltages. To perform the conversion simply, a common voltage divider can be used. Using a pair of resistors to divide the power source into two output voltages with a common ground is probably the most convenient approach and works well if wasted power is of no concern and it usually is, especially with battery operation.

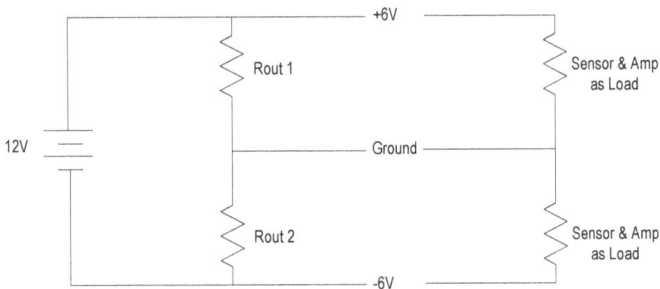

Figure 6 – The resistors on the right represent the sensor and amplifier as loads to the bipolar power supply. The resistors on the left are required to divide the twelve-volt source into two bipolar six-volt sources with a common ground.

99

However, the resistors used to divide the source voltage will consume and dissipate power as heat, which will assist in depleting the charge of a battery. In addition, the regulation capabilities of the voltage divider are very poor. Should the value of the load change, the supply voltage will also change. In other words, the supply voltages can be expected to vary for each unique value of power (current) demanded by the load. So for each different value of output from the op amp circuit the load placed upon the power source (the battery and resistors) is different and as might be expected, the value of the supply voltages will be different This will be especially true if the amount of available power (current) from the originating source is small which usually implies a high internal resistance within the source. Larger supplies like automotive batteries will exhibit little change while smaller consumer batteries may exhibit a considerable variation.

Other more efficient and better-regulated approaches are also available however. Zener or solid-state varieties of regulators work well in dividing unipolar supplies. The following diagram demonstrates the Zener diode arrangement.

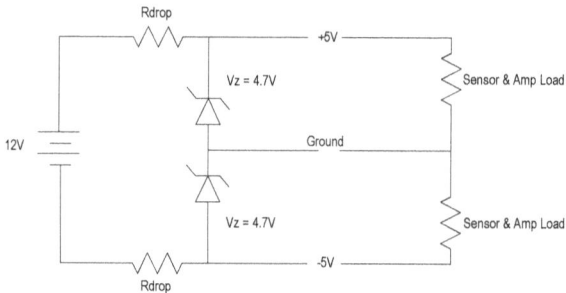

Figure 7 – Two Zener Diodes are used to divide an applied source of 12V into two 5V (5.1V or 4.7V are standard Zener values) supplies with a common ground.

Two grounds may actually exist in this circuit and in the previous resistive circuit since the battery/alternator will probably have its negative side grounded to a chassis or frame. The battery ground would be considered the chassis ground

100

and the power supply/amp ground is considered the signal ground. Each would be given different symbols on a schematic.

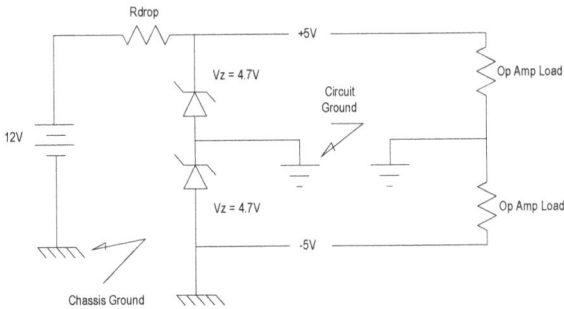

Figure 8 – Two grounds are obvious in this diagram. The "Rake" represents the chassis ground connection of the originating power supply as in the case of an automobile. The circuit ground symbol represents the zero-volt reference location of the signal and amplifier circuit.

Solid-state regulators provide an efficient means of performing the voltage division function as long as the load demands under an amp or two of current. The following diagram demonstrates how two 5-volt regulators perform the single to bipolar conversion.

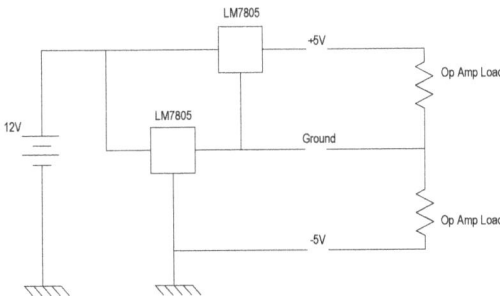

Figure 9 – Two solid-state (LM7805) regulators convert a unipolar supply into bipolar supplies.

Inputs and Outputs

As an electronic circuit, the op amp is a two-port (single input/single output) device with an additional power supply input. The operational amplifier contains two bipolar (positive and negative) inputs and an individual single-ended output (a single pin output lead referenced to ground). Every op amp configured as an amplifier requires two separate input connections, a signal input to be amplified or conditioned, and a power supply input from which the op amp will derive its output voltage and power. The two inputs are easily distinguished since the input signal is always operating within a span or range of values and is a relatively low voltage value whereas power supply voltages are always fixed and constant. Sensor output voltages, radio frequency (RF) communication or modulation data and industrial control signals are common examples of op amp input signals. The power supply inputs should never change or vary, and are equal but opposite in polarity. Common power supply voltages are $\pm$ 5V, $\pm$ 9V, $\pm$ 12V, $\pm$ 15V, $\pm$ 18V and $\pm$ 24V. Since the power supplies are of equal voltage or "balanced," no ground connection to the op amp is necessary.

Similar to every other active component such as vacuum tubes and transistors, the op amp operates as a valve regulating the amount of the power supply voltage that is allowed to appear in the op amp output. The output voltage is allowed to assume any value between the plus and minus supply voltage extremes. The extreme supply voltage values are occasionally referred to as the "rails" of the amplifier's operation, designating the limits of the output. The varying input signal span is operating in the capacity of a controlling signal by dictating the amount of the supply to appear at Vout.

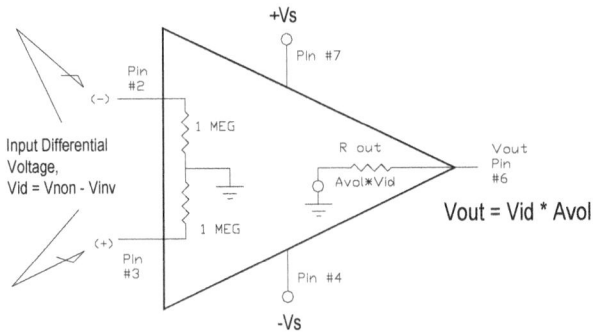

Figure 10 – The concept of the amplifier is similar to a common industrial control valve. The difference of voltage applied to the two inputs terminals (Vid) is amplified by the internal voltage gain (Avol) of the op amp. The resulting output voltage is drawn from the positive or negative power supply and appears at pin #6. Where the control valve utilises a controlling input to determine the amount of material from the supply to pass to the output, the op amp utilises the difference in voltage between the inputs to determine the amount of supply voltage to place at the output.

The gain and offset of the amplifier establish the input/output relationship, or Y = MX + B characteristics. For a given amount of input signal span there exists a given amount of output signal span, the amplifier generates a linear and proportional output in response to the input signal span.

The focus of this text is upon design and analysis of scaling amplifiers. Scaling amplifiers are designed for the purpose of converting an input voltage span into a completely different span of values. Scaling amps rarely utilize the full "rails" of output values available between the supplies but instead utilize a span of output values that correspond to the input by the gain and offset quantities the amplifier is designed for use with. Subsequent chapters will investigate design of scaling amplifiers.

Impedance

The op amp offers a high amount of resistance to input signals. The output terminal of the op amp, being connected to a power amp stage, must have an

103

associated low resistance for current drive characteristics, even though the current drive capability of the op amp is rarely over 20 milliamps. The characteristics of input and output resistances are often referred to as input and output impedance. Ideally, impedance includes resistance and frequency dependent quantities of reactance. Yet input and especially output impedance contain little reactance and as a result is usually listed in op amp specification sheets as input resistance or output resistance. Should these characteristics appear as resistances, input and output capacitance values are also provided. Inductive characteristics within operational amplifiers are negligible.

The op amp closely resembles an ideal amplifier in many ways, including open-loop voltage gain (Avol), input impedance (Zin) and output impedance (Zout). Regardless of the resistance or impedance designation, in an ideal (perfect) amplifier, input impedance of infinity is necessary to acquire maximum voltage from a signal source of any internal resistance. At the output of an ideal amplifier a resistance approaching zero ohms is desired to provide the amount of current demanded by any sized load. As a result, high input and low output impedance are understood to be characteristics associated with the ideal amplifier. The voltage divider figures in the chapter 3 diagrams reduces an op amp amplifier to its equivalent input and output voltage divider circuits. All cascaded signal coupling stages can be represented as a simple series circuit or voltage divider circuit for analytical purposes. In this manner the voltage or power effects of larger or smaller, source or load resistances can be determined. To summarize, Hi input resistance and low output resistance are required for maximum signal (voltage) transfer from stage to stage as is exhibited in the series voltage divider circuit diagrams in chapter 3. Maximum signal transfer should not be confused with maximum power transfer. To facilitate maximum power transfer the input and output resistances (impedance) must be equal and usually approximate 75 ohms.

Differential Inputs

Most conventional amplifiers contain single-ended inputs and outputs. A single-ended input provides a single lead-wire input to apply the signal for amplification. The input signal is applied in conjunction with a connection to ground eventually requiring two separate connections, two wires. The single-ended output is similar with the output being available at a single lead-wire. The output is then measured or taken for further processing with respect to, and in conjunction with a ground connection. Again, to utilise the single-ended output signal two connections are required, one wire to the output and one wire to ground.

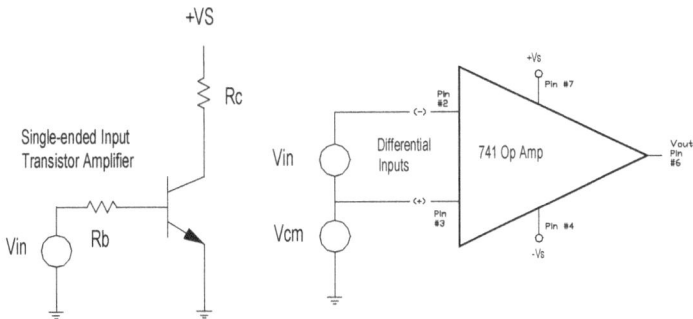

Figure 11 – Comparison of differential inputs (right) and single-ended inputs (left) as found with transistor amplifiers. Note the input voltage symbol on the op amp contains two designations. Vin is a bipolar differential input, Vcm is any AC or DC voltage common to both inputs, as in the case of a bridge circuit where each bridge output is nulled initially, and subtracted. The two input designations represent the cancellation of voltages common to both inputs (Vcm, or common-mode voltages) and the amplification of input differentials (Vin, or Vid).

Differential inputs do not require a ground connection. Although always available, a ground connection is not necessary to reference the signal. The signal voltage is available as a difference in voltage between to non-zero voltages available between two non-grounded leadwires. Differential voltages are commonly available at Wheatstone resistive bridge outputs and center-tap grounded transformer outputs. Differential voltage sources will always be bi-polar with one lead-wire labeled positive and the other lead-wire labeled negative. Differential amplifiers amplify the difference, or effectively subtract the voltages available from the two lead-wires. A common voltmeter is an example of a differential measurement since all voltmeters indicate the difference in voltage between two points.

Operational amplifier inputs are labeled as inverting (negative) and non-inverting (positive). These designations represent the polarity of the inputs, or the phase relationship of each input with the output. As examples, if an increasing (positive going) voltage is applied to the inverting (-) input, the output will move in a decreasing (negative going) direction. An increasing voltage applied to the non-inverting (+) input will cause the output to move in an increasing (positive going) direction.

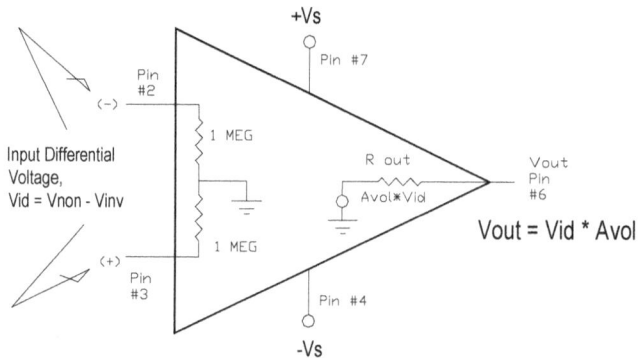

Figure 12 – The input differential voltage, Vid, is amplified by the internal voltage gain, Avol, to generate Vout.

An op amp's output voltage will always be equal to the voltage difference between the inverting and non-inverting inputs (Vid) multiplied by the open-loop gain, Avol. In equation form,

$$Vout = Vid * Avol$$

or,

$$Vid = Vout/Avol$$

Where, Vid represents the voltage difference between the two inputs.

As an example and assuming an Avol of 200,000, an output voltage of +5 Volts is created by an input difference of +25 microvolts.

$$Vid = +5V/200,000$$
$$Vid = +. 000025V$$

This concept may appear to be of little consequence initially but will prove to be a valuable troubleshooting tool with negative feedback amplifiers. Since the op amp's output voltage will always be between the rails (power supply values of $\pm12V$, $\pm15V$, $\pm24V$, etc.) of the op amp, the input differential voltage (Vid) will always be microvolts. Measuring the voltage between the two inputs (Vid) of an op amp amplifier should yield about zero volts (microvolts) or something is improperly connected or failed.

Gain * Bandwidth Product

Operational amplifiers are available under a wide variety of part numbers and specifications. The "741" is considered a general-purpose operational amplifier and is manufactured by over a dozen integrated circuit manufacturers. As a general-purpose op amp, the 741 exhibits an open-loop gain of 200,000 or greater, an input resistance of at least 1 Megohm from either input to ground (referred to as single-ended), at least two Megohms differential input resistance (between the two inputs), and 100 ohms or less of output resistance. These specifications are the most common and may vary slightly from one IC manufacturer to another but it is generally understood that a 741 will exhibit these specifications as minimum quantities.

A category of op amp exists to amplify high frequency or rapidly changing (high slew rate) signals such as square waves. These operational amplifiers are classified as high-speed op amps and are generally somewhat more expensive than a common 741. Op amp frequency rating is usually combined with gain. Every amplifier possesses a maximum frequency that can be processed for a given gain or, a maximum gain that can be developed at some maximum frequency. In other words, when attempting to optimize gain or frequency in any amplifier (including op amps) frequency and gain are always a trade-off, one is increased at the expense of the other. For this reason, the two specifications are combined into Gain*Bandwidth Product.

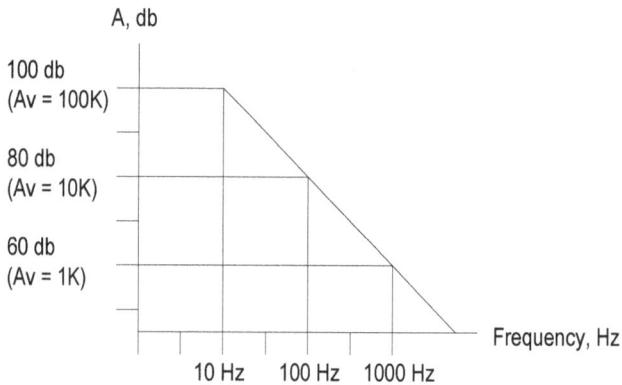

Figure 13 – Graphical representation of a 20db/decade "roll-off" characteristic used to determine the stability and gain-bandwidth product of an operational amplifier. The product of the frequency and voltage gain at any point along the curve is 1,000,000 or less. Any "compensated" operational amplifier will exhibit a similar A*BW characteristic. The 741 contains a single implanted capacitor internally to establish the 20db/decade roll-off characteristic.

Gain*Bandwidth Product, or A*BW is the result of multiplying the desired amplifier operating gain times the maximum frequency to be amplified. The resulting quantity, or product, should always be less than the op amp's A*BW rating. As an example, the 741 A*BW specification is one million. This means the 741 could be used to amplify frequencies up to 20-Khz at a maximum gain of 50 since;

Max Gain * Max Frequency = Gain Bandwidth (A*BW) Product

or,

$$A_{Vmax} * F_{max} = A*BW \text{ Product}$$
$$50 * 20\text{-KHz} = 1,000,000$$

If the amplifier were assembled with a closed-loop, negative feedback gain (Av) of 50 and a 20 Khz signal applied, an output of 20 Khz fifty times larger than the input would be measured. If the gain were fixed at fifty and the input frequency increased above 20 Khz, the resulting output amplitude would decrease to maintain the one million gain*bandwidth product. Conversely, if the gain were increased above fifty and the frequency remained at 20 Khz, the output amplitude would not increase but would remain constant at an effective gain of fifty to again satisfy the one million gain*bandwidth product.

Gain*Bandwidth Product is a characteristic usually associated with internally frequency compensated operational amplifiers. Frequency compensated operational amplifiers, such as the 741, contain a capacitor implanted within the op amp to limit the amplifier's "roll-off" or frequency vs. gain operating characteristics to -20 decibels per each decade of frequency. This results in a gain decrease of 20 decibels for each increase in frequency of 10:1. A roll-off of -20db/decade is considered inherently stable meaning the amplifier will not oscillate or create significant self-generated distortion when presented with a rapidly changing or unstable input. Uncompensated op amps are susceptible to self-oscillation unless provided with external compensating circuitry. Additional pins are provided for compensating such uncompensated types of op amps. Resistor-capacitor combinations are assembled on the op amp according the diagrams provided on the particular op amps' specification sheet to provide the desired roll-off characteristic.

Op Amp Comparator Circuit
As mentioned previously the internal or open loop gain (Avol) of the op amp is too large to be practical. Negative feedback will be utilized to effectively reduce

the gain to a useable value. Concepts of negative feedback amplifiers are discussed in a subsequent chapter.

There exists a single application for the open-loop op amp however. Since an input differential of only microvolts is required to drive the output of an op amp to its full rail (± Vsupply) voltage, the op amp's output can be used to indicate which input is greater, or more positive. This circuit is only rarely used and is referred to as a comparator.

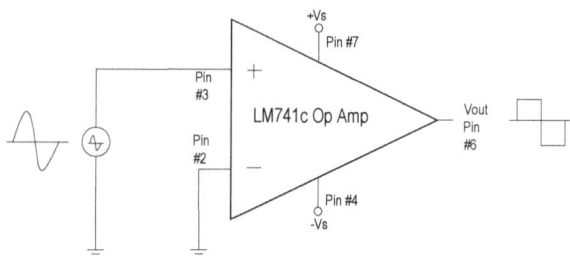

+Vs
Pin #7

Pin #3

+

LM741c Op Amp

Vout
Pin #6

Pin #2

−

Pin #4
-Vs

Figure 14 – Op Amp comparator application with input and output waveshapes displayed. In this application, a varying input voltage alternates moving positive and negative a few fractions of a volt. The open loop gain will attempt to amplify the input by 200,000 or more causing the output to reach the positive and negative extremes of the output, approximately equal to the plus and minus supply voltages, or output "Rails."

As an example, assume the inverting input is connected to ground and the non-inverting input is connected to a low frequency (say about 1 cycle every 20-30 seconds), .20-volt peak-to-peak, zero referenced (or zero offset, meaning the wave peaks at -.1 and +.1 volts) sinusoidal waveshape. If this is difficult to visualize, imagine someone slowing varying a variable DC voltage source between plus .1 volts and minus .1 volts. AC voltages can always be substituted by a varying DC voltage for analytical, and in many situations application reasons.

With the inverting input referencing the op amp comparator circuit at a ground potential (zero volts), the output will switch whenever the non-inverting input gets within microvolts of the ground reference voltage. When the input moves from zero in a positive direction, the output will immediately go to the positive rail, ideally the positive supply voltage value. Realistically the output will approach the supply but will be slightly less. The output will remain at the positive rail (also called saturation) value until the input moves through zero into the negative half cycle.

With the input moving negative from zero volts, the op amp amplifies the input with its full Avol of 200,000 causing the output to switch rapidly within microvolts of negative Vin to negative saturation. The large gain will also distort the input sine wave causing the output to appear as a square wave. This specific application of the comparator is referred to as a zero-crossing detector, so called because of the circuit's ability to switch Vout as the input swings through zero volts. Since the input was applied to the non-inverting or positive input of the op amp, the output will be in phase with the input. With the input applied in this manner, the circuit is also referred to as a direct-acting zero crossing detector. Direct-action means the output directly follows the input or the output goes up as the input goes up. Switching the input to the inverting terminal and the reference voltage (ground in this example) to the non-inverting input will result in a reverse acting or 180 degree out of phase input/output characteristic. In the reverse acting example the out will move negative as the input moves positive causing a phase inversion.

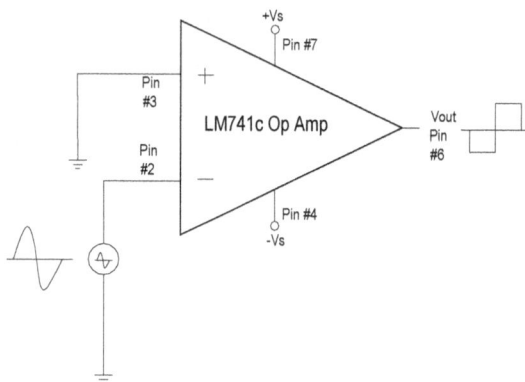

Figure 15 – The op amp comparator as an inverting, or reverse acting zero crossing detector. Note the out of phase output and input signals.

Several variations on the basic comparator exist. Among the more common involve utilizing a non-zero reference voltage as the output switching point. By providing a DC voltage as the reference, the output will switch states whenever the input passes through the reference DC voltage value. This type of circuit is referred to as a reference comparator and operates in an identical manner to the circuit described previously except the output will no longer switch at zero volts but rather at the value of the reference DC voltage. In automation applications, reference comparators provide a simple means of On/Off control. The varying input voltage would be provided by a sensor output, with the DC reference voltage applied to the op amp representing a Set Point of desired system (process) operation. The output from the op amp would be used to eventually activate a valve, alarm or other type of process correction mechanism. The varying output state would cause the valve to open or close depending on the sensed value. Ultimately, the sensor output (from the process) would cycle around the reference value applied to the op amp.

Unfortunately, the op amp is rarely used in Comparator applications. This is due to a number of reasons but primarily because the op amp was specifically designed for linear, amplification purposes and not non-linear comparison functions. If comparator functions are desired one should consult the databooks for comparator IC's such as the LM311 and similar comparator IC's.

The databooks should always be consulted before applying any IC. Besides the available operating specifications most databooks provide a wealth of typical application circuits that can result in considerable savings of design and development time.

Negative Feedback

The term "Negative Feedback" is a reference to the manner with which an output signal is returned, or fed-back to the input of an operational amplifier. Negative feedback indicates the input signal and the feedback signal are summed "out-of-phase," or the signals are applied to the inputs differentially, requiring some form of subtracting function. Application of negative feedback is not exclusive to operational amplifiers, virtually every component in automatic control systems contain or are contained within a negative feedback system. As will be demonstrated, when applied properly negative feedback is capable of maintaining a variable constant under varying external conditions.

The amplifier output signal can be returned to the input in one of four methods. Series voltage, series current, parallel voltage and parallel current are the four techniques to return an output, or portion of an output to the input for feedback purposes. The feedback signal is always returned to the negative, inverting input when linear, proportional amplification is desired.

Positive feedback, although rare, is performed when discontinuous switching functions (as with hysteresis switches) is desired. Positive feedback applies a portion of the output to the input in a manner that results in an overall larger output signal. As the output increases, the feedback increases and the output increases further. The obvious conclusion is that positive feedback will result in an output that is always close to being a supply voltage value, plus or minus. Positive feedback is accomplished by placing a resistive divider from the op amp output to ground, and "feeding-back" a portion of the output to the positive, non-inverting input terminal. Positive feedback results in a Hysteresis or Schmitt Trigger type of characteristic when used with an op amp. Operational amplifiers make poor hysteresis switches since the original intent of the op amp was for use as a linear device, whereas the Hysteresis switch is a non-linear characteristic. A class of op amp derivatives called Comparator IC's has been developed to accommodate the demand for a switching output IC component. A positive feedback, op amp circuit and Hysteresis characteristic follow.

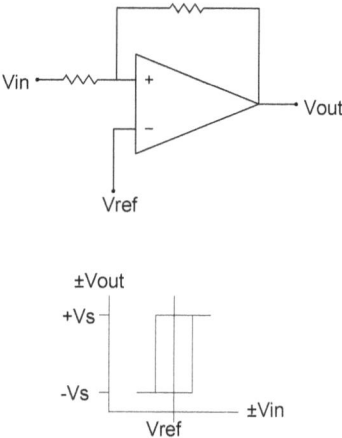

Figure 16 – An operational amplifier with <u>positive feedback</u> becomes a Hysteresis switch.

Positive and negative feedback result in entirely different concepts and equations, and should not be confused. *Op amps configured and applied as amplifiers always utilize negative feedback.* This text is devoted to negative feedback amplifiers and will not investigate positive feedback, op amp applications. The vast majority of all op amp applications will utilize negative feedback.

114

Of the four methods available to create negative feedback with operational amplifiers, two types (series current and parallel voltage) appear most often. These two feedback techniques are applied in the inverting and non-inverting amplifiers respectively. The inverting and non-inverting amplifiers form the basis of all op amp amplifier circuits. If the operation and equations associated with these two amplifier types are well understood, any negative feedback op amp application can be determined at a glance.

As mentioned previously, feedback is the manner with which an output signal is provided back to an input. Since the signal is returned to the inverting (negative) input, the feedback is understood to be negative. Negative feedback can be seen as comparing the input signal with the output, or a portion of the output. Seen another way, the input and output are algebraically summed out of phase in an attempt to generate a difference, or error, Vid. Using a high gain (Avol $\geq$ 200,000) op amp with negative feedback, an input signal will be applied, amplified and an output generated. A feedback circuit, usually composed of a resistive divider, will voltage divide the output and provide the reduced output signal (or "feedback" signal) back to the input for comparison against the input signal. As the feedback and the input voltages are compared (subtracted), the resulting error signal (Vid) becomes less, ideally becoming within plus or minus 75 microvolts, ($\pm$ Vsat/Avol = $\pm$15V/200,000).

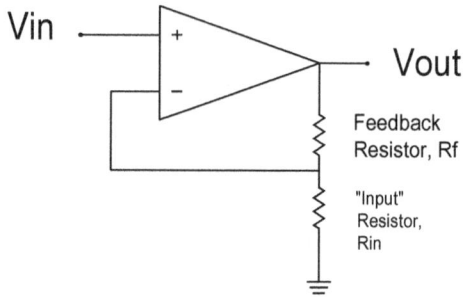

Figure 17 – Op amp with negative feedback. A voltage divider reduces the output voltage and provides the "Feedback" voltage to the inverting input for comparison against Vin.

With a negative feedback op amp amplifier the realistic difference between the input and feedback signals will be very close to zero, within millivolts. Any larger than required difference between the two signals (Vin and Vfb) is considered an "error signal" and will be amplified by the op amp's Avol to adjust the output and the subsequent feedback signal to the correct value. This provides a valuable troubleshooting tool, the voltage on each input should be virtually the same when measured with respect to ground. The difference between the inputs when measured differentially should be within a few millivolts of 0.00 volts. If this is not the case, the op amp *may* be faulty. Excessively high gain, excessively large input voltage or improper circuit connection will yield the same symptoms however.

The amount of feedback will determine the overall *closed-loop voltage gain, Av*, of the amplifier circuit. If an output is 100% fedback to the input, the amplifier gain will be limited to unity. As the amount of feedback decreases, the gain of the overall amplifier circuit increases. The closed-loop voltage gain can be proven in the following derivation of gain (Av).

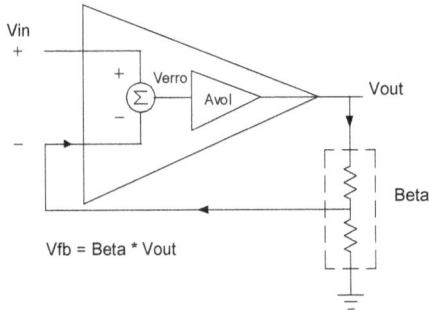

Figure 18 – A model for an analysis of negative feedback amplifiers. The arrows indicate the feedback signal flow path. The Beta (voltage divider) network provides a portion of the output (Vfb, the feedback voltage) to the "Summing junction" for comparison against Vin (performed by the differential inputs). Any difference between the feedback and input voltages will be amplified by the open-loop voltage gain to generate an output voltage. Negative feedback is inherently self-corrective, resulting in a single proportional value of Vout for each unique value of input voltage and closed-loop voltage gain, Av.

Defining the terms in the diagram;

Beta = Vfb / Vout   (% of Vout returned (fedback) to the input)

Vfb = Beta * Vout   (feedback voltage)

Verror = Vin-Vfb   (difference between Vin and Vfb, also called Vid)

Vout = Avol * Verror   (Vout is the amplified difference

between Vin and Vfb)

Deriving the equation for closed-loop voltage gain, Av yields;

$$Vout = Avol * Verror$$

$$Verror = Vin - Vfb$$

$$Vfb = Beta * Vout$$

$$Vout = Avol * (Vin - Vfb)$$

$$Vout = Avol * (Vin - (Beta*Vout))$$

$$Vout = (Avol*Vin) - (Avol*Beta*Vout)$$

$$Vout + (Avol*Beta*Vout) = Avol*Vin$$

$$Vout (1+Avol*Beta) = Avol*Vin$$

$$Vout = Avol*Vin/(1+Avol*Beta)$$

117

Since,   Av = Vout/Vin

                          Vout/Vin = Avol/(1+Avol*Beta)

Assuming, Avol*Beta is approximately equal to 1+Avol*Beta,

                          Vout/Vin = Avol/(Avol*Beta)

Canceling Avol yields,

                          Av $\cong$ 1/Beta

or, Av closely approximates the <u>inverse</u> of Beta.

In conclusion, the closed-loop voltage gain, Av, (the relationship between Vout and Vin) is approximately equal to the inverse of Beta. Where Beta is the feedback factor, or the amount of the reduction of Vout. If Vout is reduced to one-half, Av becomes the inverse of one-half, and Av equals two. Vout is twice Vin.

The Non-Inverting Amplifier

As can be seen in the previous derivation, the Av equation is an approximation. However, even as an approximation the results are quite accurate, certainly within the tolerance of the resistors used to assemble the feedback circuit.

Av is always as easy to compute as 1/Beta. However this equation and the circuit application holds for parallel voltage feedback only. For any op amp amplifier one simply needs to recognize the circuit type, determine the fraction of Vout fedback to the input, and inversion determines the amplifier gain. As an example, assume a circuit is assembled as in figure 11, where Rf is 9K ohms and Rin is 1K ohms. The two resistors form a series voltage divider where 10% of Vout is applied at the negative input as Vfb. The feedback factor Beta is therefore 10% or .1, and the voltage gain, Av is the inverse of .1, or 10.

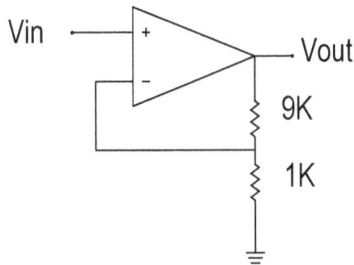

Figure 19 – A non-inverting amplifier with a feedback factor (Beta) of .1, and an Av of 10.

With a voltage gain of 10, any input voltage will appear ten times larger in the output. If an input of 1 volt AC or DC is applied, 10 volts AC or DC will appear in the output. The output can never exceed the amount of the supply voltage however. With ±15 volts as a supply and 2 volts DC applied to the input of the previous circuit, the computed Vout would be 20 volts. Since the "rails" of the amplifier are at plus and minus 15 volts, the output cannot acquire the computed 20 volts DC. Vout will be limited to slightly under 15 volts. Note also the input is applied to the positive or non-inverting input terminal. When the input "sees," or is applied to the non-inverting terminal, the amplifier is direct acting or non-inverting, meaning the output varies in phase with the input.

Amplifier Circuit Analysis

Of greater importance in assessing circuit operation is analysis of the circuit's operation in terms of voltage and currents. When input is applied to the positive (non-inverting) terminal, negative feedback will "drive" the negative (inverting) terminal to become within millivolts of the positive terminal voltage. In this manner, whatever voltage is placed on the positive terminal effectively appears on the negative terminal, and the input differential voltage (Vid) is zero. Therefore, according to the circuit connection and since Vid equals zero volts, the input voltage appears across the resistor labeled Rin. The voltage across Rin will create a current Iin, or I_{Rin}. Since the input resistance of the op amp is large

119

(1 Megohm) the current established through Rin continues through Rf and into the op amp output. Vout becomes the sum of all voltages between the op amp output and ground, V_{Rf} and V_{Rin}. The current in Rin and the current in Rf are equal.

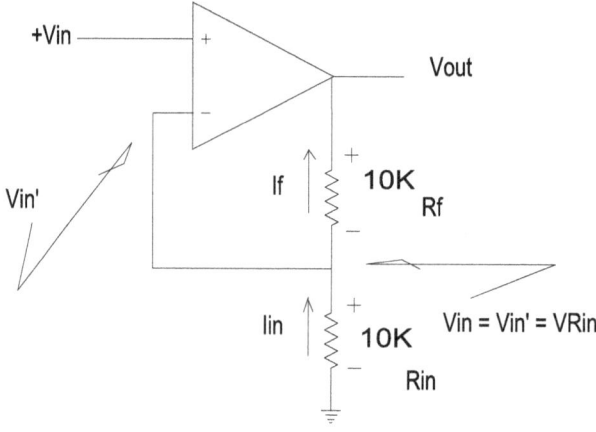

Figure 20 – The non-inverting amplifier. Vin is applied to the non-inverting input, since Vid = 0V (within millivolts), Vin is also measured at the inverting input and across Rin as indicated. Iin is equal to Vin divided by Rin, If equals Iin, the voltage across Rf equals If multiplied by Rf, and Vout equals the sum of voltages across Rf and Rin. This operation applies to *all* negative feedback op amp circuits. Vin' is also labelled as Vfb, the feedback voltage.

Two concepts mentioned in the previous description hold for all negative feedback op amp applications. The two inputs will always be of equal voltage (the input differential equals zero), and the currents through the individual resistors will be of equal value. In addition, all negative feedback op amp amplifier circuits work in an identical manner. As will be seen, the circuit analysis technique described previous holds for non-inverting and inverting amplifiers. This single analytic approach allows the student to focus upon applications of the op amp rather than memorization of circuit configurations and equations. Two gain equations and a single description of circuit operation will support the vast majority of op amp amplifier applications.

The Inverting Amplifier

Since the operation of all negative feedback operational amplifiers is identical that described previously, the inverting amplifier distributes its voltages and currents in a manner similar to the non-inverting. As an example, consider the diagram following. This amplifier is considered inverting since the input signal is applied to the inverting input. As can be seen from the diagram, the input signal is applied to Rin, which is connected to the inverting input directly. Since Vin is applied to the inverting input, this circuit configuration will cause a 180° phase inversion between the input signal and output voltage, which means the output will move downward (negative) as the input moves up (positive). In the case of the unbiased (no supplemental offset voltage) inverting amplifier, a positive DC input will cause a negative DC output. The inverting circuit configuration is also referred to as the reverse acting amplifier.

In analyzing the operation of the circuit, two concepts must be assumed. Since the internal voltage gain is large, the amount of input difference to generate any output between the rails (±Vs) is ideally microvolts, or essentially nothing. For this reason the input differential voltage is assumed to be zero, or Vid = 0V. This concept allows a determination of the voltage on each input terminal. Given the circuit configuration with the grounded non-inverting (positive) input and the feedback connection to the inverting (negative) input, each input will assume the potential placed upon the non-inverting terminal. In this case ground potential, or zero volts.

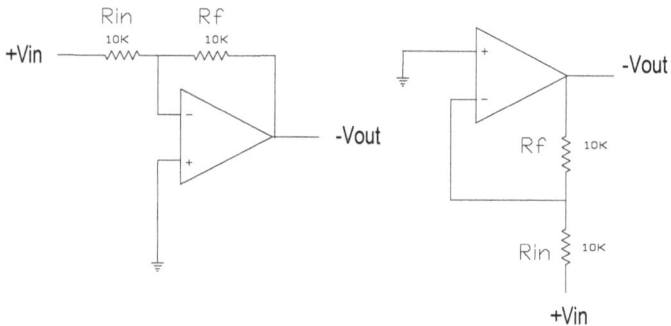

Figure 21 – Inverting amplifiers can be represented in two types of circuit configurations. Notice both exhibit negative feedback and an input voltage placed to the inverting input. When an op amp symbol is recognized on a schematic diagram, always verify negative feedback first, and note where the input signal is applied next. The circuit at the right will be used for circuit analysis purposes in the following description since it is drawn in a manner to be consistent with the Non-inverting amplifier.

With the inverting input being connected to the output (albeit indirectly), any difference between the inputs (Vin and Vfb) will be amplified by the open-loop voltage gain (Avol) and used to change the output. As the output changes the feedback signal provided back to the inverting input also changes. In the inverting amplifier the feedback signal is a series current from the output back to the input voltage passing through Rf and Rin. As the current passes through Rin a voltage is established across Rin. The voltage across Rin, in conjunction with the input voltage Vin will algebraically sum to equal the voltage at the non-inverting input terminal, at least within millivolts. With the non-inverting input grounded, Vin and V_{Rin} will sum to zero or the op amp will again amplify the difference, vary Vout, which varies the current through Rin, which will again attempt to force the two inputs to be approximately equal. With negative feedback the process becomes inherently self-corrective, the inverting input will be driven to equal the non-inverting input.

122

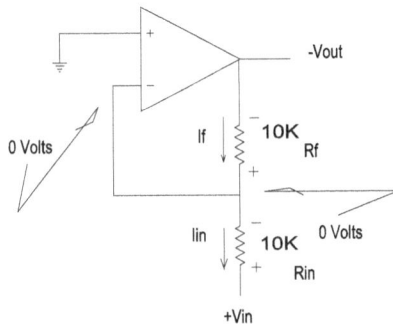

Figure 22 – An Inverting amplifier places the input signal at the inverting input and grounds the non-inverting input. The voltage on the non-inverting input is zero volts, the voltage on the inverting input is zero volts (since, Vid = 0V), and the voltage at the junction of Rf and Rin is zero volts. Iin equals Vin divided by Rin, If equals Iin, the voltage across Rf is If multiplied by Rf, and Vout is the sum of all voltages from the output to ground. This circuit exhibits the virtual ground effect, caused by grounding the non-inverting input, where the output achieves a value that results in both inputs being held close to ground potential. Inverting circuits of this type (non-inverting input grounded) will always result in Vin and Rin voltages canceling to equal zero volts at the junction of Rf and Rin.

Finding the voltages and currents around the inverting amplifier is as rapid as with the non-inverting amplifier. To begin, note the potential placed at the non-inverting input. Again, in this case zero volts. This will also be the voltage at the inverting input, zero volts. Knowing the input voltage at Vin, the voltage across Rin can be determined. Once the voltage across Rin is known, the current through Rin can be established. The same current passes through the feedback resistor, Rf, between the output and the input (hence the term series-current feedback). Finding Iin allows determination of If, the current through the input resistance equals the current through the feedback resistance. Knowing the feedback resistor current through Rf allows computation of the voltage across Rf. Writing a Kirchoff voltage loop, the output voltage is the algebraic sum of all the voltages from the output to any ground connection.

The following equations are utilized in the previous analysis.

$$V^{\text{inverting input}} = V^{\text{non-inverting input}}$$

$$V_{Rin} = V^{\text{inverting input}} - Vin$$

$$Iin = V_{Rin}/R_{in}$$

$$If = Iin$$

$$VRf = If * Rf = Iin * Rf$$
$$Vout = VRf + Vid, (Vid = 0)$$

$$\text{or, } Vout = VRf + VRin + Vin$$

With the non-inverting input grounded, the output voltage will always equal the voltage across Rf, as can be seen in the previous equations. This is due to Vid being very small, practically zero volts.

Upon examination the inverting and non-inverting amplifiers are very nearly identical. Each contain input and feedback resistors (Rin, Rf), and each has feedback from the output to the inverting input through the feedback resistor, Rf. Unique to each amplifier is the location of the input voltage signal. Since the operation of each circuit configuration is the same, there exists only a single method for analyzing all negative feedback op amp applications. This principle cannot be over emphasized. Though the input and feedback components may vary, and the input signal may be applied at the inverting or non-inverting locations, all negative feedback op amp circuits work according to the manner described previously.

Simplified Gain Equations

In analyzing schematics with a single op amp circuit, performing the previously mentioned technique to find the resistive voltage drops of VRf and VRin is fairly rapid and accurate. However most schematic diagrams of commercial or industrial systems utilize numerous op amp circuits where component level voltage and current analysis is not practical. With exposure and experience the efficient technologist will apply an abbreviated approach to analyzing and designing op amp signal conditioning circuits.

As mentioned previously the non-inverting amplifier exhibits a voltage gain that closely approximates the inverse of the feedback factor created with Rin and Rf. Expressed differently, the non-inverting amplifier gain equals the inverse of the feedback factor, or

$$Av = 1/Beta$$
$$Since, Beta = Rin / (Rin + Rf)$$
$$Av = (Rin + Rf) / Rin$$
$$Algebraically\ rearranging,$$
$$Av = 1 + (Rf / Rin)$$

This simple equation for Av provides a rapid, single step analysis to determining a non-inverting amplifier's gain. As technicians and designers of op amp signal conditioning circuits, a single linear equation independent of characteristic curves and conditions provides an efficient means of applying and analyzing operational amplifiers.

A similar yet slightly different equation can be derived for the inverting amplifier. Due to the grounded non-inverting input on the inverting op amp circuit configuration, the voltage measured from the inverting input to ground equals zero volts since Vid is driven to equal zero volts with negative feedback. Likewise the output voltage from the inverting amplifier is equal to the voltage appearing across Rf, since one side of Rf is connected directly to the output pin and the other side of Rf is connected to *the virtually grounded* voltage at the inverting terminal. Therefore, Vout = VRf with an inverting amplifier. Deriving an Av equation for the inverting amplifier becomes,

$$Vout = VRf$$

Including polarities,

$$Vout = -VRf$$
$$VRf = If * Rf$$

Or,

$$Vout = -If\ Rf$$
$$If = Iin$$

Likewise,

$$Vout = -IinRf$$

Since,

$$Iin = Vin\ /\ Rin$$

Substituting,

$$Vout = -(Vin/Rin)\ Rf$$

Solving for Av = Vout / Vin,

$$Vout/Vin = -Rf/Rin$$

And finally,

$$Av = -Rf\ /\ Rin$$

As a result, the inverting amplifier can also be reduced to a single equation representative of the amplifier's gain. Again, with exposure and experience the inverting amplifier will be observed, as a circuit performing the –Rf/Rin equation and the component level analysis of voltages and currents will become less frequently used. Associating the two Av equations with the appropriate circuits would be time well spent since subsequent chapters will assume the student is familiar with these equations.

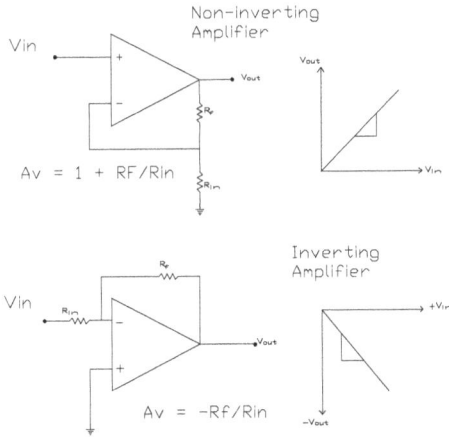

Figure 23 – Inverting and Non-inverting amplifiers with voltage gain equations and corresponding characteristic curves.

The negative sign associated with the Av equation indicates the inverting, reverse acting nature of the circuit. As the input signal moves in one direction, the output moves in an opposite direction. Or expressed another way, should the input be positive going, the output will be negative going. In DC applications with no additional circuitry applied to a purely inverting amplifier (without an applied offset voltage), a positive DC input will create a negative DC output voltage. However this is a single case of the inverting amplifier. As will be demonstrated in the following chapters, an inverting amplifier is capable of generating a positive output with a positive input. Again, the important concept to retain with the inverting amplifier is the directions of the input and output, and not the polarity of the individual input and output voltages.

Negative Feedback and Inherent Self-Correction

An earlier description referenced inherent self-correction with negative feedback. The concept is employed in numerous applications other than operational amplifiers and becomes beneficial when two variables are to be equated. The following provides an alternative description using an operational amplifier model.

Inherent self-correction describes the process resulting from two signals being "summed" out of phase. Described another way, when two signals of the same polarity are subtracted the resulting differential (Vid) or "error" signal becomes less than either of the two original signals. The resulting differential signal is then amplified and appears in the operational amplifier output. The output voltage or a portion thereof is then applied back to the inverting input as feedback, for continued comparison against the signal on the non-inverting input. Should the feedback voltage be incorrect for the given input signal, the difference will again be compared (subtracted) and amplified to form a different Vout value. The cycle continues until the correct amount of feedback voltage results, and the corresponding amplified difference between the input signal and feedback voltage becomes the output voltage, Vout.

The correct amount of feedback voltage is contingent upon the input signal at the non-inverting pin, and the internal, open-loop voltage gain (Avol) of the amplifier. In the case of a general-purpose op amp, the internal voltage gain is typically 200,000. With this value of Avol, the difference between the input and feedback voltages will be within a hundred microvolts. As an example, refer to the following diagram.

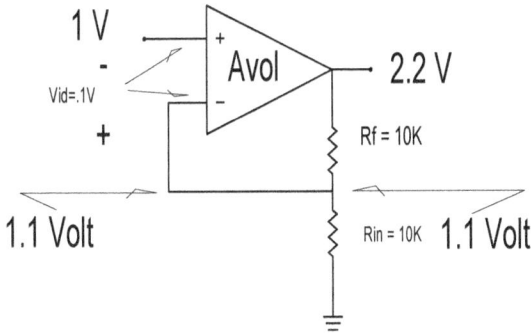

Figure 24 – An example of inherent self-correction with negative feedback.

In the diagram, an input signal of 1-Volt DC is applied to a non-inverting amplifier with a non-inverting gain of +2. This should result in an output voltage of 2 Volts but momentarily assume the voltage at the output moves to 2.2 volts. Given the feedback and input resistors are equal, one-half the output voltage will appear at the inverting terminal as a feedback voltage. The resulting 1.1 volts at the inverting input will be compared with the 1.0 volts from the input source, and a difference or error of .1 volts will be available for amplification by the amplifiers' internal open-loop gain, Avol. Note the polarity of the voltages available at each input in the diagram. The amplifier will attempt to amplify the .1 volt Vid "error" voltage by 200,000 times, *in the direction of the polarity of the input differential* applied to the inverting and non-inverting inputs. The amplifier will drive the output in a negative (reducing from 2.2 volts) direction due to the input differential polarity, in a attempt to "correct," or reduce the input differential voltage. Given the high internal amplifier gain and the relatively large amount of error, or input differential voltage (Vid), the amplifier output will reduce from 2.2 volts (moving negative) in a large way. As Vout reduces, the feedback voltage will also reduce moving negative from 1.1 volts.

Momentarily assume the amplifier output is at 1.8 volts. Again, the amplifier output should reside at 2.0 volts but to demonstrate the inherent self-correction provided by negative feedback, assume the amplifier output moves negative from 2.2 to 1.8 volts. The feedback voltage resulting from an output of 1.8 volts and equal feedback and input resistances would be one-half of 1.8 volts, or .9 volts. The following diagram represents the current condition.

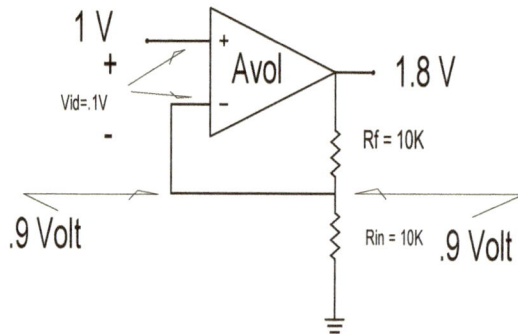

1 V

+

Vid=.1V

Avol

−

1.8 V

Rf = 10K

.9 Volt

Rin = 10K   .9 Volt

Figure 25 – An op amp uses the difference of voltage between the inputs and the polarity of the voltage difference to generate output and feedback voltages.

In this condition the amplifiers' input differential voltage is again .1 volts, but of an opposite polarity to the previous condition. Again, given the internal op amp voltage gain, amount of error and especially the polarity of the error, the output will be driven in a positive direction from where it currently resides. As the output rises from 1.8 volts, the feedback voltage also rises. As the feedback rises, the differential or error voltage is reduced. As should be obvious from the previous examples, if the output varies too much the feedback will over-correct and again become too great. Should the feedback become too large, the differential input polarity will reverse causing the output too move in the opposite direction.

The outputs' directional reversal, dictated by the input differential error polarity and the relationship of the feedback voltages' inverting input with the output becomes the essence of negative feedback. Inherent self-correction is the manner with which the inputs are compared and the output is driven to cause one of the inputs to effectively match the other. The final circuit condition is approximated in the following diagram.

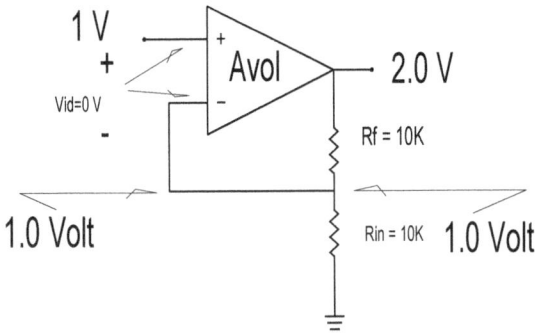

Figure 26 – The product of inherent self-correction resulting from negative feedback and high gain, the input and feedback voltages are approximately equal, and the output equals the closed-loop gain, Av, multiplied by the input signal.

But the diagram and the resulting voltages are somewhat misleading. For if the inputs were truly equal, the difference between the feedback and input voltages would be zero, resulting in an output of zero volts. In reality the inputs are almost equal and the differential is almost zero, but not quite. Since the output voltage equals the product of Vid and Avol, rearranging and solving for Vid yields the correct amount of input differential for the given amplifier gain and output voltage conditions

Since,   Vout = Vid * Avol

                          Vid = Vout / Avol

                          Vid ≈ 2V / 200,000
                          Vid ≈ .00001V
Or,   Vid ≈ 10 microvolts

131

From the previous computation, if Vin is exactly 1.000000 volts, the feedback voltage is 10 microvolts away at .99999 volts, and Vout is actually 1.99998 volts. However, internal op amp errors will cover and distort the precision of the high open-loop voltage gain causing Vout to deviate somewhat from the six figure quantities indicated. But in practice the amount of Vid measured in the previous circuit example should be well within the millivolt range. In addition, since the inverting terminal is connected to the output to provide the feedback path, the voltage on the inverting terminal will be "driven" to equal the voltage on the non-inverting terminal.

If the measured input differential, Vid, is ever over a few millivolts and the input voltage(s) and amplifier closed-loop gain are compatible (the computed Vout is within the amplifiers' rails), and the circuit is connected correctly, an excessive Vid is a key troubleshooting tool to locating a bad op amp. When an op amp goes bad an internal stage will short (or occasionally open) and no longer amplify. The loss of a single stage internally may not necessarily cause the op amp to cease operations. It will cause a reduction in the internal open-loop voltage gain (Avol), which will reduce Vout, which reduces the feedback voltage, which increases the amount of measurable Vid between the inputs. Thus, measuring Vid becomes a rapid means to determining the "health" of an operational amplifier.

Chapter 3 Homework

#1
Assume a non-inverting amplifier contains an Rin of 10K and an Rf of 91K, with an applied voltage of .4 Volts DC. Find Vout, VRf, and VRin

#2
Assume the component values and input voltage is configured in an inverting amplifier. Find Vout, VRf and VRin.

#3
From the LM741 spec. sheet following this homework problem set, locate and determine the following items under "Absolute Maximum Ratings." A search of available specifications should always precede an application experiment with a new integrated component.
  a. supply voltage range,
  b. supply current,
  c. output short circuit duration,
  d. open-loop voltage gain (aka, large signal voltage gain),
  e. input resistance,
  f. output resistance,
  g. slew rate,
  h. bandwidth,
  i. pin connections for an 8-pin package,
  j. input bias current
  k. input offset current

  Author's note – the 741 spec sheets following this problem set were converted on-line from an Adobe pdf document. As a result, some of the units associated with the requested items in this problem aren't quite accurate. As an example, the units associated with input resistance should be MΩ instead of M&.

#4
A non-inverting amplifier contains an Rin of 100K and an Rf of 1K. If Vin is applied at 1 Volt, determine the input and feedback resistor voltages, and Vout. What would the Av equal for this amplifier?

#5
An inverting amplifier with an Rin of 10K and an Rf of 100K is connected to 15 Volt supplies. Assuming the maximum output (saturation) voltage will equal the supply voltages, what is the maximum DC input voltage that can be applied without saturating the output? What is the maximum RMS input voltage that can be applied without saturating the output voltage?

#6
Assume the input of problem #2 is connected to the output of problem #1. If a .1 Volt DC signal is applied to the input of the first amplifier, what is the output voltage? What is the total gain of both amplifiers in cascade?

#7
Assume .05 Vpp, 1 Khz sine wave is applied to the circuit of problem #4. What is the amplitude, frequency and shape of the output signal?

N *National* Semiconductor

August 2000

LM741

Operational Amplifier

General Description

The LM741 series are general purpose operational amplifiers which feature improved performance over industry standards like the LM709. They are direct, plug-in replacements for the 709C, LM201, MC1439 and 748 in most applications.

The amplifiers offer many features which make their application nearly foolproof: overload protection on the input and output, no latch-up when the common mode range is exceeded, as well as freedom from oscillations.

The LM741C is identical to the LM741/LM741A except that the LM741C has their performance guaranteed over a 0°C to +70°C temperature range, instead of −55°C to +125°C.

Connection Diagrams

Metal Can Package

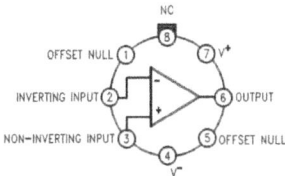

DS009341-2

Note 1: LM741H is available per JM38510/10101

Order Number LM741H, LM741H/883 (Note 1),
LM741AH/883 or LM741CH
See NS Package Number H08C

Dual-In-Line or S.O. Package

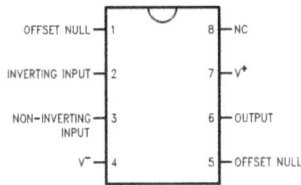

DS009341-3

Order Number LM741J, LM741J/883, LM741CN
See NS Package Number J08A, M08A or N08E

Ceramic Flatpak

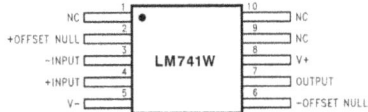

DS009341-6

Order Number LM741W/883
See NS Package Number W10A

Typical Application

Offset Nulling Circuit

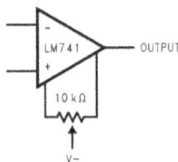

DS009341-7

135

Absolute Maximum Ratings (Note 2)

If Military/Aerospace specified devices are required, please contact the National Semiconductor Sales Office/ Distributors for availability and specifications.

(Note 7)

	LM741A	LM741	LM741C
Supply Voltage	±22V	±22V	±18V
Power Dissipation (Note 3)	500 mW	500 mW	500 mW
Differential Input Voltage	±30V	±30V	±30V
Input Voltage (Note 4)	±15V	±15V	

±15V Output Short Circuit Duration

Continuous   Continuous

Continuous Operating Temperature Range   −55°C

to +125°C	−55°C to +125°C	0°C to +70°C Storage Temperature
Range	−65°C to +150°C	−65°C to +150°C −65°C
to +150°C Junction Temperature	150°C	150°C

100°C Soldering Information

N-Package (10 seconds)	260°C	260°C	260°C
J- or H-Package (10 seconds)	300°C	300°C	300°C M
Vapor Phase (60 seconds)	215°C	215°C	215°C
Infrared (15 seconds)	215°C	215°C	215°C Se
ESD Tolerance (Note 8)	400V	400V	400V

Electrical Characteristics (Note 5)

Parameter	Conditions	LM741A			LM741			LM741C			Units
		Min	Typ	Max	Min	Typ	Max	Min	Typ	Max	
Input Offset Voltage	$T_A =$ 25°C R_S $\leq$ 10 kΩ					1.0	5.0		2.0	6.0	mV
			0.8	3.0							mV
	$T_{AMIN} \leq T_A \leq$ T_{AMAX} $R_S \leq 50\Omega$			4.0							mV
	$R_S \leq 10$ kΩ						6.0			7.5	mV
Average Input Offset Voltage Drift				15							µV/°C
Input Offset Voltage Adjustment Range	$T_A = 25°C$, $V_S = ±20V$	±10				±15			±15		mV
Input Offset Current	$T_A = 25°C$		3.0	30		20	200		20	200	nA
	$T_{AMIN} \leq T_A \leq T_{AMAX}$			70		85	500			300	nA
Average Input Offset Current Drift				0.5							nA/°C
Input Bias Current	$T_A = 25°C$		30	8		80	500		80	500	nA
	$T_{AMIN} \leq T_A \leq T_{AMAX}$			0.210			1.5			0.8	µA
Input Resistance	$T_A = 25°C$, $V_S = ±20V$	1.0	6.0		0.3	2.0		0.3	2.0		MΩ
	$T_{AMIN} \leq T_A \leq$ T_{AMAX}, $V_S = ±20V$	0.5									MΩ
Input Voltage Range	$T_A = 25°C$							±12	±13		V
	$T_{AMIN} \leq T_A \leq T_{AMAX}$				±12	±13					V

Electrical Characteristics (Note 5) (Continued)

Parameter	Conditions	LM741A			LM741			LM741C			Units
		Min	Typ	Max	Min	Typ	Max	Min	Typ	Max	
Large Signal Voltage Gain	$T_A = 25°C$, $R_L \geq 2$ kΩ										
	$V_S = \pm20V$, $V_O = \pm15V$	50									V/mV
	$V_S = \pm15V$, $V_O = \pm10V$				50	200		20	200		V/mV
	$T_{AMIN} \leq T_A \leq T_{AMAX}$,										
	$R_L \geq 2$ kΩ,										
	$V_S = \pm20V$, $V_O = \pm15V$	32									V/mV
	$V_S = \pm15V$, $V_O = \pm10V$				25			15			V/mV
	$V_S = \pm5V$, $V_O = \pm2V$	10									V/mV
Output Voltage Swing	$V_S = \pm20V$										
	$R_L \geq 10$ kΩ	±16									V
	$R_L \geq 2$ kΩ	±15									V
	$V_S = \pm15V$										
	$R_L \geq 10$ kΩ				±12	±14		±12	±14		V
	$R_L \geq 2$ kΩ				±10	±13		±10	±13		V
Output Short Circuit Current	$T_A = 25°C$	10	25	35		25			25		mA
	$T_{AMIN} \leq T_A \leq T_{AMAX}$	10		40							mA
Common-Mode Rejection Ratio	$T_{AMIN} \leq T_A \leq T_{AMAX}$										
	$R_S \leq 10$ kΩ, $V_{CM} = \pm12V$				70	90		70	90		dB
	$R_S \leq 50\Omega$, $V_{CM} = \pm12V$	80	95								dB
Supply Voltage Rejection Ratio	$T_{AMIN} \leq T_A \leq T_{AMAX}$,										
	$V_S = \pm20V$ to $V_S = \pm5V$										
	$R_S \leq 50\Omega$	86	96								dB
	$R_S \leq 10$ kΩ				77	96		77	96		dB
Transient Response	$T_A = 25°C$, Unity Gain										
Rise Time			0.25	0.8		0.3			0.3		μs
Overshoot			6.0	20		5			5		%
Bandwidth (Note 6)	$T_A = 25°C$	0.437	1.5								MHz
Slew Rate	$T_A = 25°C$, Unity Gain	0.3	0.7			0.5			0.5		V/μs
Supply Current	$T_A = 25°C$					1.7	2.8		1.7	2.8	mA
Power Consumption	$T_A = 25°C$										
	$V_S = \pm20V$		80	150							mW
	$V_S = \pm15V$					50	85		50	85	mW
LM741A	$V_S = \pm20V$										
	$T_A = T_{AMIN}$			165							mW
	$T_A = T_{AMAX}$			135							mW
LM741	$V_S = \pm15V$										
	$T_A = T_{AMIN}$					60	100				mW
	$T_A = T_{AMAX}$					45	75				mW

Note 2: "Absolute Maximum Ratings" indicate limits beyond which damage to the device may occur. Operating Ratings indicate conditions for which the device is functional, but do not guarantee specific performance limits.

Electrical Characteristics (Note 5) (Continued)

Note 3: For operation at elevated temperatures, these devices must be derated based on thermal resistance, and T_j max. (listed under "Absolute Maximum Ratings"). $T_j = T_A + (\theta_{jA} P_D)$.

Thermal Resistance	Cerdip (J)	DIP (N)	HO8 (H)	SO-8 (M)
θ_{jA} (Junction to Ambient)	100°C/W	100°C/W	170°C/W	195°C/W
θ_{jC} (Junction to Case)	N/A	N/A	25°C/W	N/A

Note 4: For supply voltages less than ±15V, the absolute maximum input voltage is equal to the supply voltage.

Note 5: Unless otherwise specified, these specifications apply for $V_S = \pm15V$, $-55°C \leq T_A \leq +125°C$ (LM741/LM741A). For the LM741C/LM741E, these specifications are limited to $0°C \leq T_A \leq +70°C$.

Note 6: Calculated value from: BW (MHz) = 0.35/Rise Time(μs).

Note 7: For military specifications see RETS741X for LM741 and RETS741AX for LM741A.

Note 8: Human body model, 1.5 kΩ in series with 100 pF.

Schematic Diagram

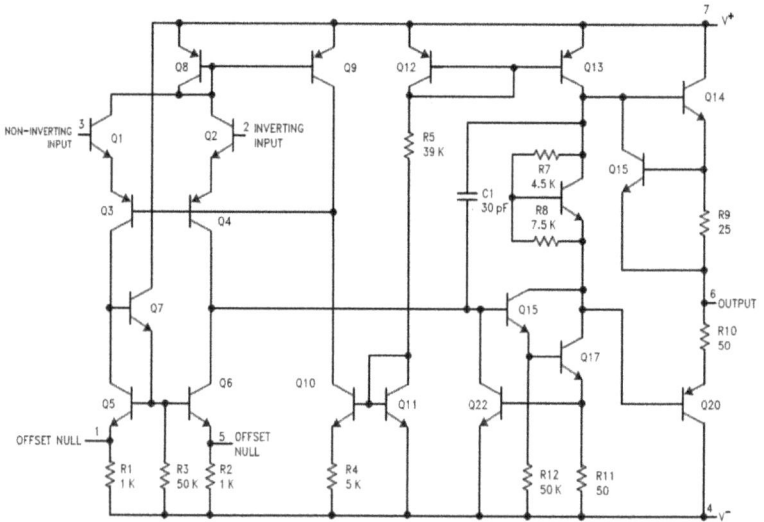

DS009341-1

Physical Dimensions inches (millimeters) unless otherwise noted

Metal Can Package (H)
Order Number LM741H, LM741H/883, LM741AH/883, LM741AH-MIL or LM741CH
NS Package Number H08C

Ceramic Dual-In-Line Package (J)
Order Number LM741J/883
NS Package Number J08A

139

Physical Dimensions inches (millimeters) unless otherwise noted (Continued)

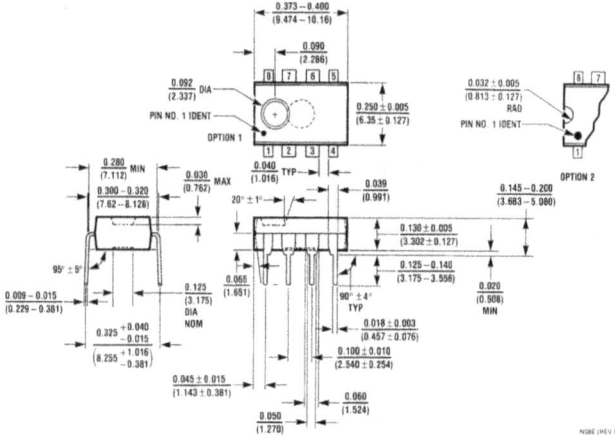

Dual-In-Line Package (N) Order Number
LM741CN NS Package Number N08E

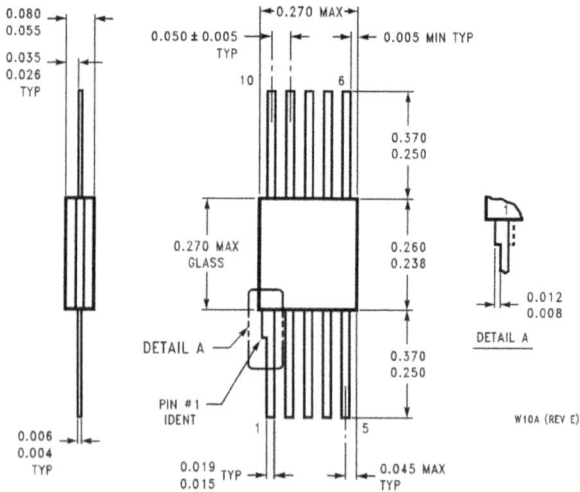

10-Lead Ceramic Flatpak (W)
Order Number LM741W/883, LM741WG-MPR or LM741WG/883
NS Package Number W10A

Chapter 4 – Output-Offset and Biasing of Operational Amplifiers

All configurations of negative feedback op amp circuits can be understood if the material in this chapter is well digested. Among the covered topics: linear Y=mX+b concepts are reviewed and applied to operational amplifier circuits, concepts of biasing and biased amplifiers are introduced, circuit analysis is continued and using previously introduced analytical concepts, circuit design is introduced. Digital panel meters are introduced. Impedance matching from sensor to amplifier and from amplifier to load is reviewed. Variable gain and bias circuit configurations are designed and component value determination is demonstrated. Calibration is reviewed.

Objectives:
Upon completion of this chapter, you should be able to:
- Explain amplifier offset biasing
- Explain the purpose and operation of general scaling amplifiers
- Provide sensor-based application examples of scaling amplifiers
- Interpret scaling amplifier circuit diagrams
- Design and assemble a scaling amplifier
- Explain the purpose and process of calibrating a scaling amplifiers
- Explain the operation of inverting and non-inverting op amp circuits

Scaling Amplifiers

It should be understood initially that the term "Amplifier" is a general one that represents virtually every circuit containing an "active" component. Active components are capable of developing gain and require an external or supplemental power supply. Amplification, or "gain," is in the form of a voltage or current increase from the amplifier's input to the circuit's output. Hence gain (symbolized with an upper case "A") is computationally equal to an amplifier's output divided by its input. As an example, consider a circuit whose output voltage is twice its input voltage, the circuits gain would be two since output/input = 2.

A similar example could be made of a circuit whose output current is increased over its input's. Usually a current amplifier is referred to as a power amplifier since the voltage gain is held close to unity (by the circuit's design) and the resulting Pout/Pin, which equals (Vout*Iout) ÷ (Vin*Iin), is greater than one. More about power gain when impedance matching and unity gain amplifiers are discussed.

Certainly a circuit whose output power is greater than its input should and would be considered a true amplifier. There exists however an entire class of circuits whose gain may commonly be less than unity, and require a circuit to perform a mathematical computation, the operational amplifier's primary function. These types of circuits are referred to as signal conditioners and perform filtering, linearization, isolation, calibration and/or scaling computations upon a sensors' output signal (voltage) range. Various signal-conditioning functions will be addressed in locations throughout the text. This chapter will introduce a form of sensor signal conditioning from an Y=MX+B, computational standpoint.

The Y=MX+B model allows an investigation of proportional input/output relations. For our current purpose, Y=MX+B is representative of an amplifier performing a *scaling* function. Scaling involves amplification (input signal voltage amplitude change) and offsetting (from an initial output value of zero) a signal by an amplifier. A sensor measuring some physical variable such as pressure, temperature, flow, stress/strain and others commonly provides such input ranges. Scaling is required to fit the sensors' output signal range to the input requirements of a display, control or other signal-processing device. The scaling amplifier receives the input from the sensor and generates a proper output for the signal processor.

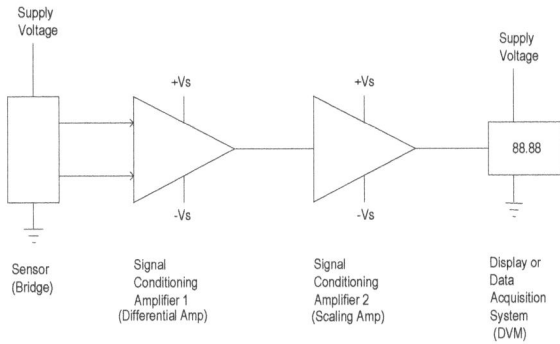

Figure 1 – Most signal conditioning is performed in multiple cascaded stages. Typically, the first amplifier stage amplifies (scales) and converts the signal from differential to single-ended. The second signal conditioning stage might perform filtering, linearization or analog to digital conversion.

Y=MX+B is used to describe many linear (proportional) applications. Amplifier scaling applications require substituting the appropriate input/output voltage quantities into the Y=MX+B equation,

Generally, for all input/output functions,

$$Output = (Gain*Input) + Offset$$

More specifically for scaling amplifiers,

$$Vout = (Gain*Vin) + Offset$$

Where,

The 'Y' quantity is Vout,

The slope of the plot, 'M', is Gain,

The 'X' quantity is Vin,

The 'B' of the equation is the Output Offset Voltage, or V_B,

The Slope (M) is the dynamic characteristic or "rise over run" of the resulting input/output plot. The Offset (B) is the point where the plot passes (or would pass in cases where the input range is elevated above zero volts) through the 'Y' axis.

The output offset, B or V_B is mathematically referred to as the Y-intercept value of the plot. The output offset represents the 'Y' (output) value at an 'X' (input) value of zero. The slope (M) is the amplifiers gain (A_V) and the Y-intercept value (B) is the required amount of output offset.

The following figure helps to visualize these quantities. For reference purposes, the appropriate scaling amplifier is also indicated.

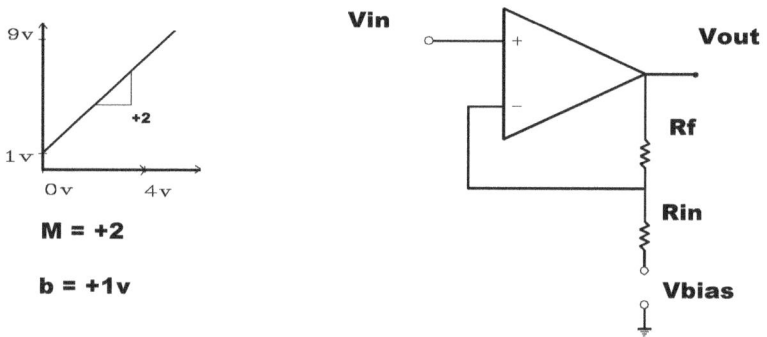

Figure 2 - Biased Non-Inverting Amplifier

The ratio of the resistors used to assemble the circuit will determine the amplifier's gain. The amount of DC voltage introduced to the amplifier's input (in conjunction with the signal) will determine the amplifier Output Offset, "B" or V_B.

From the previous figure;

A) The amplifier-input range (conceivably from a sensor) is always plotted on the x-axis of the graph,

B) The amplifier output range is always plotted on the y-axis,

C) The slope of the corresponding input/output voltage ranges represents the circuits' operational gain. The slope (or voltage gain, Av) is computed as;
Or,

$$A_v = \Delta V_{out} / \Delta V_{in}$$

D) The intersection of the input/output characteristic plot and the y-axis represents the amplifiers' output bias value, B or V_B. The offset, or "B," value is computed as "Y = MX + B" rearranged algebraically to solve for "B." Therefore,

$$V_B = V_{out} - (A_v * V_{in})$$

or, B = Y – MX

or, V_B = Vout - (Av * Vin)

E) In the general form, Y= MX + B,

In circuit form,

$$V_{out} = A_v V_{in} + V_B$$

Again, where V_B is the "output-bias" quantity computed from the input/output plot indicating the amount and direction of the output voltage ranges' displacement along the y-axis.

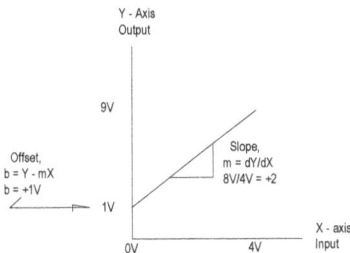

Y - Axis
Output

9V

Offset,
b = Y - mX
b = +1V

Slope,
m = dY/dX
8V/4V = +2

1V

0V

4V

X - axis
Input

Figure 3 – Scaling amplifier Input / Output characteristic. Note the output range along the Y-axis is elevated from zero volts.

To achieve the characteristic plot shift of V_B, a DC bias voltage (Vbias) will be applied to the amplifier to offset the output range's vertical displacement up (or down, when required) the Y-axis. The quantities of V_B and Vbias appear similar but are always numerically different however and should not be confused. As will be seen, Vbias is always applied to an amplifier input in the form of a DC voltage, and V_B is the amount of amplifier output at an input of zero volts, or the graphical Y-intercept value.

Bias voltages are used frequently in electronic amplifiers. In any application, a "bias" is a constant influence placed upon a device to establish a desired initial or "quiescent" (without an input signal applied) output condition. With electronic amplifiers specifically, a bias is used to determine the initial output voltage before a signal range is applied, or at an input of zero volts. The mathematical definition of the Y-intercept value (Y value at an X value of zero) is consistent with the scaling amplifier application, since B is also the output voltage value at an input of zero volts.

Again, note the difference between V_B and V_{bias};

V_B = Y-axis intercept quantity computed from the input/output plot, the output-bias quantity.

V_{bias} = DC Voltage applied to the circuit to create the above Y-axis intercept value shift.

Output Offset and Scaling Example

Assume an input voltage range of 0 to 4 Volts is to be scaled into a corresponding output voltage range of 1 to 9 Volts. From the previous equations, the resulting circuit gain is,

$$A_v = \Delta V_{out}/\Delta V_{in}$$

$$A_v = 8V/4V$$

Therefore, $$A_v = +2$$

Where, the positive polarity represents a non-inverting or direct-acting amplifier configuration. Direct-action means the amplifiers output will increase as the input increases, in a direct manner. An inverting or reverse-acting characteristic would exhibit a _negative_ gain polarity and input/output slope, (V_{in} increases, V_{out} decreases).

The output bias or y-axis intercept point is;

$$B = Y - MX$$
$$Or, V_B = V_{out} - A_v V_{in}$$

Substituting corresponding V_{out} and V_{in} values from the given input and output ranges,

$$V_B = +1 \text{ Volt}$$

Again, the output-offset voltage V_B merely represents the distance and direction of the output voltage range from 0 Volts along the y-axis. The required V_{bias} to create the output-bias V_B remains to be determined and is contingent upon the specific circuit type and the component values used to design the amplifier.

Output Equation Determination - The Biased Non-Inverting Amplifier
The previous chapter introduced the two most common op amp amplifier circuits; the inverting and non-inverting amplifiers. The vast majority of all op amp applications will utilize one of these two basic circuit configurations. If circuit requirements are such that the output voltage range is not expected to be offset

from zero volts, the inverting and non-inverting amplifier circuits will work fine as drawn in the following figure (shown with corresponding output/input plots).

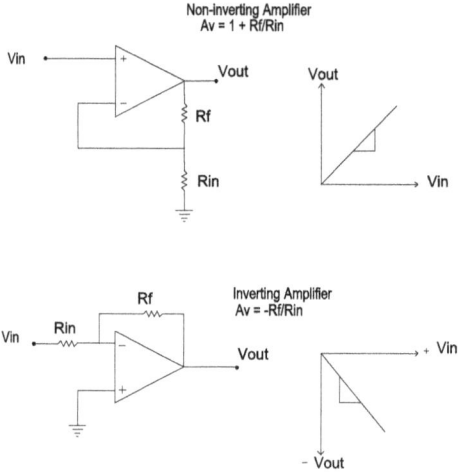

Figure 4 - Inverting and Non-inverting amplifiers without bias, and corresponding true-zero input / output ranges.

If however the output voltage range is to be offset, or vertically shifted in relation to the input, the non-inverting amplifier circuit will require a bias voltage, V_{bias}, at one of its inputs. The most commonly accepted location for the bias voltage is at the other input as shown.

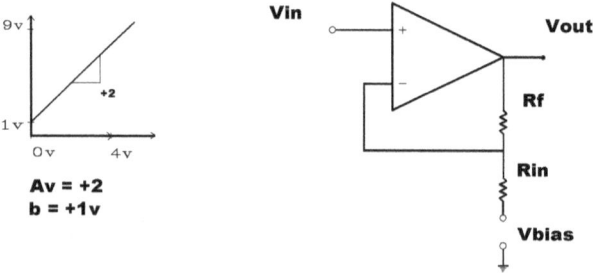

Figure 5 -Non-inverting amplifier with bias applied.

148

Referring to the figure, Vout is 1V when V_{in} is 0V, or more appropriately the entire eight volt output range (ΔV_{out}) is biased or shifted positively by 1V in relation to the four volt input span (ΔV_{in}) on the x-axis. In order for the output to establish 1V when V_{in} is zero, a separate, constant DC "bias" voltage must be applied to the amplifiers' input circuit. The DC bias input is commonly provided by a voltage divider connected to a voltage supply and is referred to as the biasing or offsetting voltage, V_{bias}. The bias voltage is differentiated from the input signal since op amp circuits are designed to operate over a range of input values (input range is 0V to 4V) and the bias input is fixed, or not capable of changing.

The bias input voltage appears to complicate the circuits operation. But if one is familiar with the operation of the inverting and non-inverting amplifier, circuit analysis and subsequent circuit design is not altogether new, but becomes an extension of existing circuit operational theories and equations. As an example, refer to the biased non-inverting amplifier of the following figure.

Biased Non-Inverting Amplifier

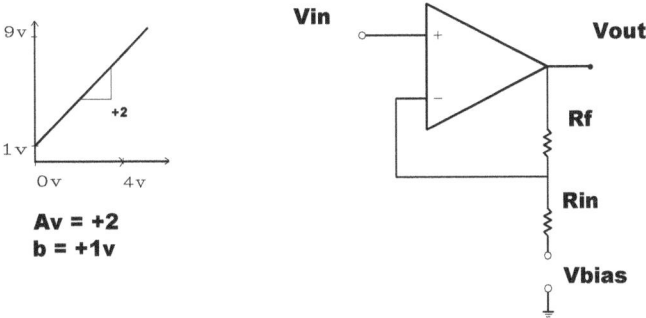

Figure 6 – The Biased Non-Inverting Amplifier. The input signal is applied to the non-inverting input terminal and the bias is applied to the inverting inputs' resistance. The output will be the sum of the individual gain* input products.

Note the input signal range (V_{in}) "sees" a non-inverting amplifier yet the bias voltage (V_{bias}) "sees" an inverting amplifier. The circuit is redrawn in the following figure to better demonstrate this concept. Similarly, the circuits V_{out} equation reflects this configuration, V_{out} equals V_{in} multiplied by the gain "seen" by V_{in} (A_{Vin}) summed with the product of V_{bias} and its appropriate gain (A_{Vbias}). If V_{in} and V_{bias} are individually assumed to be zero and equivalent circuits are envisioned the combined Vout equation becomes apparent.

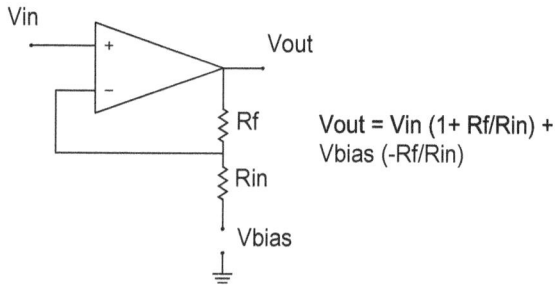

$$V_{out} = V_{in}(1+ Rf/Rin) + V_{bias}(-Rf/Rin)$$

Figure 7 - Biased Non-inverting amplifier. When using both inputs on an op amp, the resulting Vout is the sum of the input voltages and their respective voltage gain products.

Using operational concepts developed in chapter three, the V_{out} equation of the previous figure can be derived:

KVL equation from Vout to ground yields;

$$V_{out} = V_{Rf} + V_{id} + V_{in}$$

Where,

$$V_{Rf} = I_f R_f = I_{in} R_f, \text{ Since } I_{in} = I_f$$

And,

$$I_{in} = (V_{in} - V_{bias})/R_{in}$$

Then,

$$V_{Rf} = [(V_{in} - V_{bias})/R_{in}] * R_f$$

Therefore,

$$V_{out} = [(V_{in}-V_{bias})/R_{in}]*R_f + V_{in}$$

Multiplying,

$$V_{out} = V_{in}(R_f/R_{in}) - V_{bias}(R_f/R_{in}) + V_{in}$$

Recombining,

$$V_{out} = V_{in} + V_{in}(R_f/R_{in}) - V_{bias}(R_f/R_{in})$$

Finally,

$$V_{out} = V_{in}(1+R_f/R_{in}) + V_{bias}(-R_f/R_{in})$$

Or,

$$V_{out} = V_{in}(A_v) + V_{bias}(A_{Vbias})$$

Where, A_v is the gain "seen" by the input range and A_{Vbias} is the gain "seen" by the bias voltage.

Using the above approach to determine the output range yields;

Finding V_{out} at V_{in} = 0V and V_{bias} = -1V,

$$V_{out} = (0V)(1+10K/10K) + (-1)(-10K/10K)$$

$$V_{out} = (0V)(2) + (-1V)(-1)$$

$$V_{out} = 1V \text{ at } Vin = 0V$$

Finding maximum V_{out} at V_{in} = 4V,

$$V_{out} = (4V)(2) + (-1V)(-1)$$

$$V_{out} = 9V \text{ at } Vin = 4V$$

Therefore, the resultant output range is 1 to 9 Volts for an applied input voltage range of 0 to 4 Volts, and a V_{bias} of -1 Volt.

Again,

The circuit's output will equal the algebraic sum
of each respective input's voltage * gain product.

Or,

$$V_{out} = V_{in}*(A_{Vin}) + V_{bias}*(A_{Vbias})$$

Where, each <u>term</u> of the equation represents an op amp input.

The previous Vout equation derivation may seem unimportant or irrelevant to the novice engineering student, yet it should be remembered that contained within all equation derivations is an explanation of circuit operation and corresponding design and trouble-shooting criteria.

Note also the intrinsic simplicity of this analytic approach. If inverting and non-inverting signal inputs can be recognized, an output equation and subsequent computation can be made. This allows rapid approximations of the output voltage without sophisticated calculator computation. The importance of this circuits' operation cannot be overemphasized <u>since all resistive negative feedback op amp circuits operate in precisely the same manner</u>. This becomes the real benefit of using "linear," and integrated components. Hence a little time invested in circuit operation here will pay benefits over the remaining chapters.

Circuit Analysis - Biased Non-Inverting Amplifier

In analyzing the biased non-inverting amplifier, a technique will be used which is consistent to that presented in the previous chapter. The following circuit shows V_{out} is composed of a Kirchoff Voltage Loop from the output to either ground. Any loop from the output to ground can be summed, the following loop is chosen only as a matter of convenience.

To begin the operational analysis, since V_{id} = 0V, V_{out} is created by the voltage across R_f which is summed algebraically with V_{in}, and/or the other input quantities of V_{Rin} + V_{bias}. The voltage across R_f is caused by I_f which is established by the difference between V_{in} and V_{bias} across R_{in}, since I_f and I_{in} are the same current.

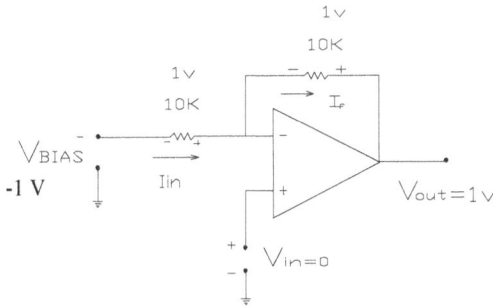

Figure 8 - Biased Non-inverting amplifier circuit analysis at an input of zero volts. Although the circuit assumes an inverting appearance, the input is applied to the non-inverting terminal.

Using a Kirchoff Loop (KVL) analysis to verify an op amps' output voltage range requires computing the quantities of I_{in}, I_f, V_{Rf} and writing the voltage loop equation(s). The author assumes electron flow in assigning polarities (negative in/positive out as the electron flows through components) across voltage drops, and acquires the first polarity observed when writing KVL statements from V_{out} around a loop. As if a voltmeter were placed at the output to measure the algebraic sum of all voltages to ground.

From the amplifier circuit diagram, verifying V_{out} minimum at $V_{in} = 0V$,

Since

$$V_{id} = 0 \text{ Volts, } I_{in} = (0-1V)/10K$$
$$I_{in} = .1 \text{ mA}$$

Since,

$$I_{in} = I_f,$$
$$I_f = .1 \text{ mA}$$

Since,

$$V_{Rf} = I_f * R_f$$
$$V_{Rf} = (.1 \text{ mA})(10K) = 1 \text{ Volt}$$

Writing the KVL equations from the output through both input paths,

$$\text{KVL 1,} \qquad V_{out} = V_{Rf} + V_{id} + V_{in}$$
$$V_{out} = 1V + 0V + 0V$$
$$V_{out} = 1 \text{ Volt at } V_{in} = 0 \text{ Volt}$$

$$\text{KVL 2,} \qquad V_{out} = V_{Rf} + V_{Rin} + V_{bias}$$
$$V_{out} = 1V + 1V + (-1V)$$
$$V_{out} = 1 \text{ Volt at Vin} = 0 \text{ Volt}$$

Verifying V_{out} maximum at $V_{in} = 4V$ using the same technique,

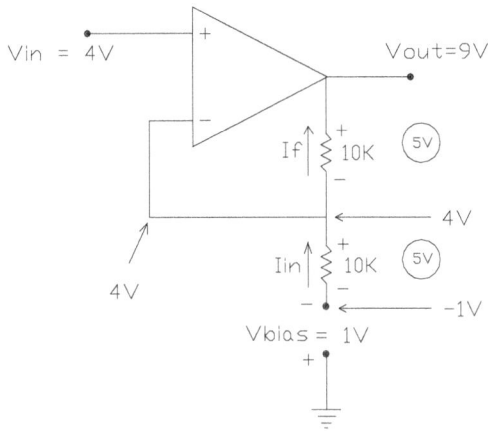

Figure 9 - Biased Non-inverting amplifier circuit analysis at an input of four volts. Again, the input is applied to the non-inverting terminal. Note: voltages in circles are measured differentially (across the component), voltages with arrows are measured single-ended (in refernce to ground).

Since,

$$V_{id} = 0V, I_{in} = (4-\text{-}1V)/10K$$

$$I_{in} = .5 \text{ mA}$$

Since,

$$I_{in} = I_f,$$

$$I_f = .5 \text{ mA}$$

$$V_{Rf} = I_f * R_f$$

$$V_{Rf} = (.5 \text{ mA})(10K) = 5V$$

Writing both KVL equations,

$$\text{KVL 1,} \qquad V_{out} = V_{Rf} + V_{id} + V_{in}$$

$$V_{out} = 5V + 0V + 4V$$

$$V_{out} = 9 \text{ Volts at Vin} = 4 \text{ Volts}$$

$$\text{KVL 2,} \qquad V_{out} = V_{Rf} + V_{Rin} + V_{bias}$$
$$V_{out} = 5V + 5V + (-1V)$$
$$V_{out} = 9 \text{ Volts at Vin} = 4 \text{ Volts}$$

The previous assumptions about the output range are verified.

In summary, the location of V_{bias} between R_{in} and ground can be seen as elevating (or suppressing) the operation of the entire circuit depending upon the V_{bias} and A_{Vbias} polarity. Previously a zero volt, ground potential reference had "biased" the op amp circuits. The resulting output had been a voltage range with a zero volt output offset, a true zero signal range. Note that the amount of output elevation or suppression is not dependent solely upon the amount of V_{bias}, but on the gain (A_{Vbias}) seen by V_{bias} as well.

Output Equation Determination - The Biased Inverting Amplifier

Non-inverting (direct acting) and inverting (reverse acting) circuit characteristics do not necessarily refer to the static input/output polarities but instead are references to the dynamic, periodic or varying input/output voltage range relation. A *biased* inverting amplifier is capable of exhibiting a positive output voltage even when a positive input is applied. As the input changes in a positive direction however, the output will be moving in a negative direction. Simply stated, a reverse acting or inverting circuit characteristic means; as the input moves in a positive direction, the output moves in a negative direction. A direct or non-inverting characteristic means; as V_{in} goes up, V_{out} goes up. With biasing, the input and output ranges may consist of all positive, or all negative, or a combination of values. In the future one should consider any references to direct and reverse circuit action as a relation between input and output voltage ranges and directions, rather than static input/output voltage point value polarities.

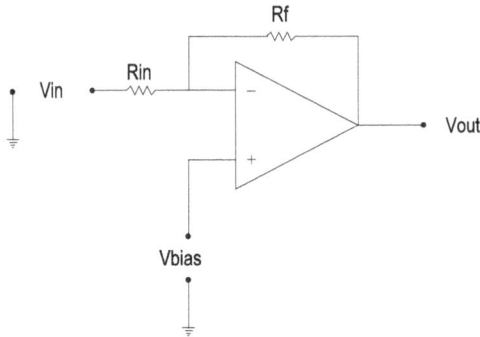

$$Vout = Vin\ (-Rf/Rin) + Vbias\ (1+Rf/Rin)$$

Figure 10 – A biased Inverting Amplifier. The input signal, Vin, "sees" an inverting amplifier, the bias voltage "sees" a non-inverting amp.

Biasing the inverting amplifier is a matter of applying the input signals' voltage range, ΔV_{in}, to the inverting inputs' resistance and providing the required V_{bias} to the non-inverting input terminal. Or switching V_{in} and V_{bias} in the biased non-inverting amplifier circuit results in the biased inverting amplifier.

Again, the corresponding V_{out} equation is the algebraic sum of the V_{in} and V_{bias} inputs multiplied by their respective gains, therefore;

$$V_{out} = V_{in}\ (A_v) + V_{bias}\ (A_{Vbias})$$

Or,

$$V_{out} = V_{in}\ (-R_f/R_{in}) + V_{bias}\ (1+R_f/R_{in})$$

Using the previously presented circuit analysis technique, the Vout equation can be verified through derivation. Deriving the V_{out} equation for the biased inverting amplifier follows;

KVL,

$$V_{out} = V_{Rf} + V_{id} + V_{bias}$$

Since,

$$V_{Rf} = I_f R_f,$$

Since,

$$V_{id} = 0V$$
$$V_{out} = I_f R_f + V_{bias},$$

Since,

$$I_f = I_{in},$$
$$V_{out} = I_{in} R_f + V_{bias}$$

Subbing for I_{in},

$$V_{out} = [(V_{bias}-V_{in})/R_{in}]^*R_f + V_{bias}$$

Multiplying,

$$V_{out} = V_{bias}(R_f/R_{in}) - V_{in}(R_f/R_{in}) + V_{bias}$$

Rearranging,

$$V_{out} = -V_{in}(R_f/R_{in}) + V_{bias} + V_{bias}(R_f/R_{in})$$

Or,

$$V_{out} = V_{in}(-R_f/R_{in}) + V_{bias}(1+R_f/R_{in})$$

As with the biased non-inverting amplifier, the biased inverting amplifiers' output voltage equation is simply <u>the algebraic sum of the individual applied input and bias voltages multiplied by their respective gains</u>.

As an example, consider the following diagram where Rf is 20K, Rin is 10K, Vin is a range of values between zero and four volts, and Vbias is 3 volts. Determining the output voltage range at the input voltage extremes produces the following results.

Biased Inverting Amplifier

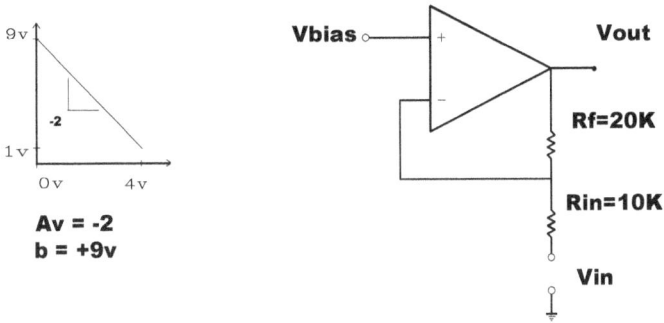

Figure 11 – The Biased Inverting Amplifier and input/output characteristic, Vin sees an inverting amplifier and Vbias sees a non-inverting amplifier. Biased inverting amps require a large amount of output bias voltage if the input and output ranges are to reside in the first quadrant. Note the resemblance in appearance to the non-inverting amplifier, the circuit analysis will be identical to the previous example.

Finding the maximum V_{out} at the minimum V_{in},

$$V_{out} = V_{in}(-R_f/R_{in}) + V_{bias}(1+R_f/R_{in})$$

$$V_{out} = 0V(-20K/10K) + 3V(1+20K/10K)$$

$$V_{out} = 0V(-2) + 3V(3)$$

$$V_{out} = +9V \text{ at } V_{in} = 0V$$

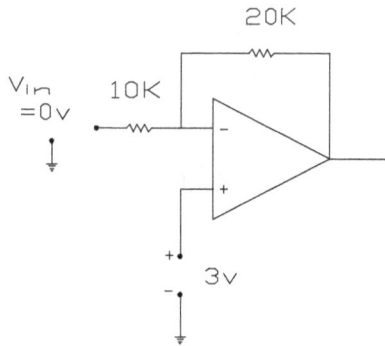

Figure 12 – A Biased Inverting Amplifier with an input of 0V, and a bias of 3V.

Finding the minimum V_{out} at the maximum V_{in},

$$V_{out} = 4V(-2) + 3V(3)$$

$$V_{out} = 1V, \text{ at } V_{in} = +4V$$

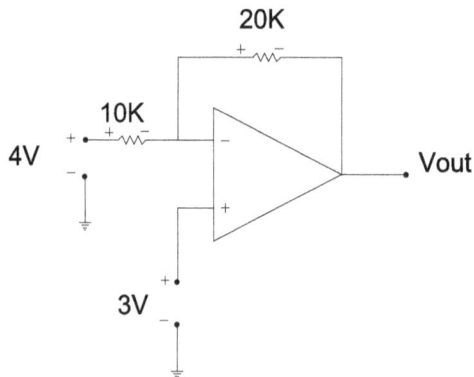

Figure 13 – The Biased Inverting Amplifier at an input of 4 volts and a bias of 3 volts. Vout is determined to be +1V.

Circuit Analysis - Biased Inverting Amplifier

Determining the circuits' operational specifics is no different for the biased inverting amplifier than for the biased non-inverting amplifier. Verifying the operation of the biased inverting amplifier using the previous example, the input voltage range is 0 to 4 Volts, V_{bias} is a positive 3 Volts, R_{in}=10K and R_f=20K.

Using the operational description provided in the previous chapter and with the biased non-inverting amp, V_{bias} would be felt at both op amp input terminals since V_{id} = 0 Volts. The difference between V_{in} and V_{bias} across R_{in} will establish I_{in}, which becomes I_f. As electron flow, I_f, passes through R_f, and V_{Rf} is given the associated polarity. Determining V_{out} is a matter of assuming an appropriate V_{in} value, computing I_{in}, I_f and V_{Rf}, and writing a Kirchhoff Voltage Loop (KVL) from the output.

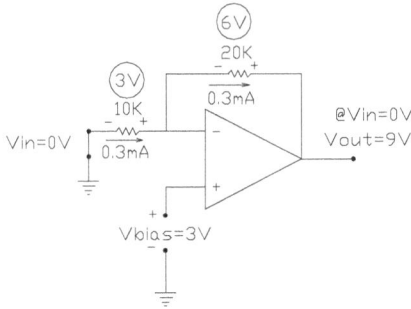

Figure 14 – A biased inverting amplifier. Vin is set to zero volts, the voltage applied at the non-inverting input is observed at the inverting input due to negative feedback and Vid = 0V. Therefore, the voltage across Rin equals the bias voltage when Vin is zero.

Given the provided figure, assume V_{in} is 0 Volts. The voltage drops across R_{in} and R_f and the corresponding polarities are the result of electron flow from the lowest or most negative, to the highest or most positive potentials. Since Vid is zero, the voltage on the inverting terminal will equal positive three volts. With the input set to zero volts, the input resistor current will be the difference between the three volts of Vbias on the right and zero volts on it's left.

Therefore,

$$I_{in} = (0V-3V)/10K$$
$$I_{in} = .3 \text{ mA}$$

Since,

$$I_f = I_{in} = .3 \text{ mA}$$
$$V_{Rf} = (.3 \text{ mA})(20K)$$
$$V_{Rf} = 6 \text{ Volts}$$

And,

$$V_{out} = V_{Rf} + V_{id} + V_{bias}$$
$$V_{out} = 6V + 0V + 3V$$
$$V_{out} = 9V, \text{ at } V_{in} = 0V$$

At an input of zero volts, V_{out} is developed solely by V_{bias} and it's associated gain, A_{vbias}. Using the same technique and assuming V_{in} equals +4 Volts yields;

$$I_{in} = (4V-3V)/10K$$
$$I_{in} = .1 \text{ mA}$$
$$V_{Rf} = (.1mA)(20K)$$
$$V_{Rf} = 2 \text{ Volts}$$
$$V_{out} = -2V + 0V + 3V$$
$$V_{out} = 1V, \text{ at } V_{in} = +4V$$

Note the reverse-acting input/output plot that results when the corresponding points are placed and joined on a graph. As mentioned previously, inverting amplification or reverse-acting signal conditioning, is represented by a negative slope and a negative voltage gain to V_{in}. Note also that +3 Volts of V_{bias} resulted in a V_b, Y-axis intercept/output-offset of +9V. Again the influence of V_{bias} on the output range is determined by A_{vbias}.

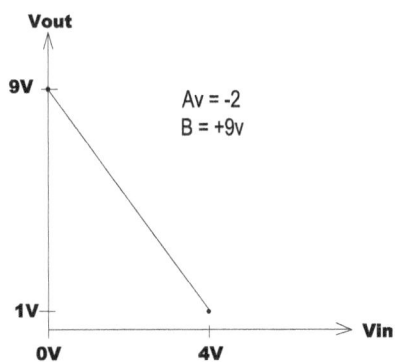

Figure 15 - An Inverting or Reverse-Acting characteristic. Note the large amount of output offset required for first quadrant operation.

An alternative Kirchoff Voltage Loop could have been written through R_{in} and V_{in} instead of V_{id} and V_{bias}. The KVL equation would appear as;

$$V_{out} = V_{Rf} + V_{Rin} + V_{in}$$

Algebraic substitution of circuit voltage values into the above equation yields;

$$V_{out} = 6V + 3V + 0V$$

$$V_{out} = 9V \text{ at } V_{in} = 0V.$$

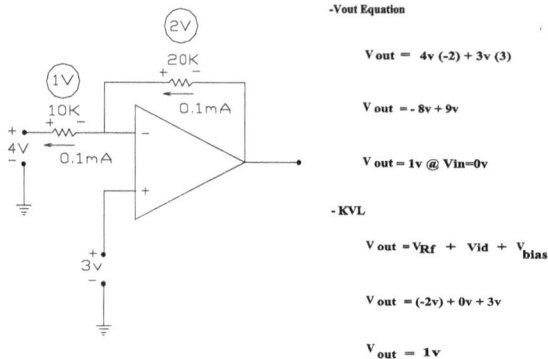

Figure 16 - The Biased Inverting Amplifier at Vin = 4 volts and Vbias = 3 volts. The currents and equations are also indicated.

Solving for Vout produces;

$$V_{out} = -2V + -1V + 4V$$

$$V_{out} = 1V \text{ at } V_{in} = 4V.$$

Obviously these results must agree with the previous reverse-acting circuit computations or something would not be correct.

Input Biasing

As a final note on biased operational amplifiers, the resulting V_{out} equations are performing Y=MX+B in a functional sense. However, since the applied bias voltage is amplified by its respective gain the general form of the output equation is;

$$Y= M_1(X) + M_2 (B)$$

This form exhibits two slope (gain) values. M_1 is the gain seen by the input voltage range, V_{in}, M_2 is the gain seen by V_{bias}, or A_{Vbias}. Circuits that operate according to this description are referred to as "input-biased" circuits, since the bias voltage is applied at the op amps' input, upstream of the gain. As a result, the Offset Voltage (Y-intercept) value from the graph becomes equal to the product of the V_{bias} and A_{Vbias}.

Or,   Offset, $V_B = V_{bias} * A_{Vbias}$

Circuits that operate according to the standard Y=MX+B equation (where $V_B = V_{bias}$) are referred to as output-biased. In output-biased circuits the applied bias voltage is independent of any voltage gain and the V_B value computed from the input/output plot is also the V_{bias} applied to the circuit.

Chapter 4 - Homework Problems

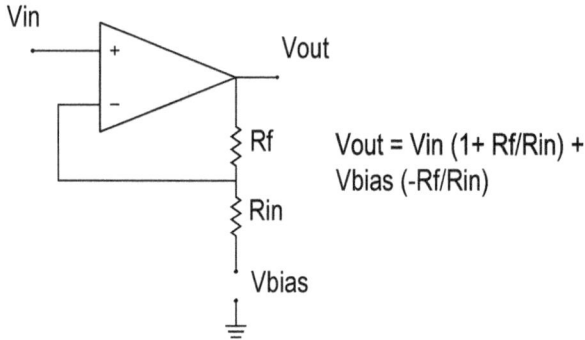

Vin

Vout

$$V_{out} = V_{in} (1+ R_f/R_{in}) + V_{bias} (-R_f/R_{in})$$

Rf

Rin

Vbias

Figure 17 – A Biased Non-Inverting Amplifier and corresponding Vout equation.

#1 – Determine the output voltage range of a biased non-inverting amplifier given the following:
ΔV_{IN} = 1 to 5 Volts
V_{Bias} = 2 Volts
R_F = 100K
R_{IN} = 100K
In Y=MX+B form {or, Vout = (Gain*Vin)$\pm$Vb}, determine the circuits' operational equation.

#2 – Determine the output voltage range of a biased non-inverting amplifier given the following:
ΔV_{IN} = -1 to +1 Volts
V_{Bias} = -2.5 Volts
R_F = 100K
R_{IN} = 200K
In Y=MX+B form {or, Vout = (Gain*Vin)$\pm$Vb}, determine the circuits' operational equation.

#3 – Determine the output voltage range of a biased non-inverting amplifier given the following:
ΔV_{IN} = -.5 to +.5 Volts
V_{Bias} = -.55 Volt
R_F = 900K
R_{IN} = 100K
In Y=MX+B form {or, Vout = (Gain*Vin)$\pm$Vb}, determine the circuits' operational equation.

#4 – Determine the output voltage range of a biased non-inverting amplifier given the following:
ΔV_{IN} = -1 to -5 Volts
V_{Bias} = -4.5 Volts
R_F = 100K
R_{IN} = 200K
In Y=MX+B form {or, Vout = (Gain*Vin)±Vb}, determine the circuits' operational equation.

#5 – Determine the output voltage range of a biased non-inverting amplifier given the following:
ΔV_{IN} = 1 to 3 Volts
V_{Bias} = 1Volt
R_F = 100K
R_{IN} = 100K
In Y=MX+B form {or, Vout = (Gain*Vin)±Vb}, determine the circuits' operational equation.

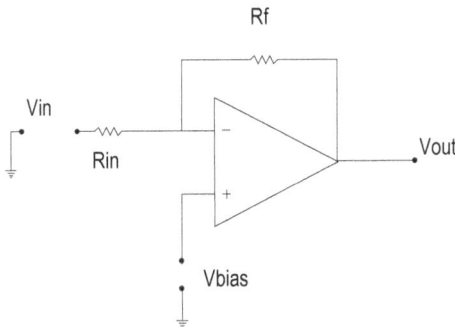

Vout = Vin (-Rf/Rin) + Vbias(1+Rf/Rin)

Figure 18 – A Biased Inverting Amplifier with corresponding Vout equation.

#6 – Determine the output voltage range of a biased inverting amplifier given the following:
ΔV_{IN} = 1 to 5 Volts
V_{Bias} = .333 Volts
R_F = 200K
R_{IN} = 100K
In Y=MX+B form {or, Vout = (Gain*Vin)±Vb}, determine the circuits' operational equation.

#7 – Determine the output voltage range of a biased inverting amplifier given the following:
ΔV_{IN} = 1 to 5 Volts
V_{Bias} = 3.66 Volts
R_F = 200K
R_{IN} = 100K
In Y=MX+B form {or, Vout = (Gain*Vin)±Vb}, determine the circuits' operational equation.

#8 – Determine the output voltage range of a biased inverting amplifier given the following:
ΔV_{IN} = -5 to +5 Volts
V_{Bias} = 2.5 Volts
R_F = 100K
R_{IN} = 100K
In Y=MX+B form {or, Vout = (Gain*Vin)±Vb}, determine the circuits' operational equation.

#9 – Determine the output voltage range of a biased inverting amplifier given the following:
ΔV_{IN} = -.1 to +.4 Volts
V_{Bias} = .5 Volts
R_F = 100K
R_{IN} = 1Meg
In Y=MX+B form {or, Vout = (Gain*Vin)±Vb}, determine the circuits' operational equation.

#10 – Determine the output voltage range of a biased inverting amplifier given the following:
ΔV_{IN} = 1 to 3 Volts
V_{Bias} = 2.33 Volts
R_F = 100K
R_{IN} = 200K
In Y=MX+B form {or, Vout = (Gain*Vin)±Vb}, determine the circuits' operational equation.

Circuit Analysis Problems

#11 – Given the circuit of the following figure, determine the input and feedback resistor voltages, the input and feedback resistor currents, the output voltage and the operational equation performed by the circuit.

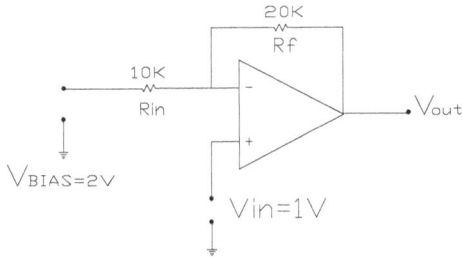

Figure 19 – Circuit for problem #11.

#12 – In the following circuit, determine the input and feedback resistor voltages, the input and feedback resistor currents, the output voltage and the operational equation performed by the circuit.

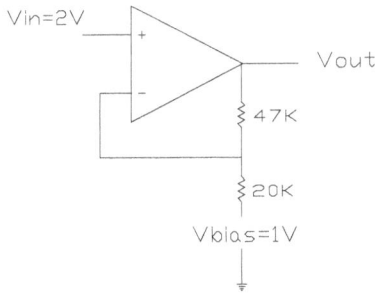

Figure 20 – Circuit diagram for problem #12.

#13 – In the following circuit determine the input and feedback resistor voltages, the input and feedback resistor currents, the output voltage and the operational equation performed by the circuit.

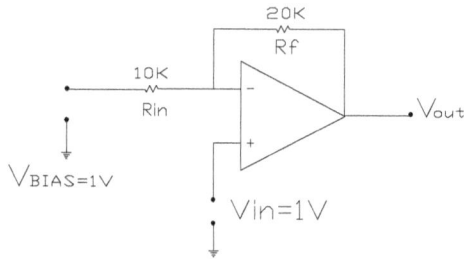

Figure 21 – Circuit for problem #13. Note the output voltage for this problem is _not_ zero.

#14, 15, 16 – Determine the operational Y=MX+B {or, Vout = (Gain*Vin) ± Vb}, output equations for each operational amplifier stage (V_C, V_F, and V_R) of the following circuit.

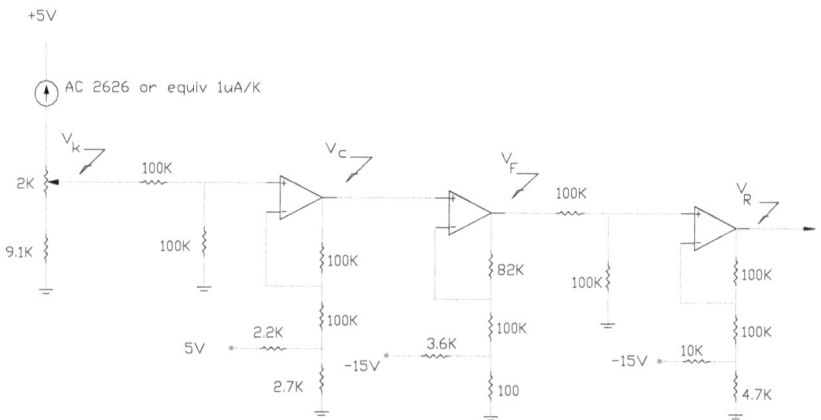

Figure 22 – Circuit for problem #14 through 16

Chapter 5 –Scaling Amplifier Design

This chapter introduces the student to concepts and procedures of scaling amplifier design. Although the intent is not to produce a circuit designer, the op amp being a linear device allows rapid design and development of sensor signal conditioning circuits. The chapter will expose the student to circuit design considerations and introductory scaling amplifier design. Subsequent laboratory assignments will require the design mechanics introduced in this chapter.

Objectives:
Upon completion of this chapter, you should be able to:
- Explain the process of scaling amplifier design
- Perform a biased direct-acting amplifier design
- Perform a biased reverse-acting amplifier design
- Address impedance matching concerns
- Design and assemble a scaling amplifier
- Interpret digital panel meter specifications
- Design and assemble a sensor measurement system
- Calibrate a sensor-based measurement system

Introduction

The proliferation of linear integrated circuits and functional modules into engineering technology has provided many opportunities to potential circuit and system designers for the two and four year engineering technology student. If linear concepts are well understood, designing scaling amplifiers for the purpose of Y=MX+B signal conditioning is quicker and easier than performing the circuit analyses.

In addition to determining the required gain and bias values, designing Y=MX+B "scaling" circuits requires an understanding of impedance matching at the amplifier input and at the amplifier output. Input and output resistance (occasionally given as impedance) values are available in the operational amplifier specifications and need to be considered between each stage of any electronic measurement system. From the signal source (commonly a sensor) to the amplifier, and from the amplifier to the load (DPM, analog input to a data acquisition system, chart recorder, etc) impedances must be "Matched" or the system will not operate as expected. In conjunction with impedance matching, additional considerations would be non-linearity, drift, frequency response, slewing rate, noise rejection, offset compensations, isolation and power supply conditions. These specifications are addressed in later chapters and in the appendix.

The focus of this chapter will be on designing biased inverting and non-inverting amplifiers for the specific purpose of impedance matching, amplifying or attenuating (voltage gain of less than unity), and output-offset biasing. The end result will allow a sensor to be interfaced with recorders, displays, analog and digital panel meters, controllers, analog to digital converters and so forth. Before discussing gain and bias considerations, a brief review of impedance matching and loading effects will help to stimulate impedance matching awareness.

Impedance Matching Review

Unlike the transistor, the general purpose "741" operational amplifier is a voltage sensitive device. Therefore, when signal processing with an op amp it is desired to transfer maximum signal voltage from one stage to the next. A common rule of thumb says a 10:1 R_{load} to R_{source} ratio will transfer maximum signal. A simple example will demonstrate. Referring to the following figure, consider the source voltage and associated internal resistance as a sensor since the vast majority of sensors can be reduced to an equivalent circuit containing a Thevenin, or equivalent voltage source and an internal source resistance (or Thevenin) output resistance. The load represents the input impedance of an amplifier, display or other electronic stage.

Momentarily assuming the source and load resistances are equal, the voltage drops across the load and the internal source resistance will be equal. If a voltmeter were placed across R_{load} under this condition, it would display half of the internally generated source voltage or 2.5V at the input to the next stage (Rload).

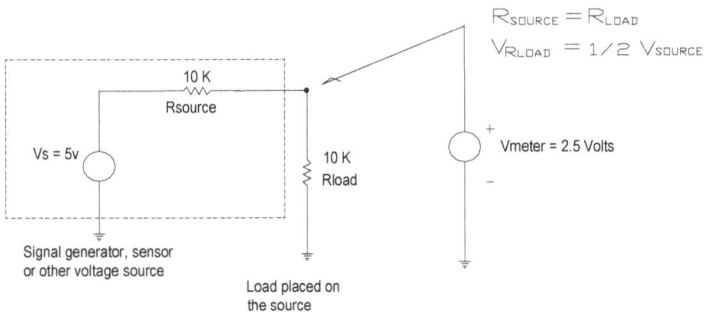

$$R_{SOURCE} = R_{LOAD}$$
$$V_{RLOAD} = 1/2 \; V_{SOURCE}$$

10 K
Rsource

Vs = 5v

10 K
Rload

Vmeter = 2.5 Volts

Signal generator, sensor
or other voltage source

Load placed on
the source

Figure 1 - Circuit diagram to represent effects of connecting a load equal to the output impedance of the source, half of the possible output signal is lost.

Unfortunately, if the load was disconnected and the voltmeter remained connected at the source's output, the voltmeter would read the full voltage from the source, or

5V. Under this condition the source is generating wasted power that is lost internally within the output resistance when the load is connected, an undesired condition. Since the output voltage increased when the load was disconnected, the 10K load is understood to be "loading" the source. "Loading" is a term used to describe the effect of a relatively large load (large load refers to power demand, and specifically current demand, large load = low resistance) placed upon a source that contains a relatively high source resistance.

The load is a relatively low resistance value that requires considerable current flow. In attempting to meet the load's current demand, the source's voltage is reduced since all sources have a finite amount of power (combined voltage and current) that can be delivered. If the load were "matched" to the source, (as with a unity-gain follower, pre-amp or other impedance matching amplifier), or if the source resistance were decreased, or if the load resistance were increased, the source would deliver more signal voltage to the load.

Simply stated, the source and load form a voltage divider. As such, the voltage distribution around the circuit will be in proportion to the resistance values. The largest resistance drops the greatest voltage. If the maximum voltage, ideally V_s, is to be transferred to the load, the load resistance must be considerably larger than the source resistance.

In the following figure, the load has been replaced with another that contains a higher input resistance. Note how the load voltage has increased from 2.5V to 4.545V. Note also that even with the required (rule of thumb) 10-to-1 load-to-source ratio, 9.1% of the signal is still lost internally at the source. For this reason it is best to make the load/source ratio as large as possible.

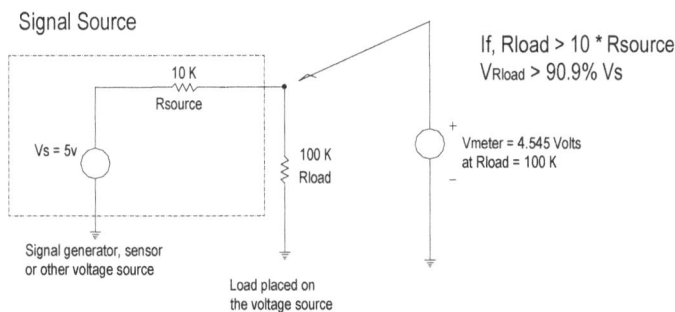

Figure 2 – For maximum signal transfer from a source to a load, the load must be at least ten times the value of the source resistance. The 10K load of the previous diagram is replaced with a 100K load and the resulting output increases from 2.5V to 4.545V. It is suggested the student assemble and verify the presented conditions if the concepts are found to be elusive.

Impedance's are usually indicated on signal sources, function generators, component specification sheets and measurement apparatus (such as oscilloscopes) to assist in impedance matching. These principles apply to AC impedance, DC resistance and combinations thereof. However, true AC impedance (the vector sum resistance and reactance) matching requires complimentary reactive components. As a brief example, an inductive source should be coupled to a capacitive load of equal value. The resulting series-resonant circuit will cause the reactors to cancel leaving a purely resistive source/load combination. Again, the load resistance should be considerably larger than the source resistance to transfer maximum signal.

If maximum *power* (current and voltage) transfer were desired rather than maximum signal transfer, the load and source resistances would be required to be equal. Equal source and load resistances pass an optimum combination of both voltage and current from the source to the load. This is not the desired condition with operational amplifiers however since op amp circuits require maximum *voltage* transfer from stage to stage. Maximum power transfer is common in computer networks, video and radio frequency (RF) communication circuits where all impedances (source, load and interconnecting cabling) are 50, 75 or 100 Ohms.

Of all the fundamental concepts in electronics, the concept of impedance matching is the most overlooked, least understood and simplest of important basic principles. It must be understood before proceeding or be assured; the amplifier you design will not operate according to specification.

Measurement and Display Systems

The first step in designing any circuit involves drawing a block diagram of the proposed measurement system, and input/output graphs of the individual stages involved in the system. Reviewing the sensor, amplifier and associated component input/output characteristics provides the designer with an intuitive understanding of system operation. To begin, assume a pressure sensor provides 1V to 5V over a 0 to 10-psi pressure span, and exhibits an internal or output resistance of 5K ohms. To make the sensor compatible with a display (often called a digital panel meter, or DPM), a signal-conditioning amplifier (or scaling amplifier) is required to convert the 1 to 5 Volts signal span into the required voltage span of the DPM. The amplifier will output 0V to 10V while simultaneously impedance matching between the indicated 5K of sensor output resistance and 10K of DPM input resistance. The following diagram illustrates.

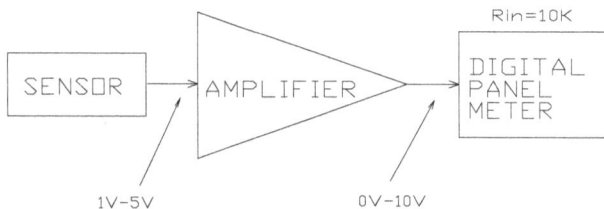

Figure 3 – Measurement system design example.

A few words about Digital Panel Meters (DPM) before proceeding, DPM's are used as digital indicators to display measured physical variables in engineering units.

Typically displayed measured variables are pressure, level, flow, temperature, stress, strain, pH, position, velocity, acceleration and so forth. The digital panel meter is simply a digital voltmeter with a fixed input voltage span, variable decimal point location and input resistance value that is usually quite large (typically 1 Megohm, even though our assumed R_{in} is a wimpy 10K). Many types of DPM's are available in varying degrees of sophistication but all require an input voltage range that numerically represents the measured variable to be displayed.

Figure 4 – A common Digital Panel Meter, or DPM with the plastic bezel removed. Notice the Gain and Zero adjustments provided for fine-tuning of the displayed quantity. The input and supply connections are made on the back of the instrument.

For our purposes the input to the DPM should be 0V to 10V which when displayed would represent the sensor input pressure in psi. With the appropriate amplifier design and decimal location, the DPM should indicate 0.00 at 0 psi and 10.0 (actually 9.99, most DPM's contain 3 digit displays) at 10psi (9.99psi). DPM's usually have self-contained gain and bias controls for the purpose of fine-tuning calibration, but these controls have a very limited adjustment range and cannot be expected to perform the roll of our op amp signal conditioning circuit. The appendix contains a typical DPM specification sheet.

The following presents the sensor to DPM, and sensor to amplifier to DPM diagrams. The block diagram figures are used to demonstrate the importance of impedance matching in transferring maximum signal voltage from stage to stage. From the corresponding voltage divider equations, connecting the sensor directly to the DPM would result in a signal transfer of only 66.6%, a third of the signal would never be felt or displayed at the DPM.

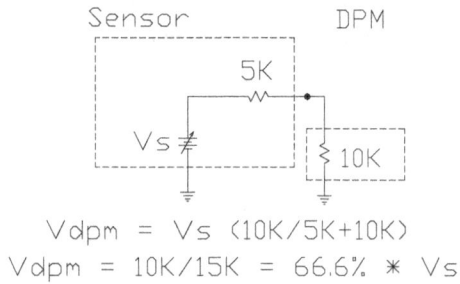

$$V_{dpm} = V_s \, (10K/5K+10K)$$
$$V_{dpm} = 10K/15K = 66.6\% * V_s$$

Figure 5 – Connecting the sensor directly to the DPM would transfer 66% of the sensor output voltage. If connected in this manner the DPM would display 0 to 3.3 psi (66% of 5V) instead of the desired 0 to 10.0 psi.

Given the operational amplifier input resistance is 1-Megohm and the output resistance is 100 Ohms or less, the impedance matched op amp circuit would pass 98.8% losing only 1.2%. No circuit can transfer 100% of an input signal since that would require an infinite input resistance and zero output resistance, ideal quantities. Note also the amplifier impedance values are the "worst-case", open-loop values taken from the specifications for a general purpose 741 op amp.

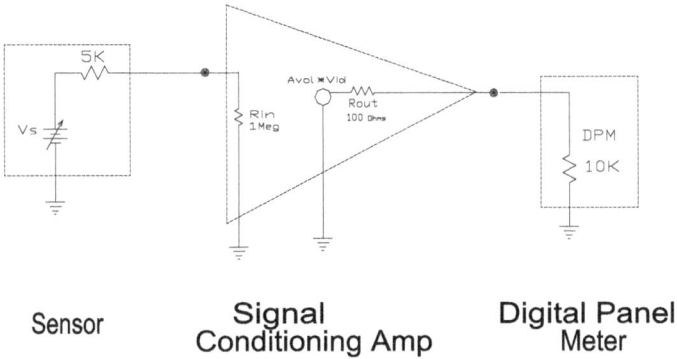

Figure 6 – Diagram of the impedance matched and scaled sensor to amplifier to display measurement circuit. The amplifier performs the functions of scaling the 1-5V span into 0-10V and impedance matching the 5K Ohm source to the 10K Ohm display.

In reality, the actual input impedance would be larger and the output impedance lower since negative feedback enhances these values. When designing however, worst case values should be assumed if one wishes to avoid occasional minor disappointments. Also, designing potentiometers into the circuit for variable gain and bias provides fine-tuning of the circuit's operational parameters. In effect, potentiometers are a variable form of compensation against the unexpected. The designs included in this chapter will be performed using fixed resistance values and modified in subsequent chapters to include potentiometers to counteract rounding error, component tolerances and other unpredictable undesirables, and to facilitate more precise results.

The two following examples will introduce circuit design of unbiased inverting and non-inverting amplifiers. The previous sensor, amplifier and DPM example requires a biased amplifier however since the output offset is computed as a non-zero value.

Non-Inverting/Direct-Acting Amplifier Design

As an example to begin the study of operational amplifier design, assume the following application is presented.

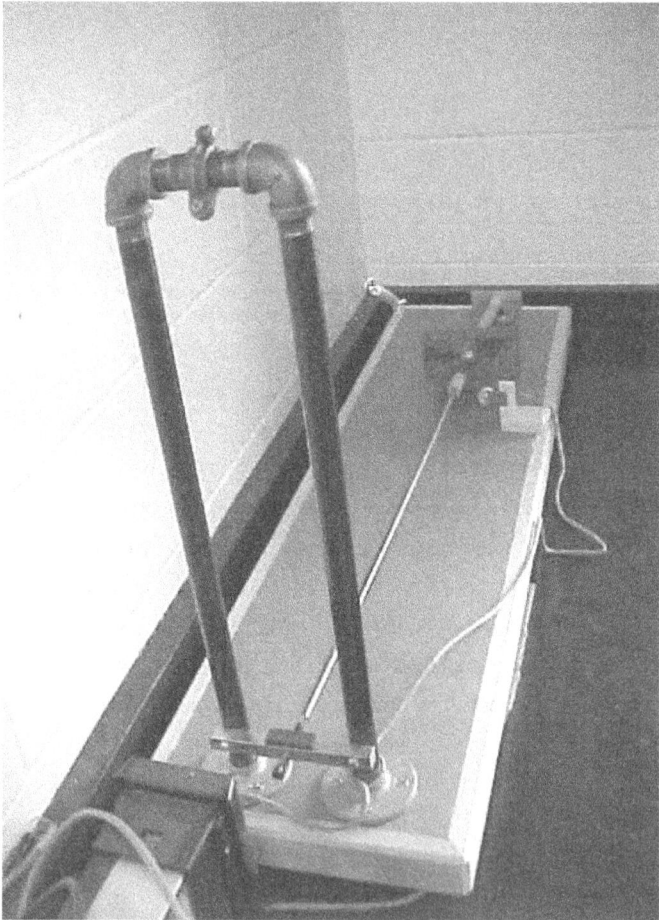

Figure 7 – A seismograph design taken from a 1979 Scientific American. The primary sensor is a coil and a permanent magnet (see following photo) located at the far end of the boom. The seismograph is a very sensitive motion detector used to sense movement within the earth caused by nuclear testing or earthquakes. Photos courtesy of Richard Harwood.

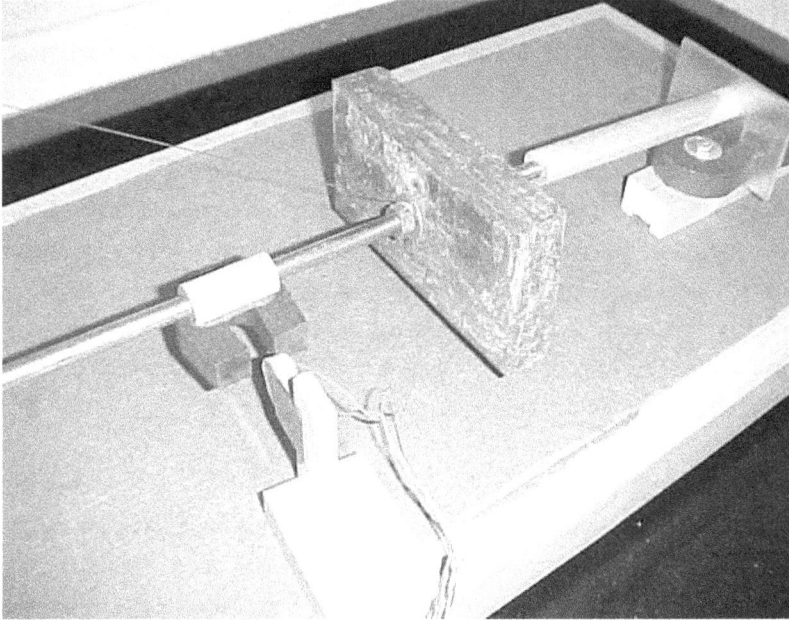

You have been asked to amplify the output of a homemade sensor. The sensor will be used to detect from a remote location, extremely large forces within the earth caused by earthquakes and/or underground nuclear testing. From the diagram provided, the hanging boom assembly will remain static (not moving) and carries a permanent magnet located directly adjacent to a coil composed of 10,000 turns of number 34 magnet wire. The coil is mounted on a platform that will be placed on the ground in very close proximity to the permanent magnet on the boom assembly.

In operation the pickup coil platform will move in a sinusoidal manner with an underground disturbance while the boom assembly remains fixed. A damper assembly located at the right end of the boom assures minimal boom motion. As the platform moves with an underground disturbance, the pickup coil will move within the magnetic field created by the permanent magnet. The sinusoidal motion of the coil will generate a corresponding sinusoidal voltage. The resulting voltage will be in the micro to millivoltage range. The task at hand is to design an amplifier that will amplify the resulting signal from the pickup and impedance match the pickup coil to an appropriate display.

As with any signal-conditioning amplifier application, as much information about the sensor and *the measurement application* must be acquired to properly design the corresponding circuitry. Unfortunately, in many experimental applications much of the required information is not known. This is after all, why the measurement system is being constructed, to acquire more information about the application. However, assume the following has been provided;

-The amplifier gain should be in the vicinity of 100, the waveforms' amplitude will be determined by the proximity of the disturbance to the measurement location. Likewise, the amplitude of the measured waveshape is unimportant. After the unit has been assembled and proven functional, relative measurements (comparison of recorded waveshapes) can be used to determine the strength of the disturbance when distance is considered.

-The phase of the measured signal taken from the pickup should not be altered, geologists have indicated the leading edge of a disturbance provides information about the nature of the disturbance,

-The pickup coil composed of 10,000 turns of #34 wire wound on a 1 inch diameter core will offer a substantial amount of source resistance, probably in the neighborhood of 10's of thousands of Ohms,

-The resulting information will be displayed on a strip chart recorder with an input span of −1 volt to +1 volt, the output voltage should be centered around zero volts during non-disturbance periods, the input resistance of the strip chart recorder is specified at 1 Megohm,

-The pickup coil platform when placed on the ground away from local disturbance sources (construction, manufacturing, lawn mowers, etc.) will be sensitive to underground disturbances occurring everywhere in the world since the disturbances are capable of propagating throughout the earth,

-The frequencies of the measured disturbance signals are typically under 10 Hertz. A simple single-order low-pass filter should be considered for passing frequencies under 10 Hz.

Given the previous criteria, a non-inverting amplifier was chosen to accurately represent phase concerns associated with the leading edge of the disturbance. Although with simple strip chart recorders the leading edge of the recorded waveshape may not be visible, if more advanced waveform recording technology is employed the first 180 degrees of the disturbance would require inspection. The amplifier gain of 100 was chosen to avoid the possibility of generating an output of substantially greater than $\pm$ 1 volt, which may damage the often-sensitive waveform-recording instrument, while providing enough amplification to differentiate between baseline noise and signal. A high input impedance input configuration was selected to best transfer the voltage from the signal source to the amplifier and eventually on to the recording apparatus. Although no amplifier type was chosen specifically for the input impedance, the non-inverting configuration allows the input to be applied directly to the non-inverting input terminal resulting in an input impedance of 1 Megohm or greater. As a side note, the input impedance of the non-inverting amplifier usually exceeds the inverting amplifier since the input resistance of the inverting amplifier is determined by the value of the input resistance itself. Or, $Zin = Rin$ (usually 100K or less) with the inverting amplifier, and $Zin \geq 1$ Megohm with the non-inverting amplifier. Fixed gain amplifier was chosen to minimize the number of variables associated with the measurement system. Since the instrument is a prototype and has never been proven functional, minimizing the number of components that are capable of changing will assist in establishing the initial functionality, operation and troubleshooting.

Again, the previous may appear as an exercise in futility since much of the provided information may not directly relate to the amplifier design. Yet many of the questions that arise during the course of application and design can be resolved, and an overall better signal conditioning and data acquisition system can be achieved if the designer is well informed initially.

The largest amount of the design has been determined in the previous application considerations. Currently, two items need to be considered, the amplifier input/output characteristic and the basic circuit configuration. Although each is fairly apparent, establishing a habitual mechanical process for the design of amplifiers will be beneficial with more demanding and sophisticated applications.

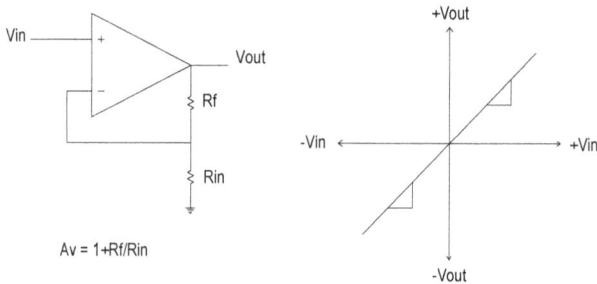

Figure 8 – The basic non-inverting or direct-acting amplifier with corresponding gain equation and input/output characteristic.

Designing the amplifier for a gain of 100 will involve specifying values for Rin and Rf. Since both are currently unknown, an assumption of Rin or Rf must be made. In later chapters, an additional resistive network will be connected to the lower portion of Rin for non-zero output voltage applications. To facilitate the additional circuitry the value of Rin must be considered. For this reason Rin will be the unknown quantity for which an assumed value will be made. Typically, Rin will be assumed at

100K Ohms. Knowing the assumed value of Rin, the voltage gain equation and the desired voltage gain quantity, Rf can be determined.

Since,

$$Rin = 100K$$
$$Av = 100$$
$$Av = 1+Rf / Rin$$
$$Rf = Rin (Av-1)$$

Rf becomes,

$$Rf = 9.9Meg$$

However a 9.9Meg Ohm resistance is not commonly available, or is not a standard value. The nearest value would be a 10Megohm resistance but this would increase the gain to 101. Again, since the value of output voltage is not critical, the corresponding gain isn't critical and the difference is acceptable. The finished circuit and I/O characteristic is shown following.

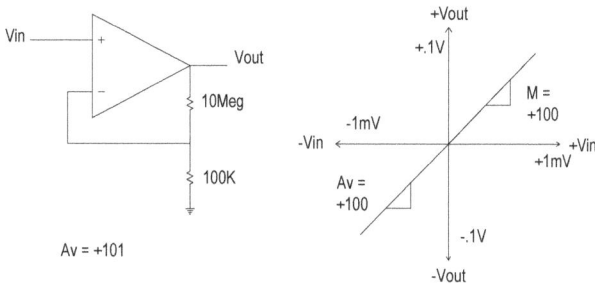

Figure 9 – The completed seismograph amplifier with component values and I/O characteristic plot indicated.

A final note about resistor value determination - later chapters will present non-ideal operational amplifier application considerations and compensations. Among the indicated non-ideal specifications will be an output voltage error condition called

Offset Voltage. Quite simply, this error condition occurs when the input is zero volts and the output voltage is anything but zero volts. Offset voltage is an unintentional shifting of the output away from zero volts when the input condition is demanding an output of zero volts. Although easily remedied, circuit designers attempt to minimize any opportunity for the occurrence of offset voltage error. One of the methods of reducing the probability of offset voltage error is to keep the value of the feedback resistor Rf, under 10 Megohms. Large values of resistance have a tendency to establish undesired voltages created from parasitic currents flowing around the amplifier circuit. Although 10-Megohms of Rf probably would not significantly alter the output voltage span of the current amplifier, values of Rf greater than 10 Meg should be considered lethal in this regard.

Inverting/Reverse-Acting Amplifier Design

Inverting amplifiers are required in sensing applications to reverse the span associated with a sensor or other signal conditioning stage, while providing amplification and occasionally offset biasing provisions as well. The following design example will assume an output offset of zero, or a true-zero voltage span, and will focus upon impedance matching, amplification and voltage span reversal.

As an application example, assume a negative-going voltage span is available at the output of a measurement instrument. Although somewhat uncommon, occasionally inverting amplifiers are required to convert a span of negative voltage into a span of positive voltage, or to convert a sensor output voltage span that moves negative with an increasing applied physical variable. Assume the following is given;

-A voltage span of 0V to –2 volts DC is available at the output of a circular recorder used to record temperatures over an eight hour period.

-The 0V to –2V span represents a temperature variation of 0° to 100°C, where 0V occurs at 0° and –2V is provided at 100°C.

-The indicated resistance at the recorder output is 1000 Ohms.

-It is desired to display temperature on a 1-volt DPM where 0V corresponds to 0° and 1V (.999V) will represent 100°C.

To prepare for the amplifier design, a graph of the existing condition is assembled and appears following.

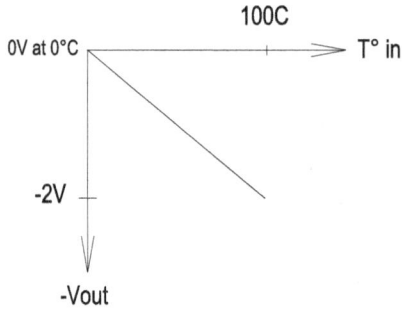

Figure 10 – The available output voltage span is a negative-going 0V to -2V from zero to one hundred degrees Centigrade.

To be properly displayed using a DPM, an amplifier is required to reverse the span to a positive-going voltage, and to convert the voltage span from 0V to –2V into 0V to +1.0V. When complete, the amplifier will output a zero to one volt span from zero to one hundred degrees Centigrade as in the following plot.

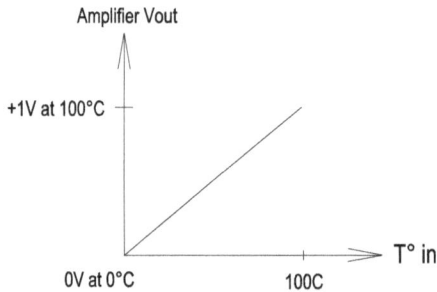

Figure 11 – The required amplifier output of a positive-going 0V to 1V from 0 to 100 degrees Centigrade.

The previous graphs were assembled to acquire a better feel for the application and could be considered optional if the designer is comfortable with reverse-acting spans. Knowing the available signal from the temperature recorder and the required voltage for the DPM, and the output resistance of the recorder, the amplifier I/O characteristic graph can be assembled.

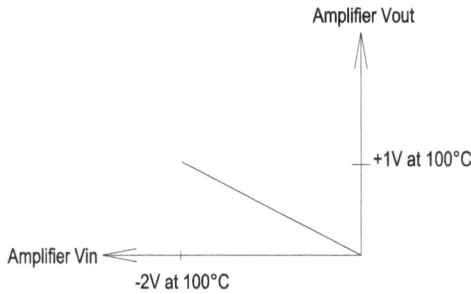

Figure 12 – The inverting amplifier I/O characteristic.

From the amplifier plot, the gain (Av) and offset (B) can be determined.

$$Av = 1.0V / -2V$$
$$Av = -.5$$
$$B = Vout - (Av * Vin)$$
$$B = +1V - (-.5 * -2V)$$
$$B = +1V - (1.0V)$$
$$B = 0V$$

Although not required in this example, the output-offset (B or V_B) was computed to verify a true-zero output span. From the computations, an inverting amplifier with a gain of one-half is required. One might ask why an amplifier is required if the necessary voltage gain is one-half. Why not use a resistive voltage divider? A resistive divider would easily reduce the input voltage span but is not capable of generating a positive output voltage from a negative input voltage. An "active" component is required for this function. The negative input voltage can be interpreted as a proportional "command" for the amplifier to place a portion of the positive supply voltage at the op amp output.

Knowing the required amplifier gain and type, an amplifier diagram can be assembled as follows.

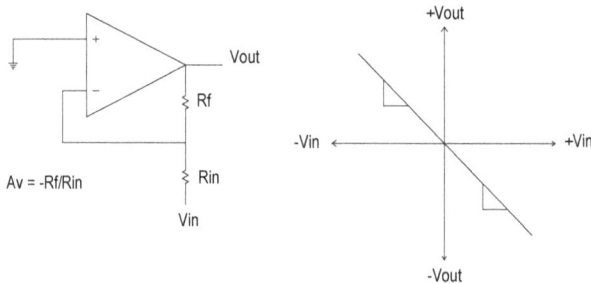

Figure 13 – A reverse-acting or Inverting amplifier with corresponding gain equation and I/O characteristic plot.

From the amplifier gain equation, the values of Rin and Rf must be determined. Knowing the desired gain is -.5, and assuming a value for Rin, Rf can be chosen. However, assuming a value for Rin requires first knowing the output resistance of the source. The inverting amplifier exhibits an input impedance quantity approximately equal to the value of Rin, and for maximum signal transfer between the signal source and the amplifier the amplifier Rin must be at least ten times the sources' output resistance. Since the source resistance was given as 1K Ohm, the operational amplifiers' Rin should be greater than or equal to 10 times 1K-Ohm, or 10K Ohms. Note the inverting amplifiers' input resistance rating is different than the Non-inverting amplifier where the input resistance is 1-Megohm or greater.

The difference in input resistance values is the result of the input being applied directly to the input terminal with the non-inverting amplifier. Whereas the inverting amplifier has Rin located between the input and the "virtual ground" or "summing junction" location. Therefore, with the non-inverting input grounded on the inverting amplifier, zero volts, or ground potential is just on the other side of the input resistance. This is demonstrated in the following diagram.

Non-inverting Amplifier

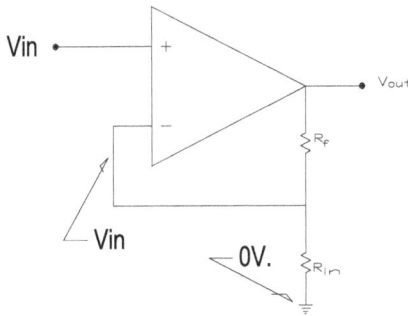

Figure 14 – In all negative feedback op amp circuits, the inverting input is "driven" to equal the voltage applied at the non-inverting input. By virtue of the inverting terminals' connection to the output, the output will adjust to whatever value is necessary to keep the two inputs within micro-volts of each other. A by-product of this concept results in the inputs signals seeing vastly different input resistances with the non-inverting amplifier configuration generating a much higher input impedance than the inverting amp. The typical input impedance of the inverting amplifier is equal to the value of Rin placed in the input circuit.

Inverting Amplifier

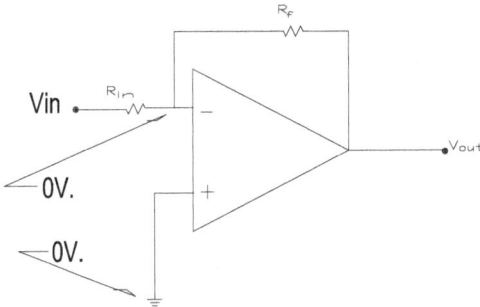

With the input applied to the non-inverting input terminal, the virtual ground or "summing junction" is at the other input. The amplifier inputs are isolated from each other, with the non-inverting applied input signal seeing a much higher resistance typically than the signal applied to the inverting input resistance. For this reason, and to maintain the greatest amount of input impedance, Rin is assumed to equal 100K instead of the previously computed 10K.

Knowing Rin and Av, Rf can be determined.
Since,

$$Av = - Rf / Rin$$

And,

$$Av = -.5$$

And,

$$Rin = 100K$$
$$-Rf = Av * Rin$$

Or,

$$Rf = -(Av * Rin)$$
$$Rf = -(-.5 * 100K)$$
$$Rf = 50K \ Ohms$$

The final circuit would appear as follows.

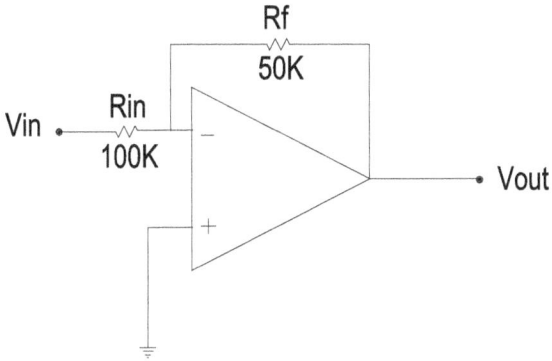

Figure 15 – The completed reverse-acting amplifier with no output-offset and a voltage gain of negative one-half.

Direct-Acting Biased Amplifier Design

As an application example of a biased non-inverting amplifier design, return to the condition provided earlier in this chapter. An amplifier is required to convert a 1-5 volt signal span from a 0 to 10-psi pressure sensor, into a 0-10 volt span for display on a Digital Panel Meter (DPM).

After reviewing the application and graphing the corresponding amplifier input and output spans, the gain and bias quantities can be computed for the given input span of 1-5V and desired output span of 0-10V.

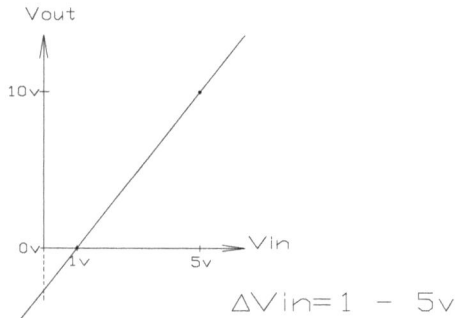

Figure 16 – Input/output characteristic of the signal-conditioning amplifier.

Determining the amplifier Gain from the graph,

$$A_v = \Delta V_{out}/\Delta V_{in}$$

$$A_v = 10V/4V$$

$$A_v = +2.5$$

The output-bias, V_B or "B" quantity is:

$$B = V_{out} - (A_v*V_{in})$$

$$B = 10V - (+2.5*5V)$$

$$B = -2.5V$$

Given the positive slope (A_v) of the input/output characteristic and the non-zero output offset (B, or occasionally V_B) value, a biased, direct-acting (non-inverting) amplifier needs to be constructed with the indicated operational values.

Before determining the resistance values the appropriate circuit diagram should be established. The following figure shows a biased non-inverting amplifier with the corresponding gain and offset circuitry. After the necessary amplifier type has been determined, the component values for proper output offset and gain can be selected. A structured sequence of events must occur when finding the resistance values of the amplifier circuit. Regardless of the amplifier type (inverting or non-inverting), if the amplifier application is scaling, the steps are as follows;

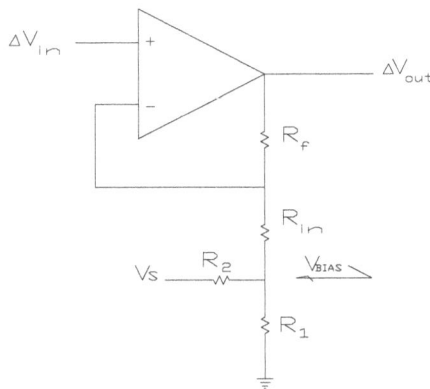

Figure 17 – A Biased Non-Inverting Amplifier required for the signal-conditioning example provided.

Step 1 - Assume R_{in} to be as large as realistically possible, typically in the hundreds of Kilohms to avoid any significant altering of the gain circuitry by the biasing circuitry. Since the offset-biasing circuit is connected to the gain determining resistance of Rin, excessive values of offset-biasing resistors will contribute to the effective value of Rin and change the gain.

For most op amp applications,

$$\text{Assume, } R_{in} = 100K$$

Step 2 - Compute R_f based upon the required A_v and the assumed R_{in}. If R_f computes to greater than 10 Megohms, reduce R_{in} and re-compute R_F. A couple Megohms is acceptable but tens or hundreds of Megs is unrealistically large. If a standard resistance this large is available and used for R_f an excessive offset error (an internally generated non-zero output error, explained in a later chapter) could result. Since the required voltage gain, A_v, is +2.5 and R_{in} = 100K,

$$A_v = 1 + R_f / R_{in}$$

$$R_f = 150K$$

Step 3 - Since R_f and R_{in} are known, an A_{Vbias} can be found. Since the output-offset biasing voltage is applied to the non-inverting input terminal, the applied bias voltage will see an inverting voltage gain $(-R_f/R_{in})$, through which the bias voltage will be processed to create the output offset of +2.5 volts. Reviewing and rearranging the direct-acting V_{out} equation yields the V_{bias} equation. The algebraic manipulation of the V_{out} equation is left to the student.

Given, $\qquad V_{out} = V_{in} (1+R_f/R_{in}) + V_{bias}(-R_f/R_{in})$

$$V_{bias} = \frac{(V_{out}-V_{in}(1+R_f/R_{in}))}{(-R_f/R_{in})}$$

Choosing corresponding input and output values from the circuits' operational specification yields;

$$V_{bias} = [(10V-(5V\ (+2.5)))\div(-1.5)]$$

$$V_{bias} = +1.67V$$

Step 4 – Although previously indicated as a battery or other DC voltage source, the biasing circuit will consist of a resistive divider connected to an available DC supply voltage. As mentioned previously, caution should be exercised when choosing bias voltage resistance values for the direct-acting amplifier. Since the biasing network is "hanging" below R_{in}, any excessively large resistance values will increase the effective R_{in} value, which will decrease the desired signal voltage gain.

The rule of thumb to avoid circuit loading should be applied here by making the gain determining resistors at least ten times, preferably a hundred times, larger than the bias voltage resistances. Since R_f and R_{in} are 100's of Kilohms, use bias resistors of 1's of Kilohms. Lower resistance values could be used as long as the biasing power supply is not providing excessive current and the biasing resistors don't become heated.

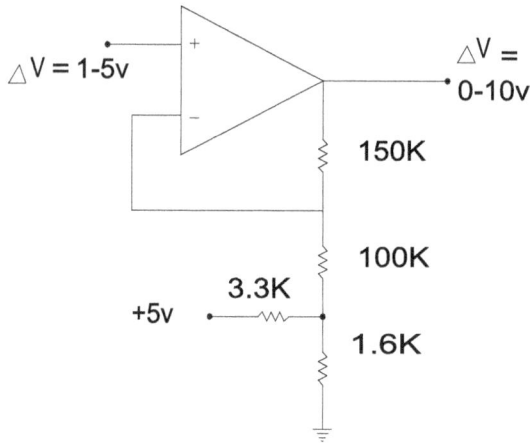

Figure 18 – The Biased Non-Inverting Amplifier with component values and Vbias computation provided.

A quick and simple method of biasing circuit resistor determination is to size the resistances according to the anticipated voltage drops across the bias divider resistors. In figure 9 the drop across R1 will be expected to provide the required V_{bias} of 1.67V. Therefore, using ones of Kilohms and the nearest available standard values,

$$R_1 = 1.6K$$

Assuming a positive five volt supply (V_s = +5V) is available, R2 will drop the difference between V_{bias} (1.67V) and V_s or 3.3V. Likewise,

$$R_2 = 3.3K$$

And the offset-bias circuit component values are determined. Not all resistance values will compute as close to the available standard resistors as these. In which case, doubling each of the resistor values or halving each value will usually result in commonly available resistors. For purposes of precision and calibration, gain and bias values are commonly made variable with potentiometers. An example of variable gain and bias circuitry design follows the inverting amplifier design example.

Biased Inverting/Reverse-Acting Amplifier Design

The Reverse-Acting or Inverting amplifier frequently finds use in experimental measurement and process control applications. Assume for the sake of example, a level sensor in an experimental application outputs a signal in proportion to the degree of empty volume within a tank. Or, as the volume increases the sensor output decreases. With some top-mounted level detectors the output signal is provided in proportion to the distance from the tank top down to the material surface contained in the vessel. Likewise, the output signal is proportional to the "empty" volume, and inversely proportional to the volume of material in the tank. To reverse the sensor output characteristic and acquire information about the volume of material in the tank, a reverse-acting or inverting amplifier is required.

Again, assume a level detector is measuring liquid volume in a storage reservoir. The sensor output varies from one to five volts between maximum and minimum level. Further assume that a DPM is to be used as a level indicator, and is desired to display zero to 100% of tank volume where zero percent represents an empty tank and 100% represents a full tank. Currently the sensor outputs minus five volts at maximum tank volume and plus five volts at minimum tank volume, and exhibits an output resistance of 100 Ohms. The task for the signal conditioner is to reverse the sensor characteristic and scale the signal span for compatibility with the DPM display. Since the DPM will require an input span that is representative of the displayed values, a 0-10 volt or 0-1 volt DPM could be used. Just for the sake of variety, assume a 0-1 volt DPM is to be used for the display, and the input resistance of the DPM is 100K Ohms.

The first step of the design is to visualize, or compose a block diagram of the proposed measurement system and of the input-output characteristic of the amplifier.

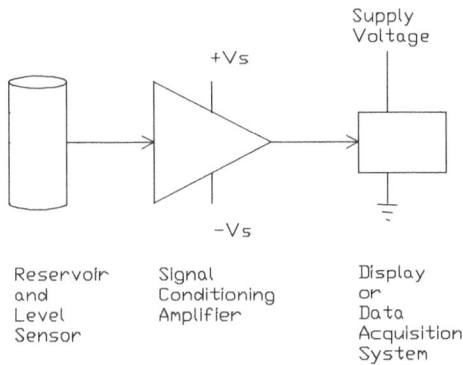

Figure 19 – Block diagram of the level system signal-conditioning example.

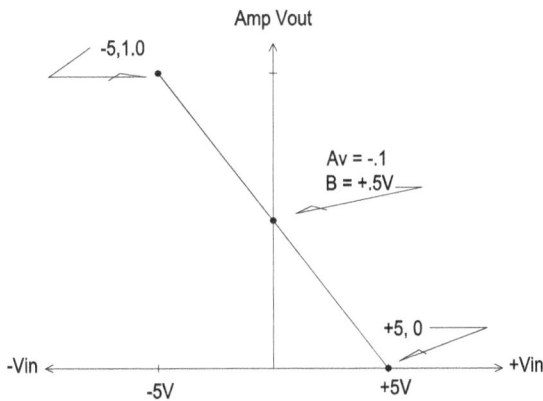

Figure 20 – Reverse-acting (inverting) amplifier I/O characteristic.

Given the previous diagrams, the amplifier circuit configuration needs to be determined. Again, since the level sensor output signal varies from +5 volts to –5 volts as the tank volume varies from minimum to maximum, and since the desired display requires a zero to one-volt signal span, an inverting amplifier is required to reverse the level sensors' signal span. In addition, from the amplifier input/output characteristic an output offset of +5 volts is obvious, and the inverting amplifier will require an offset-bias voltage to accomplish the output-offset. The following biased inverting amplifier is chosen as the appropriate amplifier type.

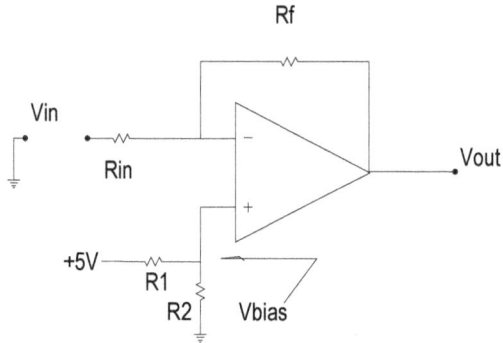

$$Vout = Vin\ (-Rf/Rin) + Vbias(1+Rf/Rin)$$

Figure 21 – A biased inverting amplifier and corresponding Vout equation.

In the final circuit configuration, a resistive divider network will be required to provide the required amount of Vbias to the op amps' non-inverting input terminal, as indicated by the diagram. The subsequent circuit design procedure is identical to that of the non-inverting amp. Assemble an input/output graph of the amplifier characteristics, determine the required impedances of all components, assume a Rin of 100K and determine Rf from the gain requirements. Determine the gain seen from the bias voltage input (1 + Rf/Rin, for the inverting amplifier), establish the required Vbias and determine the biasing circuit component values. The procedure in steps is provided following.

Step #1 – From the characteristic graph, the input and output spans were used to determine the required amplifier gain and output offset values. From the graph, the amplifier gain computed to -. 1 and the offset is determined to be +5V.

Determining the amplifier Gain from the graph,

$$A_v = \Delta V_{out}/\Delta V_{in}$$

$$A_v = -1.0V/10V$$

Note the negative sign, representative of a reverse-acting (negative slope) amplifier characteristic.

$$A_v = -.\ 1$$

The output-bias quantity is:

$$V_B = V_{out} - (A_v{}^*V_{in})$$

$$V_B = 1.0V - (-.\ 1 * -5V)$$

$$V_B = +.\ 5V$$

Step #2 – Assume Rin = 100K. This should be the only assumption necessary in determining the component values, and for the vast majority of available sensors and display or data acquisition requirements the 100K initial value for the input resistor should be adequate. Given Rin = 100K, determine the necessary Rf from the required Av. Since the computed slope was -. 1, and Rin = 100K, Rf becomes;

$$A_v = -\ R_f/R_{in}$$

$$-R_f = A_v{}^*\ R_{in}$$

$$-R_f = -.1 * 100K$$

$$R_f = 10K$$

Step 3 - Since R_f and R_{in} are known, A_{Vbias} $(1+R_f/R_{in})$ can be found. Therefore, Vbias will equal the desired output offset divided by the gain (seen from the bias voltage input) created by the previously computed Rf and Rin values. Rearranging the reverse-acting V_{out} equation yields the V_{bias} equation. The algebraic manipulation of the V_{out} equation is again left to the student.

Given,
$$V_{out} = V_{in}(-R_f/R_{in}) + V_{bias}(1+R_f/R_{in})$$

$$V_{bias} = \frac{(V_{out}-V_{in}(-R_f/R_{in}))}{(1+R_f/R_{in})}$$

Choosing corresponding input and output values from the circuits' operational specification yields;
$$V_{bias} = [1.0V - (-5V * -. 1)] \div (1.1)$$

$$V_{bias} = .5V / (1.1)$$

$$V_{bias} = .54545V$$

The bias voltage is to be located directly on the non-inverting input terminal at the operational amplifier.

Step 4 - The biasing circuit will consist of a two-resistor voltage divider which will "see" a voltage gain (A_{Vbias}) equal to the non-inverting amplifier gain of $1+R_f/R_{in}$. The combination of A_{Vbias} and Vbias will equal the required output-offset of +.5 volts. Using the previously presented technique of sizing the resistors according to the desired voltage drops, R2 will equal .545K and R1 will equal the difference between .545 volts and the available supply of say, 5 volts. R1 will need to drop 4.455 volts and according to the procedure should be 4.455K Ohms. The following diagram illustrates.

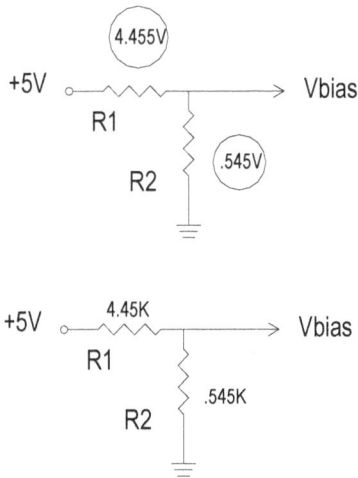

Figure 22 – Determining the biasing resistor values according to the required voltage drops. The resistor values equal the numerical values of the voltage drops.

Unfortunately, resistors of the quantities indicated do not exist. However finding available resistors is merely a matter of mathematically manipulating each resistor by the same amount. As an example, doubling the value of each would yield .9K for R2 and 8.9K for R1. And although these values are not commonly available either, the nearest standard values should work well. A 9.1K for R1 and a 1K for R2 should suffice. For the perfectionist, halving each of the original values will provide standard values of 270 Ohms for R2 and 2.2K for R1. If additional accuracy is required, gain and offset potentiometers will be required to provide provisions for calibration. The completed circuit follows.

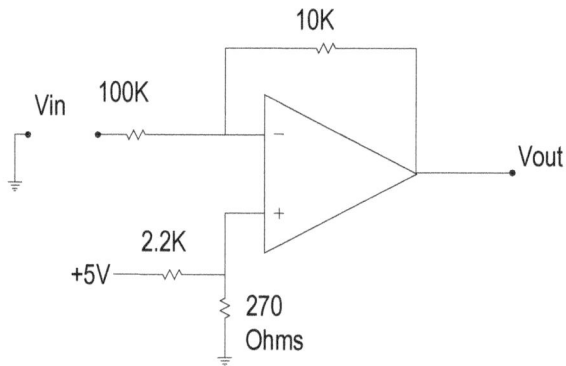

$$Vout = Vin\ (-Rf/Rin) + Vbias(1+Rf/Rin)$$

Figure 23 – The biased inverting amplifier designed for use with the level sensor and reservoir example. The positive output leadwire would be connected to the left end of the 100K Rin and the negative or ground connection from the sensor would be connected to the ground connection indicated at the input Vin.

Amplifier Design Problems

#1 - An amplifier is to convert an input signal span of 0 to 4 Volts into 1 to 9 Volts. Determine the amplifier gain (Av) and offset (b). Determine and include the circuit diagram and component values.

#2 - An amplifier is to convert an input signal span of 1 to 5 Volts into 1 to 9 Volts. Determine the amplifier gain (Av) and offset (b). Determine and include the circuit diagram and component values.

#3 - An amplifier is to convert an input signal span of 1 to 5 Volts into 0 to 10 Volts. Determine the amplifier gain (Av) and offset (b). Determine and include the circuit diagram and component values.

#4 - An amplifier is to convert an input signal span of .02 to .08 Volts into 0 to 10 Volts. Determine the amplifier gain (Av) and offset (b). Determine and include the circuit diagram and component values.

#5 - An amplifier is to convert an input signal span of -.05 to +.10 Volts into 1 to 9 Volts. Determine the amplifier gain (Av) and offset (b). Determine and include the circuit diagram and component values.

#6 - An amplifier is to convert an input signal span of -.02 to .09 Volts into 0 to 3 Volts. Determine the amplifier gain (Av) and offset (b). Determine and include the circuit diagram and component values.

#7 - A sensor outputs -. 02 to +. 09 Volts over an input span of 0 to 30 psi. Design an amplifier to condition the sensor output to 0 to 3 Volts.

#8 - A reverse acting sensor outputs .02 to -.01 Volts over an input span of 0 to 20 pounds. Design an amplifier to convert the sensor output into 0 to 2 Volts direct acting with force.

#9 - A reverse acting sensor outputs 80 to 10 millivolts Volts over an input span of -50 to +150 degrees centigrade. Design an amplifier to develop an output span that is direct acting with temperature and outputs -.5 to 1.5 Volts.

#10 - A sensor outputs 3.73 to 2.73 Volts over an input span of 273 to 373 degrees Kelvin. Design a signal-conditioning amplifier to display 2.73 to 3.73 Volts direct acting with temperature.

#11 - Design a multi-stage amplifier to convert the output of a temperature sensor into voltages representative of Centigrade, Fahrenheit and Rankin degrees. The temperature sensor outputs 2.73V to 3.73V over an input temperature span of 273° to 373° degrees Kelvin, or 10mV/K°.
The output voltages of all stages follow,
Sensor Outputs – Kelvin Degrees
2.73V – 3.73V
Stage 1 Outputs – Centigrade Degrees
0.00V – 1.00V
Stage 2 Outputs – Fahrenheit Degrees
.32V – 2.12V
Stage 3 Outputs – Rankin Degrees
4.92V – 6.72V

#12 – Design an amplifier to condition an SX30DN pressure sensor. The sensor provides an output range of -. 02V to +. 09V when an input pressure range of 0-psi to 30-psi is applied. Design the amplifier to convert the sensor output to 0 to .3V for use with the following 8054-ME digital panel meter (DPM). Address input voltage range, decimal location and related concerns on the following spec sheet.

8054-ME LED PANEL METER

FEATURES

200mV DC INPUT (SCALABLE 20,200,500)
100M Ω INPUT IMPEDANCE
SELECTABLE DECIMAL POINT
AUTO POLARITY
AUTO ZERO
5VDC POWER
EASY MOUNTING

SPECIFICATIONS

MAX INPUT:	199.9 mV	MAX DISPLAY:	1999 COUNTS
INPUT IMPEDANCE:	100MΩ	ACCURACY:	+-0.5% (23° +- 5° < 80% RH)
SUPPLY INPUT:	5VDC	SUPPLY CURRENT:	160mADC
	-Power & -Input are Common	CONVERSION RATE:	2-3/SEC.
CONVERSION TYPE:	DUAL SLOPE	CHARACTER HT.:	.56in
DISPLAY:	3-1/2 DIGIT RED LED	METER SIZE:	70MM X 45MM X 23MM
OVERRANGE:	"1" SHOWN ON DISPLAY		
DECIMALSELECT:	3 POS. JUMPER SELECT		

SETUP & OPERATION

A: Install Decimal point jumper as needed
B: Remove RB Jumper and install
   Scaling resistors RA/RB (Not Supplied)
   as required SEE CHART
   Scaling Resistors: 1/2W 0.5% Metal film
C: Connect 5VDC Power Supply
D: Connect source to be measured

CALIBRATION

Input a voltage from calibration source
equal to 1/2 of the full scale range.
Adjust R2 until reading equals that of
calibration source.

MAX INPUT	VOLTAGE DIVIDER	DECIMAL POINT
200mV	None Leave RA Jumper	P3
20V	Remove RA Jumper Install RA: 9.9MOhm Install RB: 100K Ohm	P2
200V	Remove RA Jumper Install RA: 9.99MOhm Install RB: 10KM Ohm	P3
500V	Remove RA Jumper Install RA: 9.999M Ohm Install RB: 1.0K Ohm	None

R-2
RB Scale Resistor
RA Scale Resistor
Meter V+
Power V-
Vin
P3 Jumper
P1 Jumper P2 Jumper

1.9.9.9
P P P
1 2 3
DECIMAL POINT LOCATION

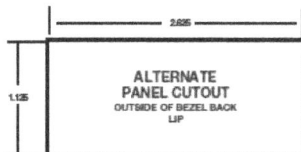

0.25   1.75
0.775
PANEL CUTOUT
SAME SIZE AS BEZEL WINDOW
0.3875
0.125
2 PL

2.625
1.125
ALTERNATE
PANEL CUTOUT
OUTSIDE OF BEZEL BACK LIP

MARLIN P. JONES & ASSOC., INC.

P.O. Box 530400 Lake Park, Fl 33403
800-652-6733 FAX 561-844-8764
WWW.MPJA.COM

210

Chapter 6 – Additional Amplifier Design Examples

This chapter provides techniques for designing inverting and non-inverting amplifiers with variable gain and offset provisions to assist with amplifier calibration. The intent is to introduce the student to the effects and operations of circuits with variable resistances and principles of calibration.

Objectives:
Upon completion of this chapter, you should be able to:
- Explain the process of scaling amplifier design

- Perform a biased direct-acting amplifier design

- Perform a biased reverse-acting amplifier design

- Address impedance matching concerns

- Design and assemble a scaling amplifier

- Address variable gain requirements

- Address variable offset requirements

Introduction
Frequently in the design of scaling amplifiers it is desired to provide a means of accommodating a number of sensor types with similar outputs, or to "fine-tune" the amplifier output. To do this, amplifiers with variable gain and variable offset

quantities are required. If the basic operational inverting and non-inverting amplifier circuit configurations are understood, designing amplifiers with variable resistances is not only a natural extension of the previous chapter, but in most cases is simpler and more forgiving than fixed gain and offset amplifiers.

Non-inverting Amp with Variable Gain and Offset

As an application example, assume a Platinum RTD connected in series with a fixed resistor, as a single active element half-bridge requires scaling into a voltage span representative of the applied temperature. The applied temperature will be 32 to 212 Fahrenheit degrees and the desired output voltage span is to be .32 to 2.12 volts. As an aid in measurement precision, the gain and offset will be variable within predetermined limits to the adjustment ranges. The range of variation will be determined from the sensor characteristics.

To begin, a circuit diagram of the RTD and associated half-bridge components are necessary. The following is representative of the RTD circuit. If the output voltage of the bridge is to "track" or "follow" the resistance variation, the RTD will be located in the bridge circuit between the output and ground, at Rx. If the output span is to be inverted in relation to the sensor resistance variation, the sensor would be placed at R1 and the fixed resistor would be moved to the output location.

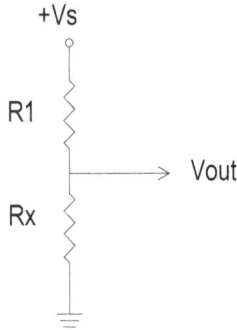

Figure 1 – A half-bridge is commonly used to convert a varying resistive sensor into a proportionally varying voltage. In this circuit the output voltage will follow the sensors' resistance variation.

Since the RTD exhibits a positive resistance-temperature coefficient (temperature increases, resistance increases), the RTD will usually appear in the location of Rx in figure 23. In this manner the bridge output voltage will vary in a direct manner with the applied temperature variation.

Knowing the sensor circuit configuration, the bridge component values can be determined and the output voltage span computed. Given the temperature span is 32° to 212° F, (or 0° to 100° C), the RTD gain is .385 Ohms per degree Centigrade, and the offset is 100 Ohms at 0 degrees Centigrade, the RTD resistance span and the fixed resistor value can be determined.

Determining RTD resistance at 0° (32 F),

$$R_{RTD} = (Gain * Temp) + Offset$$
$$R_{RTD} = (.385 * 0°) + 100 \text{ Ohms}$$
$$R_{RTD} = 100 \text{ Ohms at } 0° \text{ C } (32°F)$$

Determining RTD resistance at 100°,

$$R_{RTD} = (Gain * Temp) + Offset$$
$$R_{RTD} = (.385 * 100°) + 100 \text{ Ohms}$$

213

$$R_{RTD} = 138.5 \text{ Ohms at } 100° \text{ C } (212°F)$$

Since the fixed bridge resistance will equal the minimum value of the sensor resistance, R1 will equal 100 Ohms. The resulting half-bridge RTD temperature sensing circuit will appear as the following.

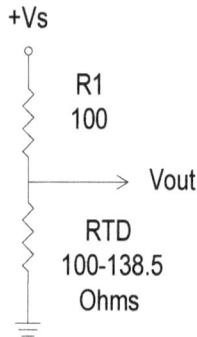

+Vs

R1
100

→ Vout

RTD
100-138.5
Ohms

Figure 2 – An RTD half-bridge with 100 Ohm fixed resistor. The RTD is placed in the output location to output a voltage span in proportion to the applied temperature.

Knowing the sensor circuit configuration, the output voltage span applied to the amplifier can be determined. From the previous figure, at minimum applied temperature (0 degrees C) the RTD will exhibit 100 Ohms. If the supply voltage, Vs, equals 5V, the output voltage at 0° C (32°F) becomes exactly half of the 5V supply, or 2.5V. At the upper temperature extreme of 100 degrees Centigrade, the RTD will appear as a 138.5-Ohm resistance and the output voltage can be determined using the voltage divider equation.

Bridge Vout at 100° C (212°F),

$$Vout = Vs \ (RTD/(R1+RTD))$$
$$Vout = 5V \ (138.5/(100+138.5)$$

$$\text{Vout} = 5V\ (138.5/238.5)$$

$$\text{Vout} = 2.90356V \text{ at } 100° \text{ C } (212°F)$$

Therefore the bridge output varies between 2.5V and 2.9V over a 0-100°C, or 32-212°F temperature span.

Knowing this, the amplifier characteristics can be determined. A graph of the input and output span follows, the input from the bridge will be the 2.5 to 2.9 volts computed previously. The amplifier output span will be the Fahrenheit represented voltage span of .32 to 2.12 volts. The input/output plot appears following.

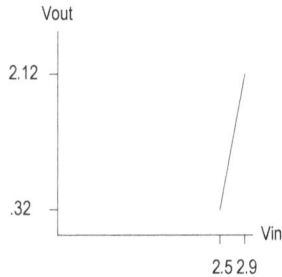

Figure 3 – Input/output characteristic of the amplifier used to condition an RTD single active element half-bridge.

From the input/output plot the amplifier gain and output offset can be determined.

$$\text{Av} = \Delta\text{Vout} / \Delta\text{Vin}$$

$$\text{Av} = 1.8V / .4V$$

$$\text{Av} = +4.5$$

$$\text{B} = \text{Vout} - (\text{Av} * \text{Vin})$$

$$\text{B} = 2.12V - (4.5 * 2.9V)$$

$$\text{B} = -10.93V$$

From the amplifier characteristics, an appropriate amplifier type can be determined. Given the positive input/output slope and non-zero output offset (Y-axis intercept value), a biased non-inverting, or direct-acting amplifier is required. In addition, to provide for a greater degree of accuracy the gain and offset quantities will be made variable. The general-purpose amplifier diagram follows.

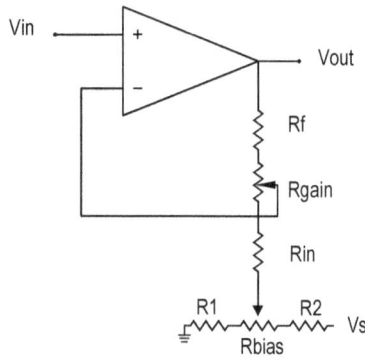

Figure 4 – A biased Non-inverting amp with variable gain and offset. This amplifier will utilise a different gain variation scheme over the previous example.

Upon inspection, the amplifier indicated in figure 18 uses a somewhat different gain variation technique as compared to the previous variable gain and offset non-inverting example. This amplifier will provide an opportunity to demonstrate the rapid approximation approach to gain and offset component value determination and should be quicker and easier to design.

The first step in the design process has been performed by graphing the amplifier I/O requirements and determining the required gain and output offset. To perform the second step of gain resistor determination, a gain variation percentage is required. Normally the amplifier gain and offset variation amount is related to the amount of gain and offset error exhibited by, or inherent within the measurement sensor. However in this case the RTD is an accurate and repeatable temperature sensor with gain and offset errors well within 10%, and probably within 1% of the published and specified temperature versus resistance tables. To accommodate the necessity of variation, a rather arbitrary gain and offset variation percentage of plus and minus 25% will be chosen. Given the accuracy of the sensor in this case, a variation span of less (say plus and minus 10%) would also be appropriate. Whatever amount of variation is chosen, the precise amount of variation *need not be adhered to*, since the ideal gain for the application will reside somewhere between the extreme values of the potentiometer adjustment. Any modification of computed circuit values to accommodate the available components will change the settings at the adjustment extremes, and not as significantly in the center. This should make sizing components to the application easier and hopefully, provide greater insight to operation and design of scaling amplifiers.

At this point the following information is known.

Amplifier Gain, Av = 4.5

Amplifier Gain variation = $\pm$ 25% (approximate)

Amplifier output Offset, B = -10.93V

Amplifier Offset variation = $\pm$ 25% (approximate)

Again, the variations are labeled as approximate to allow for the use of "Ball-park" values. Meaning the actual components placed into operation need not exactly match the computed values.

Authors Note: The design examples for determining the fixed resistance and potentiometer values in this chapter are only suggested. If one understands the operation of the basic inverting and non-inverting amplifiers, determining the component values can usually be performed without a formal, mechanical, paper and pencil computation. When designing and assembling signal-conditioning amplifiers with variable gain and offset provisions, all values can be approximated since the circuits' precise operating characteristics will require setting (calibration) before use in the intended application. As such, "ball-park" approximate resistance values often suffice in the application. It is hoped the student will eventually become adept at determining the necessary component values through an intuitive understanding of op amps and the given application, and the provided techniques will become unnecessary.

Continuing through the design process and knowing the amplifier gain adjustment span, assume Rin = 100K Ohms. Given the gain equation for the Direct Acting amplifier of 1 + Rf/Rin, and the gain variation span of 4.5 + 25% to 4.5 – 25%, the range of values for Rf can be determined.

Rf at minimum AV,

$$Rf = [(4.5 - 25\%) - 1] * Rin$$
$$Rf = (2.375) * 100K$$
$$Rf = 237.5K \text{ Ohms at min. Av}$$

Rf at maximum Av,

$$Rf = [(4.5 + 25\%) - 1] * Rin$$
$$Rf = (4.625) * 100K$$

Rf = 462.5K Ohms at max. Av

Therefore, Rf needs to vary from 237.5K to 462.5K to acquire the exact values of gain extremes. But, keep in mind EXACT values of gain extremes is not required. And in fact, a Rf variation between 200K and 500K, or between 250K and 450K, or any span of values approximating the computed Rf span will suffice. Even 100K to 600K is considered "Ball-park." The ideal gain of 4.5 will occur at a Rf value of 350K Ohms. As long as 350K is somewhere in the center of the gain variation the circuit will adjust nicely.

Rf at Ideal Av of 4.5,

$$Rf = (4.5 - 1) * Rin$$

Rf = (3.5) * 100K

Rf = 350K

The ideal Rf value is important when determining the gain seen by the bias voltage. Since the Avbias gain equation is –Rf/Rin, a value of Rf needs to be determined. For best results a value near the center of Rf should be used since this is the most likely location of the gain determining resistance setting. For this reason the ideal gain and corresponding Rf is used. Note that in the previous description, a reference to Rf includes the sum of the fixed resistance labeled as Rf in the previous circuit diagram, and the potentiometer labeled as Rgain in the diagram. Before determining the bias resistance values, the individual Rf and Rgain should be determined.

Depending upon the combination of components and specific circuit diagram utilized, the Rf and Rgain values can be determined as the sum and difference of the extreme resistance values. Referring to the circuit diagram following, at the maximum gain setting the wiper of Rgain is set to the bottom of the potentiometer yielding the maximum value of resistance. Therefore the sum of Rf and Rgain both contribute to the gain determining feedback resistance value. At minimum gain, the wiper of the gain resistance is at the upper position effectively shorting the entire resistance out of the circuit. In which case the gain determining resistance becomes only Rf. Likewise the value of Rf is the minimum computed value previous, 237.5K. And the variable resistance value will be the difference between the maximum and the minimum resistance values.

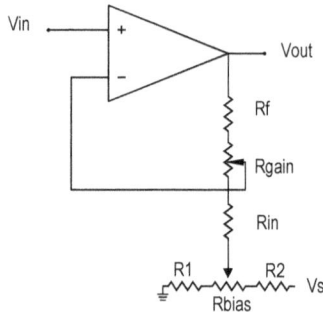

Figure 5 – The biased Non-Inverting Amplifier with variable gain and offset.

Or,

Rf = minimum computed Rf value = 237.5K Ohms

Rgain = Maximum computed Rf value minus minimum computed Rf

Rgain = 462.5K – 237.5K

Rgain = 225K Ohms

Using the nearest standard values of a single fixed Rf resistance of 237.5K (220K), and a variable Rgain resistance of 225K (200K), and an Rin resistance of 100K will result in a computed gain variation of 3.2 to 5.2. Note the ideal gain of 4.5 is located within the extremes of the variation span. Again, getting exact values of gain variation is not important, realizing the effects of using larger or smaller Rf, Rgain and Rin resistances in relation to the amplifier gain is of greatest importance. And knowing how to make subtle adjustments when required is also important. These concepts will be most easily afforded those who put the effort into preparation in the first few op amp chapters of this text.

The biasing circuit components become relatively easy after the gain circuit is completed. Again, assume the ideal gain is developed by the combination of gain determining components. In this case the ideal gain is 4.5, and would be created by an Rin of 100K, and a total Rf (with Rgain setting) of 350K. Since the gain seen by the bias voltage is inverting (-Rf/Rin), Avbias equals a minus 3.5. Or

$$A_{Vbias} = -Rf/Rin$$
$$A_{Vbias} = -350K/100K$$
$$A_{Vbias} = -3.5$$

Since,

$$\text{Output Offset, B} = Vbias * A_{Vbias}$$
$$Vbias = B/ A_{Vbias}$$
$$Vbias = -10.93V/-3.5$$

Vbias = +3.1228V

Again, the variation of plus and minus 25% was chosen arbitrarily. But even if the adjustment span were developed from sensor specification deviation, if an additional variation were added to the expected deviation circuit values could be rapidly approximated.

Using the plus and minus 25% variation yields a bias voltage span of 3.1 − 25% to 3.1 + 25%, or approximately 3 volts ± 25%, from 2.25V to 3.75V. To determine the component values the biasing circuit needs to be considered as follows.

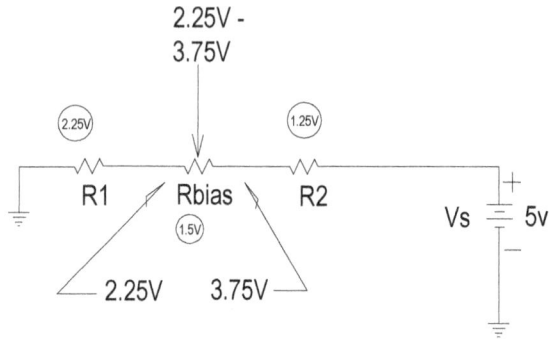

Figure 6 – Bias voltage circuit requirements for the RTD scaling amplifier example.

Once the voltage values at the extremes of the wiper location are determined, the voltage supply, the number of dropping resistors and the amount of voltage each component will need to establish can be determined. From the required Vbias and the allowed variation, the voltage at the potentiometer extremes was determined. Since the voltage span is positive and less than 5 volts, a 5-volt supply can be used. If the voltage span from the potentiometer wiper were starting or ending at exactly 5 volts or ground (zero volts), or if the computed voltage variation from the wiper were close to these values (with variable gain and offset circuits an approximation is allowed), a single, or possibly no dropping resistors (R1, R2) could be used. Knowing the voltage desired from the potentiometer wiper, the voltage across each dropping resistor is found. The circled voltages in the diagram represent the desired voltage drops across the dropping resistors. Sizing the dropping resistors (in 1's of Kilohm values to avoid loading the gain circuit) according to the voltages desired yields the resistor values. The following diagram includes the resistor values.

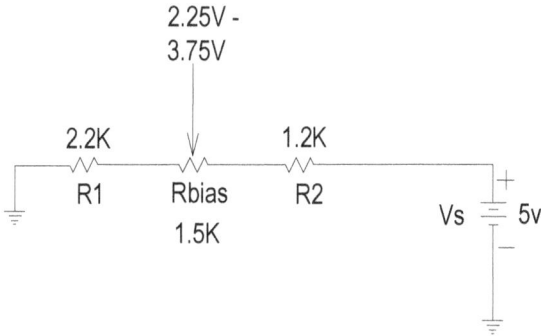

Figure 7 – Biasing circuit for the RTD scaling amplifier.

Unfortunately 1.5K potentiometers are not commonly available, but a 1K potentiometer is a common item. Reducing all indicated resistor values by one-third will result in resistors close to available standard values, and yields the finished amplifier as shown following. All indicated values are commonly available, standard resistor values. The RTD half-bridge is also included on the amplifier circuits' input.

At this point the circuit should be assembled, calibrated (gain and offset potentiometers are set for correct output voltage values) and tested for conformance to the operational specifications. The output voltage span should be set, using the two adjustments for .32V at 32 degrees Fahrenheit and 2.12V at 212 degrees F.

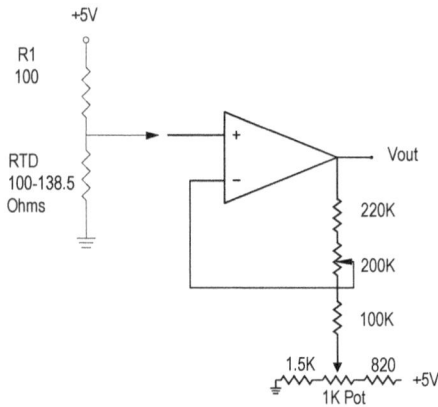

Figure 8 – The completed Biased Non-Inverting Amplifier with variable gain and offset adjustment for RTD half-bridge signal conditioning.

Inverting Amp with Variable Gain and Offset

For control purposes, assume it is desired to convert a negative-going .5V to 4.5V into a positive going 1V to 9V signal span. This type of scaling is fairly common in control system applications where a controller output requires conversion for compatibility with an electronic *servo valve*. A servo valve is a transducer, a component used to convert energy forms. Common servos convert electronic voltage or current into air pressure, hydraulic pressure, rotation, linear displacement and so forth. Servo is a general term used to describe a group of components responsible for converting an electronic signal into some other physical form. Other terms for the same components are "I to P's" for current to pressure transducer, "E to P's" for voltage to pressure, actuators, positioners and anything that implies a proportional conversion of an electronic signal into a physical variable. Servos are generally understood to provide the muscle in a control system in that they are usually expected to exhibit a positive influence on the process under control and are responsible for regulating a physical quantity responsible for returning the process to the set point should a disturbance occur.

In this example an output voltage is acquired from a controller that requires conversion for compatibility with the control system components. Often in experimental applications, a signal is acquired from a programmable device such as a Personal Computer (PC), Programmable Logic Controller or other proprietary special application device. These devices are used for purposes of cycling a test fixture, acquiring data, providing sequential test instructions, or controlling the components associated with testing processes. The output spans from programmable devices are usually fixed and may require conversion to the voltage or current span required by the servo. Functional conditioning modules to perform this type of conversion is available but can be expensive ($100 to $500 each), and in most cases if an electronics lab is available an op amp circuit is assembled to perform the required signal conversion.

Figure 9 – A signal conditioning module. This particular 120VAC unit converts a temperature sensor output into a 4-20 milliamp signal span for transmission over long distances. Similar units a available in a variety of input spans, output spans and supply voltages.

Currently, a controller output is fixed at .5V to 4.5V, and moves negative when the sensor output moves positive. It is desired to convert the output span into one that moves positive when the sensor output (and input) moves positive, and convert the voltage span into the 1V to 9V span required by the Electro-pneumatic (also called an "E to P," and designated E/P) servo valve. In order for the servo output to achieve the full 3 to 15 psi range of output, the input voltage span must achieve one to nine volts at the input. Likewise, additional conditioning is required to convert the voltage span and direction.

Figure 10 – Six "servos" or current to pressure transducers. Although difficult to observe, two front panel adjustments (span and zero) are available to calibrate the input/output relationship. These units are used to convert a 4-20 milliamp signal from six personal computers, into 3-15 psi air pressure signals to position actuators and associated valves.

Figure 11 – A voltage to pressure (E/P) servo with the cover removed. This unit converts a 1 to 9 volt input span into a 3 to 15 psi output pressure span. The pressure supply and output pressure connections are obvious at the left and the input voltage connections in the upper right. Note in this example a resistor a placed across the inputs. A 500 resistor is used to convert the available 4 to 20 milliamps signal into a 2V to 10 volt signal span, the unit also has self-contained span and zero adjustments to set the output to the required 3 to 15 psi. The "zero" adjustment effectively shifts the 2V to 10V input voltage into a 1V to 9V equivalent resulting in a 3 to 15 psi output span.

To begin designing the required signal-conditioning amplifier, a graph of the amplifier characteristics is necessary. The graph should display the desired 1V to 9V positive going output span and the .5V to 4.5V negative going input span.

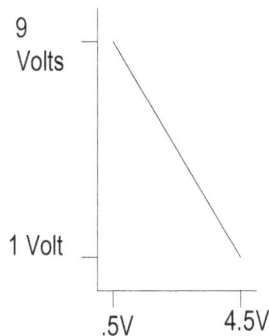

Figure 12 – I/O plot for the signal conditioning amplifier of example #5. Note the reverse-acting slope characteristic and corresponding large output offset above 9 Volts.

227

Given the required I/O characteristic, the amplifier gain and offset can be determined.

$$Av = -(8V/4V)$$
$$Av = -2$$

The negative polarity represents the negative slope of the I/O characteristic and signifies an inverting (reverse-acting) amplifier type.

$$B = Vout - (Av * Vin)$$
$$B = 9V - (-2 * .5V)$$
$$B = 9V + 1V$$

B = 10V

All quantities are taken from the characteristic plot.

From the computations, an amplifier with an inverting gain of 2 and an output offset of 10V is required.

The biased Inverting amplifier with variable gain and bias will take the appearance of the circuit in the following diagram.

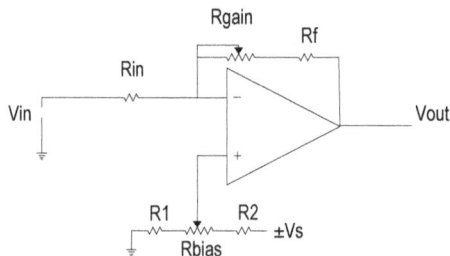

Figure 13 – A biased inverting amplifier with provisions for gain and output offset adjustment.

In determining the resistance values, Rin is assumed to have an initial value of 100K Ohms. Since Rin determines the input impedance for the inverting amplifier, the signal sources' output impedance must also be considered here. Since the .5V to 4.5V signal source is assumed to be coming from an existing controller, the output resistance will probably already be set to a minimal value. Control, testing and data acquisition equipment purchased from reputable suppliers will have impedance matching concerns already addressed. The output impedance will be low, and the input impedance high for proper voltage signal transferring concerns. For the current example, the output impedance of the signal source is assumed to be low (usually meaning 1K Ohms or less) and the corresponding op amp Rin of 100K is considered to be high. Any input impedance over 10K Ohms is considered to be high, so the signal source and the amplifier load are correctly impedance matched.

Given the desired gain is –2, and the inverting amplifier gain is –Rf/Rin, the ideal Rf would equal 200K Ohms. However variable gain and offset bias is desired in this example and since no input signal tolerances were established, an assumed gain and bias variation of plus and minus 50% will be incorporated into the design. A gain variation of plus to minus 50% will result in a gain span of –1 to –3. The ideal gain and Rf, and corresponding Rf variation can be determined from the amplifiers' Av equation. The negative signs are carried for reasons of conformity, and Rf in the following computations represents the total feedback resistance, which includes the fixed Rf and variable Rgain.

Since,

$$Av = -Rf/Rin$$
$$Rf = Av * (-Rin)$$

Rf at the ideal gain of -2,

$$Rf = Av * (-Rin)$$
$$Rf = -2 * (-100K)$$

Rf = 200K , Ideally

The ideal values are designated as a reference. It assumes the input span and all initially assumed amplifier values (such as the resistors) are available.

Max Rf at Max Av,

$$Rf = -3 * (-100K)$$
$$Rf = 300K \text{ max.}$$

Min Rf at Min Av,

$$Rf = -1 * (-100K)$$
$$Rf = 100K \text{ min.}$$

Therefore, from the computations the feedback resistance span should vary from 100K to 300K Ohms. To perform the resistance variation a fixed 100K resistance and 200K variable resistance is necessary.

If a 200K variable resistance is not available, determine the available variable resistance values and scale all existing resistors to accommodate the available variable resistance. As an example, assume a 500K variable resistance is available. The 500K is 2.5 times larger than the value determine in the current example. Increasing the size of all gain determining resistances by 2.5 will result in the same gain variation span. Likewise Rin becomes 250K, Rf becomes 250K (or as close as possible, 270K) and the Rgain variable resistance is 500K. If a 100K variable resistance is available, Rin becomes one-half of the assumed 100K or 50K (49K) and Rf becomes 50K (49K). Again, all resistors are scaled to the available variable resistance since the variable resistances are the fewest in variety typically.

At this point the circuit appears following with the initially assumed and computed component values. The output-offset biasing network will be addressed next.

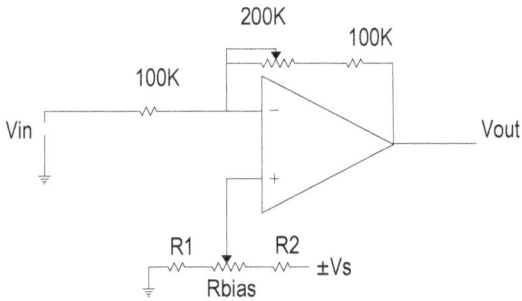

Figure 14 – Biased-inverting amplifier with gain determining values indicated.

In determining the biasing resistance values, the gain seen from the bias input must be assessed. To perform this, the ideal inverting gain must be determined. The ideal gain can be determined a number of ways. The variable gain determining resistance can be assumed to be centered and exhibiting 50% of its rated value, the original gain computation from the I/O characteristic plot can be used, or the Rf and Rin values from the ideal gain computation can be substituted into the Av_{bias} equation.

Av_{bias} looking into the non-inverting input terminal will equal the non-inverting voltage gain of 1 + Rf (ideal)/Rin.

$$Av_{bias} = 1 + (200K/100K)$$

$Av_{bias} = +3.0$

Knowing Avbias, the ideal Vbias can be determined from the previously computed output-offset taken from the original I/O characteristic plot. An abbreviated form of the amplifier circuits' Vout equation can be used to determine the required ideal Vbias..

Since,

$$Vout = Vin (-Rf/Rin) + Vbias (1+Rf/Rin)$$

And, since the input voltage value at the computed output-offset (B) is zero volts, the Vin term in the Vout equation becomes zero, leaving;

$$Vout = Vbias \ (1+Rf/Rin)$$

Again, this condition is true at the output-offset value only, since the input value is always zero at the computed Y-axis intercept.

Therefore, Vbias becomes,

$$Vbias = Vout/(1+Rf/Rin)$$

Where Vout is the computed output-offset value of "B" taken from the I/O characteristic plot. In a broader sense, this concept applies for both inverting and non-inverting scaling amplifiers, and results in the following Vbias relationship.

$$Vbias = B \ / \ Av_{bias}, \text{ for } \underline{both} \text{ amplifier types.}$$

The ideal Vbias for the current amplifier becomes,

$$Vbias = +10V/3$$
$$Vbias = 3.33V$$

Given the assumed variation span of plus and minus 50%, the actual amount of variable Vbias becomes,

$$Vbias = 3.33V \pm 50\%$$
$$\text{Minimum Vbias} = 1.67V$$
$$\text{Maximum Vbias} = 5.0V$$

Given the computed voltage variation, the biasing circuit as it appears in the previous circuit diagrams will require a slight modification. Since the desired voltage variation is positive and equal to 5 volts or less, a positive 5-volt supply can be used. In addition, since one extreme of the desired variation is at 5 volts, one side of the potentiometer will be connected directly to the five-volt supply and only a single voltage dropping fixed resistor will be required in the biasing circuit. The following diagram demonstrates.

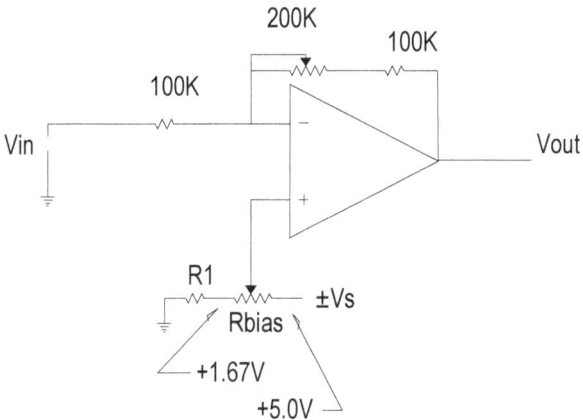

Figure 15 – Biased-inverting amplifier with bias voltage extremes labelled. Note the direct connection of the potentiometer to the supply, indicating one bias voltage extreme will be the supply voltage of 5 volts.

From the diagram, the desired bias voltage extremes are labeled at the outside of the potentiometer, and represent the voltage span that will be available on the wiper as the potentiometer is varied between the two extremes. To determine the necessary resistance values, determine the voltage drops given the indicated extreme values. Form the diagram,

$$V_{Rbias} = 5.0V - 1.67V = 3.33V$$
$$V_{R1} = 1.67V$$

233

Using the previously established technique of sizing the resistance values to the voltage drops, a 1.6K fixed resistance and a 3.3K variable will work nicely. But potentiometers are not available in 3.3K values. Variable resistances are typically available in round-number multiples of 1, 2, 5 and 10, (and the 2 multiples are somewhat scarce). Again, scaling the potentiometer to the available value and multiplying the fixed resistance by the same amount will resolve the problem. Assume a 10K pot. Is available. The 10K-potentiometer value is 3 times larger than the originally desired 3.3K, therefore the fixed resistance needs to be increased by a factor of 3, from 1.6K to 4.8K. The 4.8K is not a commonly available value but 4.9K and even 5.1K would perform well, remember precision of the established voltage(s) is not required, the gain and offset variations are compensation against any lack of precision. The final amplifier design follows.

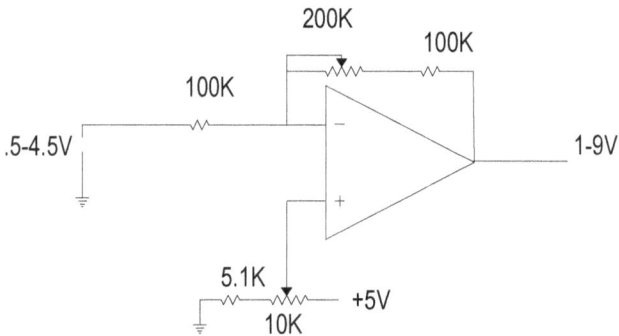

Figure 16 – The completed biased inverting amplifier with variable gain and offset provisions.

Direct-Acting Amp with Variable Gain and Offset, A Different Approach

Gain and bias variations always operate within defined limits such as +/- 20%, +/- 33%, +/- 50% or as close as reasonably possible to the required spans. The adjustment percentages are determined by the input sensor's rated gain and offset (bias) deviation. This information is provided with the sensor specifications. Full scale output (a gain equivalent) and null offset (typical distance from ideal minimum output) specifications are always indicated with typical minimum and maximum values or percent deviation from the ideal. These percentages plus another five or ten percent for non-standard resistor value computations, component tolerances and the inevitable unpredictable will determine the variation spans of the circuits gain and bias values.

The question usually arises here, why use a limited adjustment span? Why not make the gain adjustment span from the minimum to the maximum possible values, say zero to A_{vol}? Or why not make V_{bias} variable between 0 Volts and the largest available supply value? The answer is associated with a term called *resolution*, or the ability to observe small quantities clearly. Large gain and bias adjustment ranges do not yield good resolution. Sensor signal conditioning circuits often require precise (three digits) gain and bias values to display, transmit or control a measured variable accurately. A large adjustment span would be difficult to calibrate since a small potentiometer rotation would result in a large gain or bias change. Setting a specific value of gain or bias would require considerable fine-tuning. Multi-rotation, 10 or 20 turn potentiometers improve but do not remedy the condition.

Since potentiometer and circuit operating values inevitably change, environmental factors such as temperature and vibration would have serious performance effects on large gain and bias adjustment spans. For these reasons the amount of gain and bias variations is always limited to a specific percentage above and below the ideal values. The amount of percentage increase and decrease isn't quite as important when viewed with this attitude, as long as the adjustment spans provide adequate resolution and the desired gain and bias values can be easily acquired. Regardless of the percentage of adjustment span, the computed ideal gain and bias values should appear as close to the potentiometer's center of adjustment range as possible.

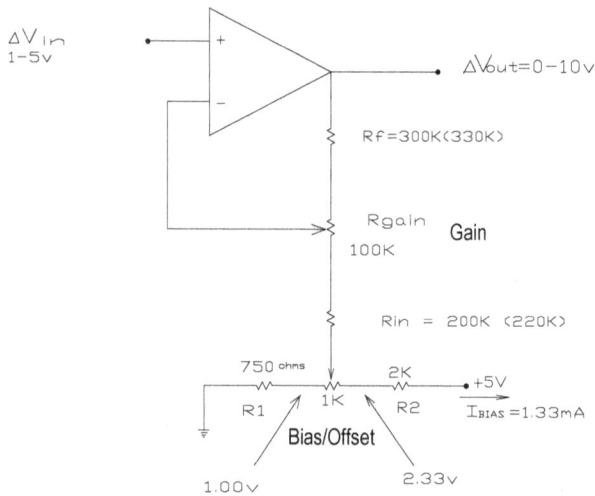

Figure 17 – The Biased Non-Inverting Amplifier with variable gain and offset provisions. This circuit represents one of many methods of performing gain and bias variation.

The previously designed direct-acting amplifier circuit is redrawn containing the gain and bias potentiometers in figure 16. Assume the gain of +2.5 is to be variable by +/- 20% and V_{bias} of 1.67V by +/- 40%. The variation percentage values are assumed to be as indicated in the sensor's specifications. Depending upon the quality of sensor, the given percentages are considerably larger than typical. In essence, the purpose behind the variable gain and variable offset amplifier is to provide an amplifier that can be "tuned" to accommodate use with sensors whose characteristics may vary over a wide range. For a given span of output voltage, a sensor with a low sensitivity would require an amplifier with a higher gain. Another sensor of the same model may exhibit a higher than average sensitivity, requiring an amplifier with a lower gain. Less expensive sensors of the same model type commonly exhibit varying sensitivity and output offset characteristics.

From the previously presented variation percentage, the variable gain span is;

$$\text{Gain span} = A_v +/- 20\%$$

$$A_v + 20\% = 2.5 + .5$$

$$A_{Vmax} = 3.0$$

$$A_v - 20\% = 2.5 - .5$$

$$A_{Vmin} = 2.0$$

The variable bias voltage span becomes;

$$V_{bias} \text{ span} = 1.67V +/- 40\%$$

$$V_{bias} + 40\% = 1.67V + .668V$$

$$V_{biasmax} = 2.33V$$

$$V_{bias} - 40\% = 1.67V - 40\%$$

$$V_{biasmin} = 1.00V$$

Gain Circuit Components

The following is one of numerous approaches to determining the necessary component values for variable gain and variable offset non-inverting (direct-acting) amplifiers. An additional, simpler technique follows.

The mathematical technique for gain resistor value determination will be applicable to only non-inverting amplifiers and should begin by choosing an available potentiometer, preferably in hundreds of kilohms. The following derivation will provide a method of finding R_f and R_{in} after a gain potentiometer, R_{gain}, has been selected. The student need not perform the derivation but should follow the reasoning behind the development of the resulting equations.

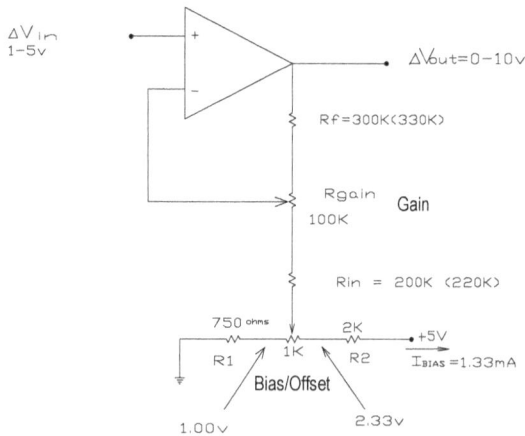

Figure 18 – The biased non-inverting amplifier with variable gain and variable output offset.

To begin, assume the wiper of the gain determining potentiometer is in the A_{Vmax}, or downward position. In this condition,

Maximum Gain Setting

Op Amp Vout

Rf

Effective Rf

$\triangle$Rg

Effective Rin   Rin

Vbias Input

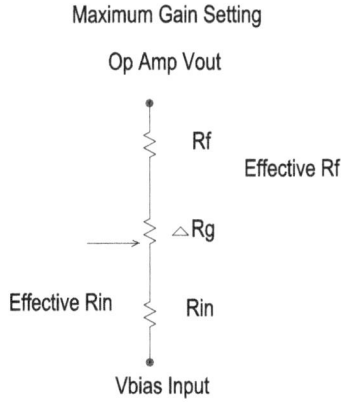

Figure 19 – The non-inverting amplifiers' gain determining resistors. The gain determining Rg
resistor is set at the maximum gain position.

$$A_{Vmax} = 1 + R_{ftotal}/R_{in}$$

Where,

$$R_{ftotal} = R_f + R_{gain}$$

Or,

$$A_{Vmax} = 1 + (R_f + R_{gain})/R_{in}$$

Therefore,

$$A_{Vmax} = (R_{in} + R_{gain} + R_f)/R_{in}$$

Since,

$$R_t = (R_{in} + R_{gain} + R_f)$$

And,

$$A_{Vmax} = R_t/R_{in}$$

Where, R_t is the sum of R_{in}, R_{gain} and R_f in series, R_{total}.

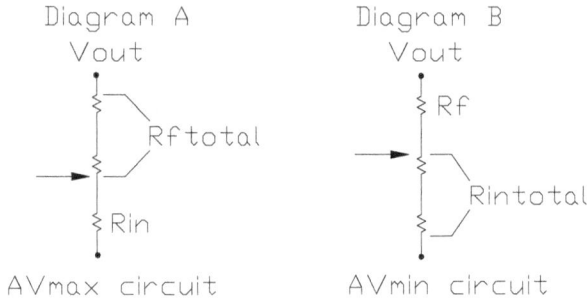

Diagram A
Vout

Rftotal

Rin

AVmax circuit

Diagram B
Vout

Rf

Rintotal

AVmin circuit

Figure 20 – Diagrams exhibiting the effective resistance values at the extremes of the variable resistance.

With the gain determining components connected as indicated in the circuit diagram, and the wiper located at an upper extreme as in Diagram B of Figure 18, the minimum voltage gain, A_{Vmin}, will be exhibited since Rf is minimum.

$$A_{Vmin} = 1 + R_f/R_{intotal}$$

Where,

$$R_{INtotal} = R_{gain} + R_{in}$$

$$A_{Vmin} = 1 + R_f/(R_{gain}+R_{in})$$

Or,

$$A_{Vmin} = (R_f+R_{gain}+R_{in})/(R_{gain}+R_{in})$$

And,

$$A_{Vmin} = R_t/(R_{gain}+R_{in})$$

Rearranging the A_{Vmax} equation to solve for R_{in} produces,

$$R_{in} = R_t/A_{Vmax}$$

Rearranging the Avmin equation to solve for $R_{gain} + R_{in}$ produces,

$$R_{gain}+R_{in} = R_t/A_{Vmin}$$

Subtracting R_{in} from the sum of $R_{gain} + R_{in}$ leaves R_{gain}.

$$R_{gain} = (R_{gain}+R_{in}) - R_{in}$$

Substituting, $\qquad R_{gain} = (R_t/A_{Vmin}) - (R_t/A_{Vmax})$

Or, $\qquad R_{gain} = R_t (1/A_{Vmin} - 1/A_{Vmax})$

Finally, $\qquad R_t = R_{gain}/(1/A_{Vmin}-1/A_{Vmax})$

The equations in this derivation will allow for the design of the gain determining circuitry around the available potentiometer values, and known A_{Vmin} and A_{Vmax} values. Once R_t is found R_{in} can be computed since,

$$A_{Vmax} = R_t/R_{in}$$

$$R_{in} = R_t/A_{Vmax}$$

In conclusion R_f can now be computed since,

$$R_t = R_f + R_{gain} + R_{in}$$

Therefore,

$$R_f = R_t - R_{gain} - R_{in}$$

All three resistance values are now known. Initially this process may appear arduous. With experience the technique will prove rapid and will also provide an improved understanding of circuit operation. To find the circuit values, assume the following are known;

$$R_{gain} = 100K \text{ potentiometer}$$
$$A_{Vmin} = 2$$
$$A_{Vmax} = 3$$

$$R_t = 100K/(1/2-1/3)$$

$$R_t = 600K \text{ (approximate)}$$

$$R_{in} = 600K/3 = \underline{200\ K}$$

$$R_f = 600K - 100K - 200K$$

$$R_f = \underline{300\ K}$$

And the given R_{gain} Potentiometer = $\underline{100\ K,}$

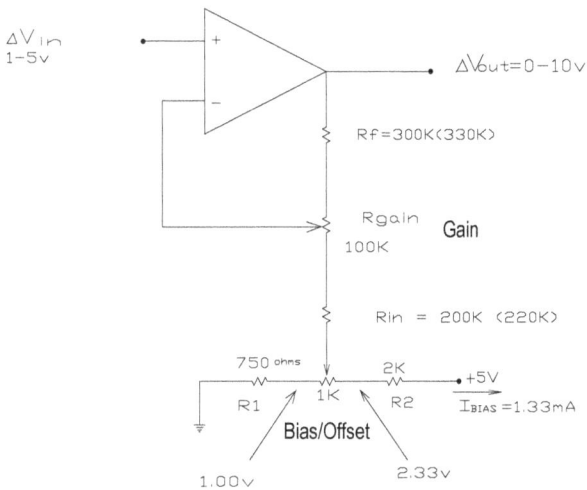

Figure 21 - The designed non-inverting amplifier with variable gain and bias.

The minimum and maximum voltage gains should be verified using the computed values of R_{in}, R_f, and R_{gain}.

$$A_{Vmin} \text{ at wiper up} = 1 + (300K/300K)$$
$$A_{Vmin} = 2$$

$$A_{Vmax} \text{ at wiper down} = 1 + (400K/200K)$$

$$A_{Vmax} = 3$$

The desired gain variation is verified.

Since the computed R_f of 300K and R_{in} of 200K are not standard value resistors, the closest available values will be used. In order to keep the ideal voltage gain of 2.5 near the center of our adjustment span, R_f and R_{in} should both be raised or lowered to the nearest available values. This may result in a slightly altered gain span than originally intended but, it should be remembered the original adjustment percentages were chosen with an additional five or ten percent tolerance to accommodate approximations and non-standard resistor computations. The additional tolerance should retain the ideal gain near the center of our adjustment span. As an exercise the student is encouraged to compute the gain spans using higher and lower standard values of R_f and R_{in}. Of greatest importance is realizing the effect of assuming the higher or lower values for R_f and R_{in} upon the circuit gain, which requires an understanding of circuit operation and applicable equations.

Bias Circuit Components
As mentioned previously, the bias circuit resistance values should be kept small (ones of kilohms) to avoid significantly altering the larger gain resistance, R_{in}. The op amp will consider the total resistance from the inverting input through any bias circuit resistance to ground as R_{in}, and will distribute the feedback voltage from V_{out} accordingly. The circuit's voltage gain is determined from the feedback voltage as discussed in previous chapters. The effective R_{in} will consist of the actual R_{in} and the Thevenin equivalent resistance from the bias potentiometer wiper to all ground connections, (directly to ground and/or through the V_{bias} voltage supply).

As an example, consider the circuit following. Assuming the bias potentiometers' wiper is located to the extreme right, the effective R_{in} will be the actual value of R_{in} and the combination of R_1 plus $R_{bias,}$ paralleled with R_2. If the biasing resistance values are tens or hundreds of kilohms, the effective R_{in} will be large and the voltage gain will be decreased considerably.

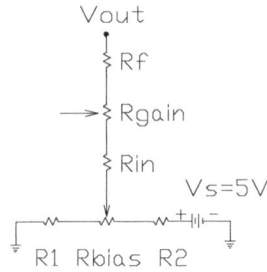

Vout

Rf

Rgain

Rin

Vs=5V

R1 Rbias R2

Bias circuit effect upon Rin

Rin effective = Rin + (R1+Rbias\\R2)

Figure 22 – Circuit to demonstrate the loading effect of the Vbias circuit upon the gain determining resistances. Any resistance placed at the bottom of Rin for biasing purposes will effectively increase Rin and decrease Av, resulting in a possible output signal error.

Since,

$$A_v = 1 + R_f/R_{in}$$

As R_{in} increases the signal voltage gain, A_v will decrease. For this reason it is best to keep bias resistances of single digit kilohms and gain resistances of hundreds of kilohms.

As with the gain determining network, the fixed biasing resistance values are computed around the available potentiometers. The biasing circuitry is determined by assuming a potentiometer value, finding the necessary voltage drops across the potentiometer and remaining resistor(s) and sizing the dropping resistor(s) according to the circuit current and corresponding voltage drops. As was done in the previous examples for our application, the desired bias voltage span is 1.0V to 2.33V. The potentiometer, two fixed resistors and a +5 Volt supply will be required.

Three resistors are essential since the bias voltage span is not starting at ground potential (0 Volts) or stopping at +5 Volts. Only two resistances, one variable and one fixed would be required if the bias span were bounded by the available supply or ground voltages. Occasionally the potentiometer alone will suffice, when the desired span of variation is close to the available connection voltages (+Vs, -Vs, or ground (0V)). Assuming a 1K potentiometer and the 1.0 to 2.33 Volt variation, the drop across the potentiometer is;

$$V_{Rbias} = V_{bias\ max} - V_{bias\ min}$$

$$= 2.33V - 1.00V$$

$$V_{Rbias} = 1.33\ Volts$$

Using a 1K linear (as compared with audio-taper logarithmic) variable resistance, the bias circuit current becomes;

$$I_{bias} = 1.33V/1K = 1.33\ mA$$

In making the gain determining resistances hundreds of Kilohms and the biasing resistances single values of Kilohms the loading effects of one network on the other are minimized. Therefore, the amount of bias circuit current lost in the gain network is insignificant. The current passing through the biasing potentiometer will also be the current through R_1 and R_2. Since the dropping resistor, R_1, drops 1V and I_{bias} is 1.33 mA, the R_1 value becomes;

$$R_1 = VR1/I_{bias}$$

$$R_1 = 1.0V/1.33\ mA$$

$$R_1 = 751.88\ Ohms\ (750\ Ohms)$$

Determining R_2 in the same manner,

$$R_2 = VR2/I_{bias}$$

$$V_{R2} = 5V - 2.33V = 2.67V$$

$$R_2 = 2.67V/1.33\ mA$$

$$R_2 = 2007.52\ Ohms\ (2000\ Ohms)$$

The closest standard values, as indicated (750, 2000 Ohms), should be used for R_1 and R_2. As a precaution the current demand of the bias circuit, I_{bias}, on the five-Volt supply should be computed.

$$I_{bias} = 5V/(R_1 + R_{pot} + R_2)$$

$$I_{bias} = 5V/(750 + 1K + 2K)$$

$$I_{bias} = 5V/3750\ Ohms$$

$$I_{bias} = 1.33\ mA$$

As expected the bias circuit current, I_{bias}, is 1.33 mA and should not place an excessive demand upon the five Volt supply. Since resistor values of kilohms were chosen, any increase in the effective value of R_{in}, and subsequent degradation of A_v created by the biasing circuit resistances will be negligible (less than a single percent).

The biased direct-acting/non-inverting amplifier is designed as displayed following with standard resistor values indicated in parenthesis.

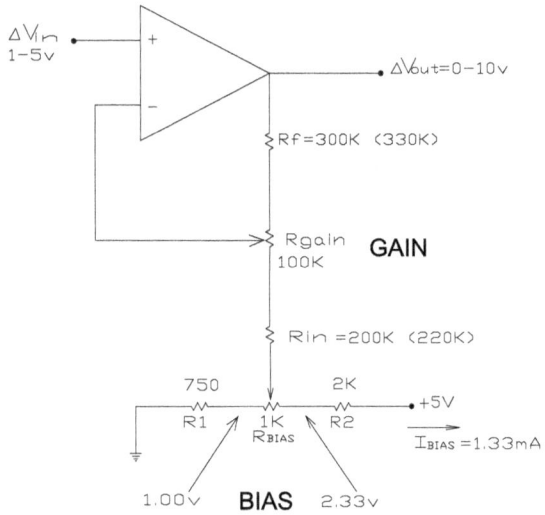

Figure 23 – The biased non-inverting amplifier with provisions for variable gain and variable offset.

Direct-Acting Amplifier with Unity Gain

A pressure sensor measures 0 to 10 psi and outputs a -. 25 to +. 75-volt span. The sensor exhibits 5K Ohms of output resistance. The output signal is to be connected to a data acquisition system for analysis, recording and informational purposes. The input voltage span of the data acquisition system is 0V to 1.0V. An input device is not required to use all of the input span, but needs to have the minimum and maximum output values within the available input span for the data acquisition system to receive all of the measured information. Any voltages outside of the 0.0 to 1.0-volt input span will not be measured by the data acquisition system. If the sensor were to be connected without any adjustment of its output signal span, roughly 25% (all of the pressures corresponding to negative output voltages) of the available information in the sensors' output would be lost. This means that all pressures responsible for generating outputs of less than zero (negative) volts would not be recorded or observed by the data acquisition system. If a test were being performed where all the input pressures resulted in less than zero (negative) volt outputs, no data would be recorded. The test would be a wash (wasted). Obviously not a good situation, as most experimental tests are expensive to run.

An operational amplifier scaling application is required to offset the sensors' output signal span for compatibility with the input requirements of the data acquisition system, while maintaining the 1V input span requirement. In short, the available output voltage span needs to be contained within the 0.0 to 1.0-volt extremes. Variable gain and offset provisions will be included, and impedance matching concerns will be addressed.

At first, this application may appear comparable to a previous non-inverting amplifier design. However a problem is created when a gain of unity or less is required of a non-inverting (direct-acting) amplifier. The problem is associated with the voltage gain equation and the required non-zero output offset. Following is the common non-inverting amplifier with variable gain and offset, Vout equation.

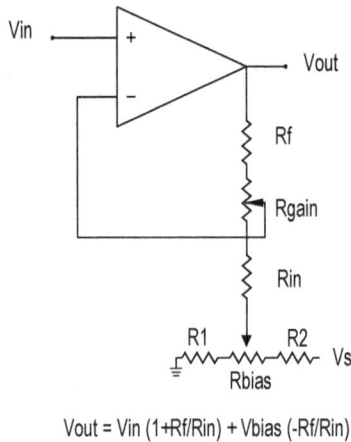

$$Vout = Vin\ (1+Rf/Rin) + Vbias\ (-Rf/Rin)$$

Figure 24 – The biased non-inverting amplifier is limited to a minimum gain if output-offset biasing is required.

Since the gain equation seen by the input is 1+Rf/Rin, the apparent minimum gain is unity. This would occur if the value of Rf is set to zero Ohms. The circuit would function with a short in place of Rf by exhibiting a gain of +1. Unfortunately the gain seen by the biasing input would equal zero in this condition since Rf has assumed a value of zero, and the Av_{bias} equation is –Rf/Rin. With an Av_{bias} of zero, biasing provisions would be eliminated. The dichotomy is this, utilize an amplifier gain of one without biasing, or utilize a higher than required amplifier gain and retain an offset biasing provision. A solution is available but unless the current situation is well understood, the reason for the solution will not be obvious. Before continuing, assure the dilemma is noticed and understood.

To resolve the problem, an additional circuit will be attached to the amplifier input to reduce the available input voltage span. By reducing the input signal span, a larger than necessary amplifier gain will be used. By maintaining an amplifier gain above unity, biasing will also be available. The following circuits assist in demonstrating the conditions.

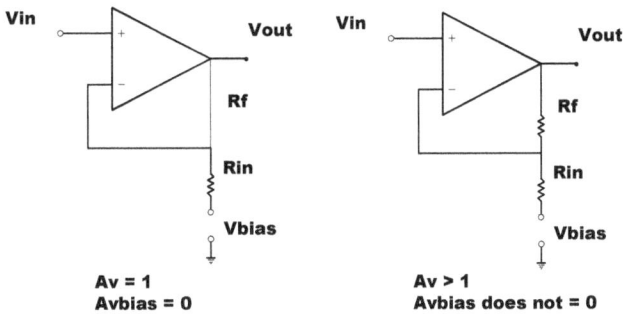

Vin   Vout   Vin   Vout
Rf   Rf
Rin   Rin
Vbias   Vbias
Av = 1   Av > 1
Avbias = 0   Avbias does not = 0

Figure 25 – The biased non-inverting amplifier dilemma, with a voltage gain of 1, biasing is not possible. For biasing to be possible, the voltage gain must be greater than one.

Continuing with the operational specifications, assume the bias and gain variation is to be plus and minus 25%, and the amplifier should be adjusted to output 0.0V at 0 psi, and 1.0V at 10 psi. In addressing the impedance concerns, the operational amplifier will need to consider impedance matching from the 5K sensor to the op amp input, and from the op amp output to the data acquisition system. Neither condition will place an undo hardship on the op amp. Recalling that the op amp input impedance is at least 1 Megohm, connecting to a 5K Ohm source should allow maximum signal transfer from the sensor to the op amp. In regards to the data acquisition system at the op amp output, all commercial data acquisition components rated for voltage inputs will exhibit 1 Megohm or greater of input impedance. Although I have not personally reviewed *all* data acquisition input impedance specifications, commercial equipment manufacturers are aware of the maximum signal transfer condition and have designed the input circuitry to conform to the high input impedance requirement for maximum signal transfer. Impedance matching from the op amp output (at 100 Ohms or less) to the data acquisition input (at 1 Megohm or greater) will not provide an obstacle.

Returning to the design procedure, a graph of the expected input/output characteristic would appear as the following.

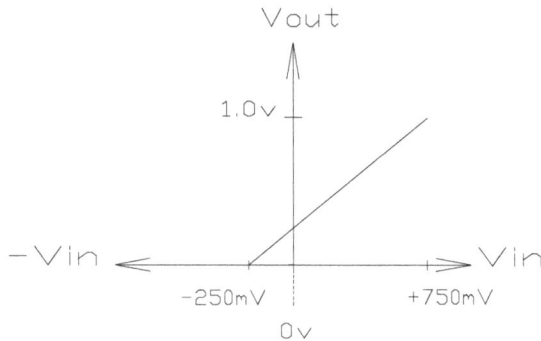

Figure 26 – Input/Output characteristic for the current example. The plot represents an input span of 125 to +75 millivolts, and a corresponding output span of 0.0 to 1.0 volts.

Given the I/O graph, the gain and offset values can be determined.

$$Av = 1.0/1.0 = +1.0$$
$$B = 1.0V - (+1.0*.75V)$$
$$B = +. 25V$$

As a review, it should be noted the output-offset (B) occurs at an input voltage of zero volts, which is obvious in the plot. The signal conditioning circuit will require an *overall* gain of 1.0, and an offset of .25V. An overall gain is the result of multiple (in this case two stages) stages combined to produce a total gain. Each stage will develop its own gain or attenuation, and will have an associated quantity representative of the signal amplification or reduction factor. The two stages will be assembled in cascade with the total gain being the product of the individual stage gains. For the circuit to function given the limitation placed on the non-inverting amplifiers' gain, the first stage will be composed of a resistive divider to reduce the signal amplitude (voltage) by one-half. The second stage

252

will need to complement the first stage signal reduction by amplifying the remaining signal (after reduction) by 2.0. The combined reduction and amplification will result in an overall gain of unity. A block diagram of the individual stages follows.

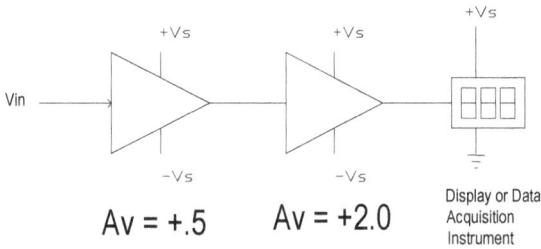

$$Av = +.5 \qquad Av = +2.0$$

Display or Data Acquisition Instrument

Figure 27 – Amplifiers may not always amplify. In the diagram, two amplifier stages are shown in cascade where the output of the first becomes the input of the second. In the current example, to accommodate the circuit requirements the first amplifier will become an attenuator since its purpose is to reduce the input voltage span (Av = .5). The second stage will generate the required gain. Although attenuation is not gain in a formal sense, both gain and attenuation or assigned the symbol designation, Av. True gain is an Av greater than 1, attenuation is a gain of less than one, a "Follower" provides a gain of exactly one or unity. Unity gain amplifiers are used to impedance match signal sources and loads.

To establish an amplification or attenuation factor of .5, an amplifier is not required. Only a series resistive voltage divider network. Amplifier circuits are occasionally used for attenuation but when the signal source requires attenuation and output-offset biasing concurrent. In the current application, a non-inverting amplifier follows a resistive divider. Any offset biasing will be performed in the amplifier stage. In fact, the overall circuit configuration has been chosen to facilitate biasing in the amplifier stage.

The second step of the circuit design process requires a slight modification, since two stages are involved. Before determining the specific component values, a strategy needs to be determined. When connected, the attenuator and amplifier will form a system with a combined gain of 1.0. To accomplish this the first stage could reduce the signal by one-half and the second stage could amplify by two, as the previous diagram indicates. Or, the first stage could reduce the signal by one-fourth, and amplify the resulting signal span by a gain of four. Or the first stage could reduce the signal to one-third and amplify with a gain of three.

Each of the proposed attenuation/amplification strategies would function. The difference between each approach is in the amount of the resulting bias voltage gain, which may influence the power supply required for Vbias. Assume the first condition is to be used, where the attenuation factor is to be .5 and the amplifier gain is to be 2.0, as in the previous diagram.

Another consideration involves the value of the individual resistors that will compose the first stage attenuation network. For an attenuation factor (also called a Beta coefficient) of one-half (B = .5), both resistors need to be of equal value. But should the individual resistors be 1K, 10K, 100K or 1Meg? The circuit diagram will appear as a resistive divider followed by a biased non-inverting amplifier as indicated in the following diagram.

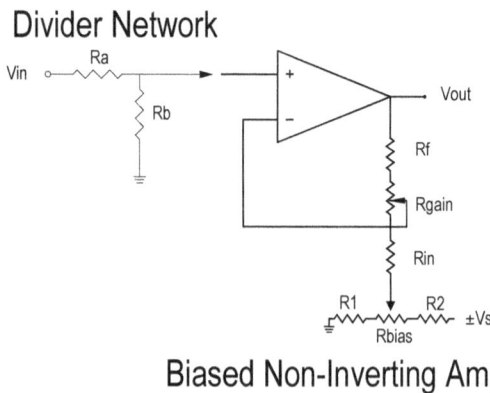

Figure 28 – In order to bias a non-inverting amplifier with a desired gain of unity or less, a resistive divider is required to reduce the signal span. The resulting amplifier gain and resistive divider "gain" are multiplied to determine the overall gain.

The divider resistors are labeled Ra and Rb in the diagram. As can be seen the resistors are connected in series, and will establish a reduction or attenuation factor equal to;

$$Beta, B = Rb/(Ra+Rb)$$

And Vout of the resistive divider will equal,

$$Vout1 = Vin * Beta, or$$

Vout1 = Vin * B, or

$$Vout1 = Vin * Rb/(Ra+Rb)$$

Since gain equals Vout divided by Vin,

Av = Vout/Vin = Rb/(Ra+Rb) = Beta, B

Therefore, the "Gain" of the resistive divider circuit equals the reduction factor or Beta. In order to develop a Beta of .5, Ra and Rb must be equal. Equal divider resistors will evenly divide the applied signal span of -. 25V to +. 75V. The resulting output from the resistive divider will be -. 125V to +. 375V. With a first stage gain of .5, the amplifier gain must equal 2.0 to establish an overall gain of 1.0.

Since Vin is originating from the pressure sensor that contains 5K of output resistance, the input resistance seen looking into the resistive divider (to ground) should be at least ten times the sensors' output resistance of 5K Ohms. The resulting divider resistance from the input to ground should equal at least 50K Ohms, or ten times the sensor output resistance. To assure maximum signal transfer from the sensor to the resistive divider circuit, the divider circuit will be composed of two 100K resistances yielding an input resistance to ground of approximately 200K Ohms. The actual input resistance to the divider would equal Ra plus Rb in parallel with the op amps input resistance of at least 1Megohm.

Input resistance of the divider circuit equals,

Ra + (Rb||Rin of Op Amp)

Given the 100K Ra and Rb resistors and the 1Meg minimum input resistance of the op amp, the resulting input resistance still equals approximately 200K Ohms, as seen from the divider input to ground. The circuit currently appears as follows.

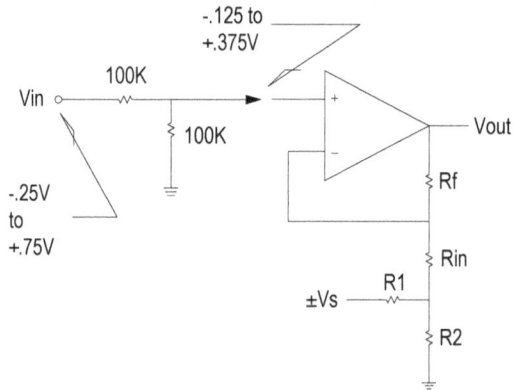

Figure 29 – An abbreviated circuit diagram of the biased non-inverting amplifier. The variable gain and bias resistances have been omitted for purposes of clarity. The resistive divider (B = .5) has been placed at the amplifiers' input and the input voltage spans have been indicated. Eventually, the overall circuit gain will become 1.0, an attenuator gain of .5 followed by an amplifier gain of 2.0.

To determine the amplifiers' resistor values, both the amplifier gain and the resistive attenuation factor (Beta) need to be considered. The signal will propagate (travel) from the input (-. 25V to +. 75V), to the amplifier input (non-inverting input terminal), and on to the amplifier output. The gain, or attenuation, of the individual stages that compose the amplifier determines the overall amplifier gain. The overall gain, Av_T, becomes the product of the individual gain or attenuation factors. For the non-inverting amplifier with a gain of unity or less,

$$Av_T = Beta * Av$$

Where,

       Av_T is the desired overall gain,

       Beta is the resistive divider reduction factor,

       Av is the amplifier voltage gain (1+Rf/Rin).

Since a desired overall gain of unity is desired, and Beta is chosen to equal .5,

$$Av = Av_T \div Beta$$

And,

$$Av = 1.0 \div .5$$
$$Av = 2.0$$

Most of this has been considered previously when values for the resistive divider reduction factor (Beta) were determined. It is performed here as a matter of formality, and indicates a non-inverting amplifier gain of two is required to provide an overall signal gain of one. Before continuing however, the gain and offset variation spans need to be addressed. From an earlier reference, a gain and offset variation of ± 25% was provided. The adjustment span was arbitrarily provided and would normally correspond to the amount of error associated with the incoming signal from the sensor output, with an additional percentage added for "slop." Where "slop" is associated with unpredictable occurrences. Given the variation percentages, the variable gain and variable offset spans follow.

$$\Delta Av = 2.0 \pm 25\%$$
$$\therefore Av \text{ min} = 1.5$$
$$And \ Av \text{ max} = 2.5$$

$$\Delta B = .25V \pm 25\%$$
$$\therefore B \text{ min} \approx .2V*$$

And B max ≈ .3V*

<u>*Note:</u> With such a small variation of the output-offset, or B, value, the span is usually opened to facilitate design and assembly using commonly available values. For these reasons, the output-offset variation span is modified to 0.0 to +.5V. Due to component tolerances, narrow offset spans in the fractional volt range can be difficult to set and frequently require modification after prototype assembly and testing. For this reason our circuits' adjustment will be widened at this point in the design. It is left to the student to determine which approach yields the most beneficial results, but experience will demonstrate that slim variation spans of offset and gain frequently require widening after circuit assembly. Only in vary rare conditions (output spans of well under a volt) should the output-offset spans be variable over a small fractional of a volt.

From the previous, The variation spans become;

$$\Delta Av = 2.0 \pm 25\%$$
$$\therefore Av\ min = 1.5$$
$$And\ Av\ max = 2.5$$

$$\Delta B = .25V \pm 100\%$$
$$\therefore\ B\ min \approx 0.0V$$
$$And\ B\ max \approx 0.5V$$

The proposed circuit diagram appears as follows.

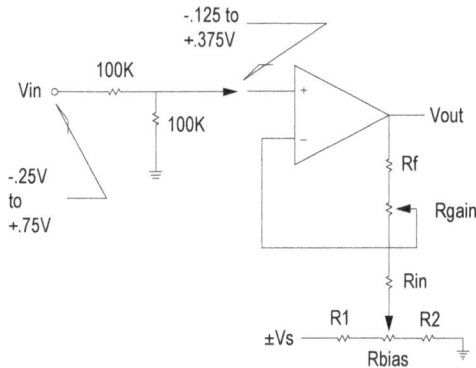

Figure 30 – A unity gain, direct-acting/non-inverting amplifier with variable gain and biasing provisions. The input voltage spans before and after the resistive divider has been labelled.

When determining the gain associated resistance values, assume a Rin of 100K Ohms. From the non-inverting amplifier gain equation, this results in a Rf range from 50K to 150K Ohms. From previous practice, a fixed 50K resistor and 100K variable resistance should suffice as the gain determining resistances in the feedback path.

Since,

$Av = 1 + Rf/Rin$

And,

$$Rin = 100K$$

$$Rf \text{ total at Av max} = Rin (2.5-1)$$
$$Rf \text{ total at Av max} = 150K$$

$$Rf \text{ total at Av min} = Rin (1.5-1)$$
$$Rf \text{ total at Av min} = 50K$$

$$Rf \text{ total at the ideal gain of } 2.0 = Rin (2.0-1)$$

Rf total at ideal gain = 100K

Since the minimum Av is determined by the minimum Rf (when the variable resistance is set to zero Ohms), the fixed Rf resistance corresponds to the minimum Av and equals 50K. The gain determining variable resistance will equal the difference between the computed Rf extremes. The ideal gain will be exhibited when the variable resistance is centered, or when set close to center.

Rgain = (Rf at Av Max) – (Rf at Av min)

Rgain = 150K – 50K = 100K

The circuit diagram including the gain determining resistances follows.

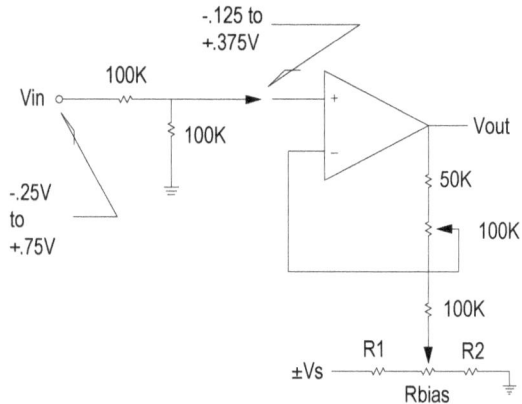

Figure 31 – A biased non-inverting amplifier of unity ideal gain, including gain determining resistance values.

Before beginning computations of the biasing component values, it should be mentioned that the addition of the divider network at the amplifier input would influence only the circuit gain, and not the output-offset value (B) or biasing voltages. This becomes obvious when one considers the input voltage condition when the output-offset (B) is determined. Since the Y-axis intercept quantity (or output-offset) is determined with an X-axis (or input voltage) of zero, the voltage divider at the input becomes insignificant in the discussion of biasing voltage considerations.

The biasing voltages and subsequent resistances can be determined by assuming the gain determining resistances are set for an ideal gain of 1. In this condition, the gain seen by the bias voltage is –Rf/Rin. Or,

$$Av_{bias} = -Rf/Rin$$
$$Av_{bias} = -100K/100K$$
$$Av_{bias} = -1.0$$

From the previous computations, the range of output-offset was re-determined to be 0.0 to .5 volts.

Since,
$$B = V_{bias} * Av_{bias}$$
$$V_{bias} = B/ Av_{bias}$$

Substituting the B range of values into the Vbias equation yields,

Vbias minimum = 0V/-1 = 0 volts
$$Vbias \ maximum = -. 5/-1 = -. 5 \ volts$$

Therefore the bias voltage span becomes 0.0 to -. 5 volts, and the biasing circuit provided in the previous amplifier diagram requires modification. With the biasing voltage having zero volts as one extreme value, a ground connection can be placed at one side of the potentiometer. The other side of the potentiometer must see the previously computed -. 5 volts. Reviewing the biasing circuit and assigning the desired voltage drops, the resistance values can be determined. Note, the biasing voltage range is negative and will require a negative supply.

Assuming Proto-Board supply voltages are used, only +5V, +15V and –15V are available. The –15 volt supply will be used as the biasing circuit supply voltage.

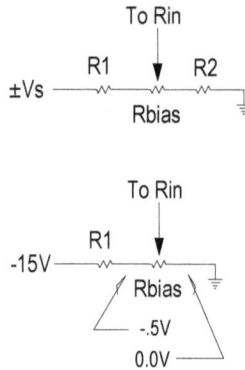

Figure 32 – The original biasing voltage divider circuit (above), and the modified voltage divider with expected voltages labelled.

Using the expected voltage drops to size the biasing circuit resistors results in the following circuit diagram.

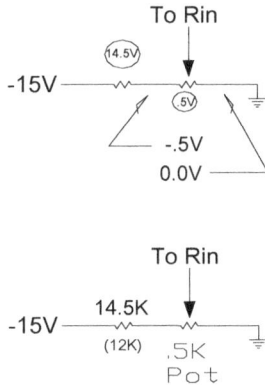

Figure 33 – The voltage drops are labelled according to the voltages desired at the potentiometer extremes (above). From the voltage drops the resistance values are assigned below, the parenthetic 12V represents an available standard resistance value. 15K is also available but was not used since a larger dropping resistor would reduce the voltage available at the potentiometer. The resulting potentiometer voltage span should be close to 0.0 to -.6V.

As was mentioned throughout this chapter, designing amplifier scaling circuits isn't only a matter of finding the correct resistance value but knowing the effect of changing any given resistance around the op amp. The dropping resistance in the final biasing voltage circuit was reduced from an unavailable value of 14.5K to allow a slightly greater amount of voltage to the potentiometer wiper.

The completed amplifier follows.

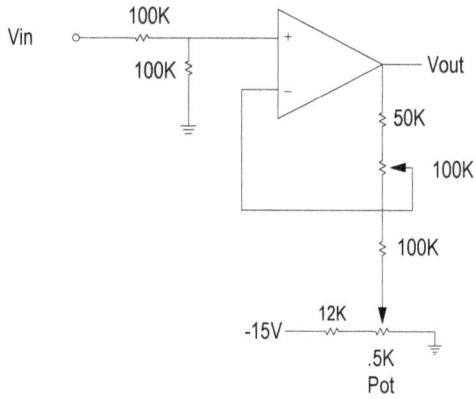

Figure 34 – The completed unity gain, biased non-inverting amplifier with variable gain and output-offset adjustments.

Amplifier Calibration

It appears appropriate at this point to comment briefly upon the required technique for setting the amplifier circuits' gain and bias potentiometers. Calibration, as defined by the Instrument Society of America (ISA) is the procedure of checking, quantifying and correcting the graduations on a measurement instrument. Since a majority of biased operational amplifiers are used for the purpose of sensor signal conditioning, the sensor, amplifier and associated output load become a measurement instrument. In order to receive accurate data from the instrument, provisions to minimize the error inherent in all instruments is provided. These provisions are the gain, usually referred to as span, and bias, or zero adjustments. An additional adjustment of linearity will occasionally be present to assist in assuring a proportional, linear and predictable relation between the physical variable at the input and the output span.

Now that a variable gain and bias amplifier has been designed, a procedure for setting the potentiometers needs to be developed. This procedure is typically referred to as "setup" or calibration, and should not be attempted on a piece of measurement apparatus without first consulting the service manual. Since our circuits are not associated with proprietary measurement apparatus, a general procedure will be developed for span and zero (gain and bias respectively) adjustments. This procedure can be applied to all span and zero adjustments regardless of whether located on a measurement instrument or not.

Span (gain) and zero (bias) are terms that relate to the graphing of straight-line, linear, Y=MX+B input/output relations. Simply stated, the zero (bias) adjustment determines the Y-axis intercept location and the span variation will establish the distance between the output end points or slope of the input/output plot. Given the input/output spans of the previously designed non-inverting amplifier, the effects of the individual span and zero adjustments are demonstrated in the following figures. As can be seen, the figure represents the ideal input/output characteristics of a previously designed amplifier. The characteristic plot at the left will be the end

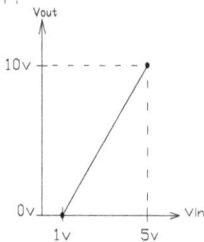

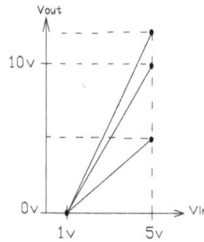

result of the calibration attempt. The graph at the center illustrates the effect of varying the span (gain). A lower span is associated with a smaller distance between output end points and lower slope (lower gain), whereas a greater span results in a larger distance between end points and a steeper slope (greater gain).

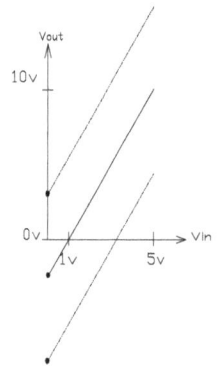

Figure 35 – Demonstration of the graphical effects of gain and output offset variation.

The effects of bias variations, or zero adjustments, become obvious when the input span is varied between 1 and 5 Volts. As is demonstrated in the graph to the right, a variable zero adjustment results in different values of the Y-axis intercept quantity, best observed if the input/output plots are projected below the 1 Volt minimum input value as shown in the diagram. A lower bias or "zero" setting is understood to suppress the output voltage span since some of the output values will be negative. This is called an "elevated-zero" span since zero is elevated within the operating span. A higher, more positive, Y-axis intercept voltage is understood to elevate the output span into a range where all output values are positive. This type of output is referred to as a "suppressed-zero" span, since zero is suppressed below the operating span. Output spans starting at zero volts and moving positive are referred to as "true-zero" spans.

To set the span and zero adjustments the input process variable, or a sensor-simulating voltage span must be applied to the amplifier input, and the individual adjustments set at the appropriate input extremes. It is highly beneficial to retain the desired input/output plot in mind and imperative that the concepts of Y-axis intercept and input/output slope be understood when performing the calibration procedure.

If a sensor is connected to the input of the amplifier undergoing calibration, the physical variable to be measured should be applied over the entire range of operation when possible. During the calibration process, the physical variable being sensed will need to be measured using an accurately known measurement standard, or the results of the calibration process will be questionable. If a sensor is not attached to the signal conditioner to be calibrated, or if varying the process variable throughout the entire operating span is not feasible, the sensor's output (amplifier's input) span must be simulated as closely as possible. In short, don't make assumptions about the applied physical variable at the sensor, verify your assumptions with a measurement standard, or by using an accepted standard procedure.

As an example, guessing at a value of applied twist in a shaft (torque) or even using a commercially available torque wrench will more than likely result in a significant amount at error in the amplifier output, DPM or data acquisition system input. However realizing torque is a force applied through a distance that attempts to create a rotating or turning effect, and that torque is equal to force multiplied by distance will allow an accepted calibration procedure by which to calibrate the measurement system at the amplifiers' span and zero adjustments. Hanging an accurately known weight at an accurately known distance from the specimen under test provides a means of determining the specific amount of torque applied to the sample. Knowing this, the zero adjustment can be set to indicate a zero condition at the zero applied torque, and the span (gain) adjustment can be used to set the amplifier output to the value of the applied torque.

As an additional example, assume a pressure sensor, signal conditioning amplifier and digital panel meter (DPM) is to be calibrated. The measurement system of Design Example 6 can be used as an application example since it contains the necessary components for the measurement system, including variable gain and bias potentiometers.

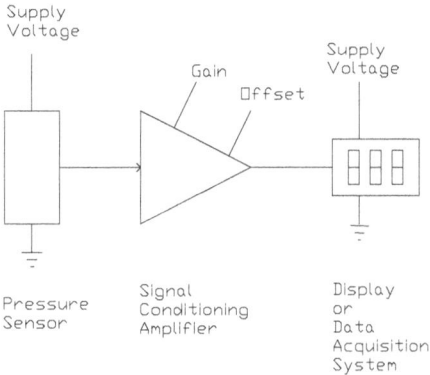

Figure 36 – Measurement system for the current calibration example. Note the location of the calibration adjustments at the amplifier. Other less influential adjustments are occasionally found at the sensor and at the display or data acquisition system

The ideal end result of the calibration process will be a 0.00 displayed when the input pressure is 0 psi, and 10.00 will be exhibited when the input pressure is 10 psi. A pressure gauge of known and documented accuracy or mercury manometer will be connected in parallel with the sensor's input to precisely measure the applied input pressure. After the system is assembled and sufficient warm-up time has passed, apply 0 psi to the sensor. Under this condition the sensor will output a negative .25 Volt to the signal-conditioning amplifier. The biasing voltage is responsible for canceling the -. 25 Volt input, and sending 0 Volts to the DPM. The panel meter will display 0 Volts as 0.00 psi. Assuming the bias and gain potentiometers are both at random positions, the zero (bias) potentiometer should be adjusted for a display as close to the desired 0.00 as possible. This adjustment will probably require resetting since the present output is determined by the value of the gain potentiometer which is about to be adjusted.

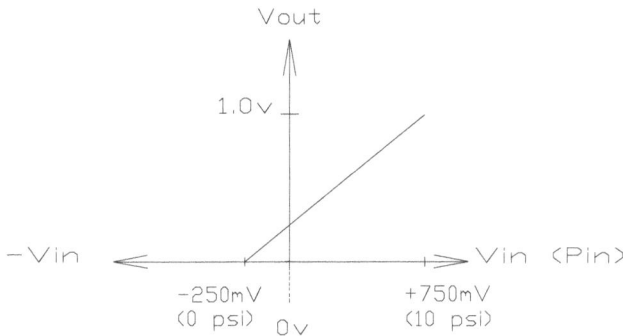

Figure 37 – Amplifier I/O plot and DPM display versus sensor output. The input pressure is 0-psi at a sensor output of -.25V, and 10-psi at a sensor output of .75V.

The provided input/output plot of input voltage (pressure) versus output voltage (and DPM displayed value) should be kept in mind while performing the calibration. The previous bias adjustment was the first step in establishing the Y-axis intercept point for the amplifiers' operation. Varying the bias adjustment should be understood to shift the input/output plot vertically without affecting the slope. If the maximum pressure of 10 psi is applied and verified by the standard instrument, the span (gain) potentiometer should be adjusted to output 1.0 volts from the amplifier, and display 10.0 psi on the DPM.

However, varying the amplifier gain will alter the output-bias since the effect of the bias is determined by its gain, A_{Vbias}, and A_{Vbias} is determined by Av, which has just been changed. Likewise, the bias adjustment will require readjusting. Reapplying 0 psi to the sensor, the bias potentiometer should again be set for a display of 0.00

psi. Raising the applied pressure to 10 psi, check and adjust for the required 10.0-psi display. After the input extremes have been calibrated, several points at various locations in between should be checked. Since provisions for linearity are not included with our signal conditioner, any error associated with the mid-values cannot be corrected.

The calibration process with operational amplifiers unfortunately requires setting the bias and gain values at least twice initially, since the gain and bias functions are interactive. The value of one is dependent upon the other. This is common with most input biased, where the bias provisions are applied to the input, signal conditioning systems

Amplifier Design Problems

Include potentiometers to vary the gain and bias in each of the following homework problems. When submitting your results show your design by including the approximate percentage of variation, the design computations and the corresponding resistance values. Submit a clearly labeled circuit diagram with the design computations.

#1 - An amplifier is to convert an input signal span of 0 to 4 Volts into 1 to 9 Volts. Determine the amplifier gain (Av) and offset (b). Determine and include the circuit diagram and component values.

#2 - An amplifier is to convert an input signal span of 1 to 5 Volts into 1 to 9 Volts. Determine the amplifier gain (Av) and offset (b). Determine and include the circuit diagram and component values.

#3 - An amplifier is to convert an input signal span of 1 to 5 Volts into 0 to 10 Volts. Determine the amplifier gain (Av) and offset (b). Determine and include the circuit diagram and component values.

#4 - An amplifier is to convert an input signal span of .02 to .08 Volts into 0 to 10 Volts. Determine the amplifier gain (Av) and offset (b). Determine and include the circuit diagram and component values.

#5 - An amplifier is to convert an input signal span of -.05 to +. 10 Volts into 1 to 9 Volts. Determine the amplifier gain (Av) and offset (b). Determine and include the circuit diagram and component values.

#6 - An amplifier is to convert an input signal span of -.02 to .09 Volts into 0 to 3 Volts. Determine the amplifier gain (Av) and offset (b). Determine and include the circuit diagram and component values.

#7 - A sensor outputs -. 02 to +. 09 Volts over an input span of 0 to 30 psi. Design an amplifier to condition the sensor output to 0 to 3 Volts.

#8 - A reverse acting sensor outputs .02 to -.01 Volts over an input span of 0 to 20 pounds. Design an amplifier to convert the sensor output into 0 to 2 Volts direct acting with force.

#9 - A reverse acting sensor outputs 80 to 10 millivolts Volts over an input span of -50 to +150 degrees centigrade. Design an amplifier to develop an output span that is direct acting with temperature and outputs -.5 to 1.5 Volts.

#10 - A sensor outputs 3.73 to 2.73 Volts over an input span of 273 to 373 degrees Kelvin. Design a signal-conditioning amplifier to display 2.73 to 3.73 Volts direct acting with temperature.

Chapter 7 – Differential Signals and Amplification

The operational concepts and equations of biased amplifiers are applied to differential amplifiers. Sensor characteristics, differential voltage sources and signal conditioning is reviewed, and differential amplification is investigated using previously introduced biased amplifier circuit analyses. Common-Mode Rejection (CMR) and testing for CMR is explained. Differential and Instrumentation amplifier operation, equations, advantages and limitations are included. Isolation amplifiers are presented.

Objectives:
Upon completion of this chapter, you should be able to:
- Explain the purpose and benefits of differential signals
- Explain differential amplification
- Explain common mode rejection
- Design differential amplifiers
- Recognize differential signal sources
- Recognize applications of differential amplifiers
- Assemble and apply multiple differential amplifier configurations

Introduction

The text discussion to this point has centered on single-ended amplifiers. A single-ended amplifier is one that requires only a single input connection, in addition to a ground connection. All of the previously introduced forms of non-inverting and inverting amplifiers are categorized as single-ended since only one of the two available amplifier inputs is connected to the signal source. The other available input is connected to ground or a bias voltage (the ground connection can be considered a bias voltage of zero volts). A differential amplifier is one that connects both inverting and non-inverting inputs to a single signal source. As a result, differential amplifiers will usually require a total of three (2 signal and 1 ground) connections, and usually will not contain output-offset provisions on the amplifier itself.

Using an operational amplifier, the circuit configuration for a differential amplifier will appear similar to a biased inverting, or biased non-inverting amplifier. The circuit analysis and equation derivations of a differential amplifier will be performed in an identical manner to any of the previous op amp circuit configurations. As such, the single op amp configured as a differential amplifier should be approached as any other biased op amp circuit.

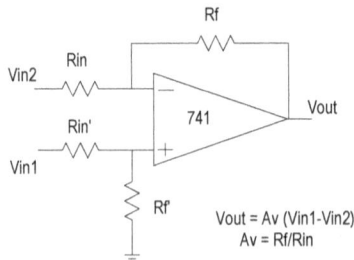

Rf

Rin

Vin2

Vin1

Rin'

741

Vout

Rf

Vout = Av (Vin1-Vin2)

Av = Rf/Rin

Figure 1 – An op amp configured as a Differential Amplifier. The circuit analysis and Vout equation derivation is performed in an identical manner to the inverting or non-inverting amplifier with bias.

Both of the inputs (Vin1 and Vin2) are received from a single differential signal source such as a bridge type sensing circuit.

Features of Experimental Sensors

Experimental sensors are commonly differential output devices. As such, experimental sensors are commonly associated with differential amplifiers. To better define the role of the differential amplifier, and to segregate experimental sensors from the larger category of industrial sensors, the following is provided. Much of the included material has been presented in previous chapters but not in elaboration.

In the broad scheme, two grades of sensors exist, industrial and experimental. Scientific sensors, used in laboratory environments are considered to be a subset of experimental sensors. Sensors can be categorized and sub-categorized in any number of ways. Among the currently available categories are output type (high level, low level), power supply requirements (AC, DC, regulated, unregulated), input type (differential, absolute, gauge), signal conditioning and compensation requirements and other operational or application driven concerns. For the purpose of differentiation and to assist with the current focus of differential signals and amplifiers, currently consider all sensors as being used for either experimental measurement, or for manufacturing measurement. Although some would argue the differences between manufacturing and experimental sensing may seem trivial and unimportant, the differences are considerable and will determine the purpose and type of subsequent signal conditioning, and the required data acquisition and analysis components.

In experimental measurement applications such as stress, strain, pressure, displacement, velocity, acceleration, vibration, force, temperature and similar measurements, precise sensors require connection to signal conditioning amplifiers. The amplifiers or amplifier modules are connected to display items, PC's or other data acquisition and data analysis devices. Occasionally the sensor output is also provided to a control device for purposes of cycling a test or controlling a specific test condition. But generally speaking experimental applications differ from manufacturing applications in the precision and quality of the sensing components, the required signal conditioning amplifiers and other components, and the nature and purpose of collecting the data.

Experimental applications are usually one of a kind, where single test fixtures (or at most two or three fixtures) are testing a specific part, design, physical variable or process. Some test fixtures may contain several devices under test for simultaneous testing. The purpose of the experimental measurement is to acquire information about the device, variable or process under test. Manufacturing and production measurement components are usually sold in mass-produced quantities, are not as accurate or precise as the experimental measurement components, usually contain on-board amplifiers or transmitters with internal or on-board calibration adjustments, and are used to measure a physical quantity for the specific purpose of *controlling the quantity*. Since the manufacturing sensor is providing an output for the purpose of control, the sensor output may not be a precise replicate of the process being measured. It doesn't need to be accurate in an absolute sense, the sensor output only needs to vary as the process being controlled varies. The experimental sensor is required to generate an output that is an absolute representative of the device, process or variable under test.

Experimental sensing components exhibit several common characteristics and are unique from their manufacturing counterparts in other ways as well. Experimental sensors are usually supplied with their own individual calibration characteristic data and rarely contain on-board amplification or calibration provisions. When an experimental sensor is purchased, a specification sheet unique to that specific sensor is shipped with each unit. The specification sheet will contain sensitivity/gain, offset, input and output resistance and other pertinent information for the designer. In this manner each sensor is slightly different, even though the model numbers are identical. In addition, most experimental sensors

276

are delicate, commonly requiring protective coatings, temperature or pressure compensation, special handling and mounting considerations.

Figure 2 – An experimental pressure sensor has been disassembled to expose _all_ internal components. The pressure sensing strain gauges are barely visible in the center of the circular section in the upper right. The bridge completion and gain related components are visible on the round printed circuit board at the lower right.

The vast majority of experimental sensors are "low-level," meaning the output voltages are usually in the range of microvolts to millivolts, and require at least one stage of subsequent and external signal conditioning. Low-level sensors are sensitive to subtle power supply variations and temperature-induced changes in the supply connection and leadwire resistance. The low-level output signal span also makes the sensor output signal susceptible to electronic noise. With low-level output voltages, Electro-magnetically induced noise sensitivity becomes greater. Long leadwire runs act as antenna inducing noise into the sensors' output signal. Experimental sensors often employ differential outputs to assist in noise immunity. And because of the previously mentioned precision and accuracy features, experimental sensors are usually expensive, often costing five hundred to a thousand US dollars or more apiece. Typically, experimental sensors are not found in industrial control applications because of the harsh environment and significantly higher expense.

Figure 3 – An industrial pressure transmitter. This unit contains the pressure sensor, amplifier and additional conditioning circuitry to convert the measured pressure range into a 4-20 milliamp current span for transmission over long distances. The expense of this unit is comparable to the expense of the experimental sensor in the previous figure.

Generally speaking, experimental sensors internally contain only those components required to measure a single physical variable. Occasionally passive compensation components required to stabilize operation over a range of temperature pressure or other operating conditions may also be included, but amplification, linearization and filtering circuitry is omitted and left to subsequent stages of signal conditioning. Given an opportunity, dissection of <u>defective</u> sensors is strongly encouraged, especially experimental sensors since these devices are functional applications of basic physical principles. Stress and strain, pressure, force and area, mass, acceleration and dynamic fluid concepts are obviously demonstrated in experimental sensors of most physical quantities.

Experimental Sensor Signal Conditioning

Since the sensor is responsible for generating an *analog* of a physical variable (an analog is a quantity which changes in proportion to another variable but which assumes a different physical form), the next stage(s) are responsible for conditioning the sensor's output signal into a useable form. Linearization, noise filtering, scaling (gain and offset) amplification, and analog-to-digital conversion are examples of sensor signal conditioning, yet any type of sensor output voltage modification qualifies as signal conditioning. The idea behind signal conditioning is to prepare the sensor output voltage span for use by the data storage, analysis, display or occasional control devices.

Many signal-conditioning functions can be performed in software currently, however the sensor output must meet the input specifications of the analog-to-digital (A/D) converter before being manipulated in software. This usually means scaling amplification is required to adjust the offset and span characteristics of the sensor output voltage to fully utilize the A/D input requirements of minus 10 volts to plus 10 Volts (typically). The trick is to modify the sensor output voltage without distorting the data contained within.

Figure 4 – A signal-conditioning amplifier from a data acquisition system. This unit was intended for use with low-level sensors.

Figure 5 – A signal conditioning "Rack" or "Front-end." As can be seen from the individual units, these signal conditioners are also intended for use with low-level experimental sensors. The gain and offset adjustments have been made obvious in two of the units. Each signal conditioner contains two outputs, one voltage, and one current for transmission over long distances. Therefore each module contains two sets of gain and offset adjustments.

Given that a typical A/D requires a twenty-volt input span and most experimental sensors generate output spans of millivolts and occasionally microvolts, the demands placed upon signal conditioning amplifiers are considerable. Gains into the hundreds, thousands and tens of thousands are not uncommon for sensor signal conditioning amplifiers. Keep in mind that any undesired Electro-magnetically-induced noise *may* also be amplified hundreds and thousands of times unless special circuit considerations are addressed.

Differential Signals and Amplification

All exposure to amplification up to this point has involved *single-ended* amplifiers. That is, amplifiers requiring two circuit connections, signal and ground. Inverting and non-inverting amplifiers are examples of single-ended amplifiers, even when biased with an offset voltage. Most conventional amplifiers contain single-ended inputs and outputs. A single-ended input provides a single lead input connection to apply the signal for amplification. The input signal is applied in conjunction with a connection to ground eventually requiring two separate connections, two wires. A single-ended output is similar with the output being available at a single lead-wire. The output is then measured or taken for further processing with respect to, and in conjunction with a ground connection.

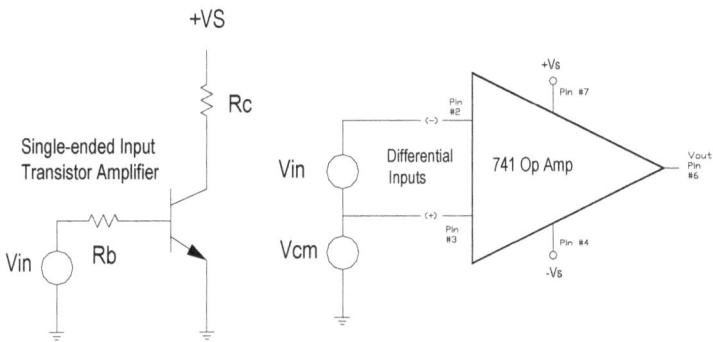

Figure 6- Differential Amplifier circuits (right) are recognized by the required dual inputs, neither of which is connected to ground. The transistor circuit at the left is an example of a single-ended input amplifier.

Again, to utilize the single-ended output signal two connections are required, one wire to the output and one wire to ground. Oscilloscope inputs are examples of single-ended instruments since one of the probe leadwires is connected to grounded through the input BNC connector. To measure a differential signal with an oscilloscope requires a dual trace scope where each channel is connected to a single output lead from the differential signal source. The scope is then placed in the "Add" mode where both channels are added, and one of the channels is inverted. The result is a single displayed trace representative of the difference between the two available output leads, the differential output voltage of the signal source. In this manner a single-ended instrument with two inputs can measure a differential signal.

A common battery operated digital voltmeter (DVM) is an example of a differential measurement instrument. All battery-operated voltmeters indicate the difference in voltage between two points. Since neither input lead requires connecting to ground, the displayed voltage is the differential voltage between the test leads.

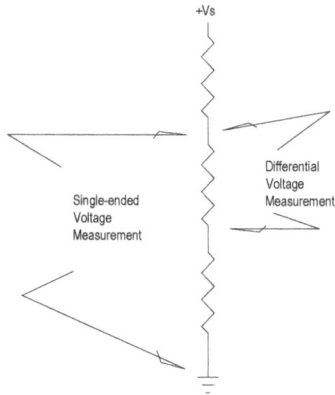

Figure 7 – A single-ended voltage measurement requires a single lead be placed, or referenced to ground. A differential measurement is measured between two non-zero voltage locations.

A similar comparison can be made with the following circuit diagram. In the circuit diagram some voltages are circled and others appear at the end of an arrowhead. The circled voltages are differential voltages, or the voltages appearing across the resistance or component. The arrowhead voltages are single-ended, meaning they are measured with respect to ground, or in reference to ground.

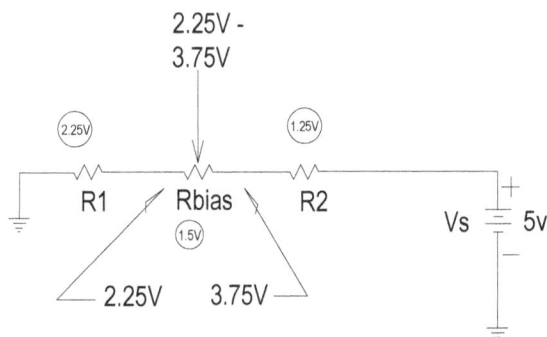

Figure 8 – The circled voltages are differential voltages, determined by placing a meter across the components, or the result of circuit analysis. The voltages appearing at the ends of arrowheads are single-ended and require a meter lead to be connected to ground. This convention of voltage measurement designation is becoming common on schematic diagrams.

Operational amplifiers are inherently differential amplifiers since the two inputs labeled as inverting (negative) and non-inverting (positive), are amplifying only the difference in voltage between them. These polarity designations represent the phase relationship of each input with respect to the output. If a positive-going voltage is applied to the inverting (-) input, the output will move in a negative-going direction. An increasing voltage applied to the non-inverting (+) input will cause the output to move in an increasing (positive going) direction. Since an op amp's output voltage is always equal to the voltage difference between the inverting and non-inverting inputs (Vid), multiplied by the open-loop gain, Avol, the operational amplifier by its design is performing differential amplification.

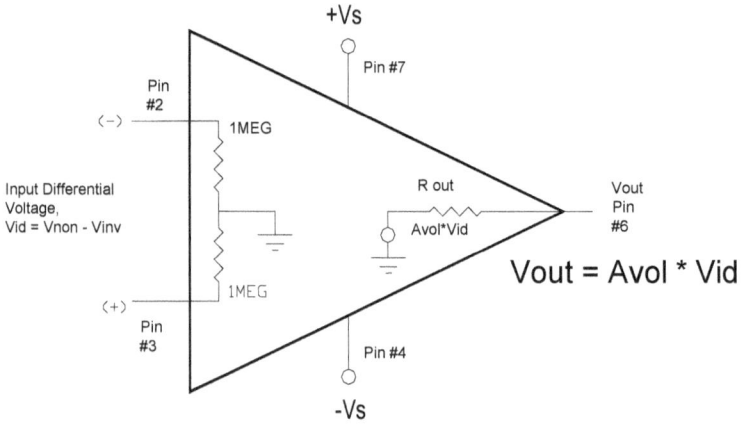

Figure 9 – A demonstration of op amp differential inputs and their contribution to the output voltage.

Differential signals may not require a ground connection. Although usually available, a ground connection is not necessary to reference the signal. The signal voltage is available as a difference in voltage between to non-zero voltages available between two non-grounded leadwires. Differential signal voltages are commonly available at resistive bridge sensor outputs, transformer outputs and although rare, some amplifiers contain differential outputs where Vout is available between two non-zero output voltage leads.

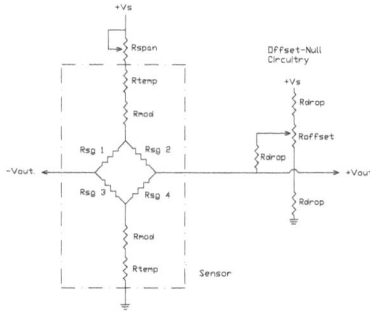

Figure 10 – Differential outputs from an experimental sensor with external gain and offset circuitry. Although neither of the two output leads are grounded, ground connections to the supply are available.

Differential voltage sources will always be bi-polar with one lead-wire labeled positive and the other lead-wire labeled negative. The polarities represent the directions the voltages will move when an input is applied. The positive lead will increase in voltage as an input is applied, and the negative lead will exhibit a negative going or decreasing voltage as an input is applied.

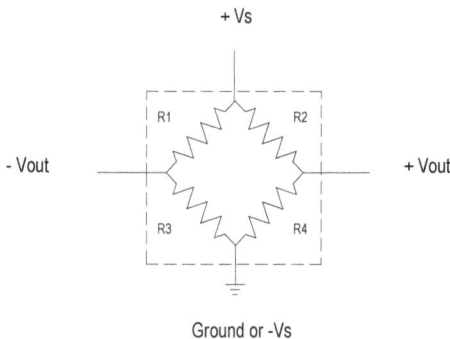

Figure 11 – Bridge circuits provide a differential output, indicated by the bipolar Vout's in the diagram. The actual output signal will be acquired from the difference of the two output voltages.

As an example of a differential signal, consider a simple resistive bridge pressure sensor similar to the previous diagram (figure 11). Bridge-type sensors contain four leadwire connections, a supply voltage (say 5 volts), a ground connection, positive and negative output leads. Initially with zero pressure applied, each output lead will exhibit half of the supply voltage, 2.5 volts. This is considered a differential output voltage of zero volts since the differential output is the difference between the two leads, or 2.5V-2.5V. As a pressure is applied the positive output lead will increase from 2.5 volts towards 5 volts, and the negative output lead will decrease towards ground from 2.5 volts. Neither output will actually reach 5V or ground, but will move in the directions of these values. As an example, should the pressure be increased to a value where the differential output voltage is .4 volts, the positive leadwire would measure 2.7 volts to ground and the negative leadwire would measure 2.3 volts to ground. Again, the differential output voltage would be 2.7V-2.3V or .4 volts. In a similar manner differential amplifiers are configured to amplify the difference (effectively subtracting) the voltages available from the two lead-wire differential voltage sources. The amount of differential amplification required, Ad, will be determined by the required amplifier output voltage and the available sensor signal span.

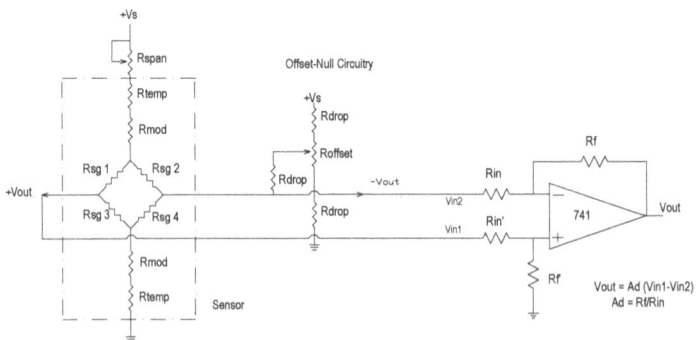

Figure 12 – A bridge circuit sensor (dashed line) followed by an op amp differential amplifier. The output from the op amp will be the amplified difference between the positive and negative output voltages from the bridge.

Common Mode Rejection

One may question the relevance and importance of belaboring the study of differential signals and amplification. Two primary purposes are apparent. First, when configuring analog I/O systems used in data acquisition and control applications, analog input modules are required to introduce the sensor-generated signals to the data acquisition, display or PC-based digital host system. An initial step in configuring the analog input module(s) is to establish the appropriate type of input circuit, single-ended or differential. If any of the signals are coming from a differential source, the input must be set-up as a differential input or half of the information contained in the signal will be lost. In other words, the type of sensor output configuration would dictate the type of analog input configuration.

The second and most important reason for using differential signal sources and amplifiers is referred to as Common Mode Rejection, or CMR. Lengthy leadwires are common to all experimental and industrial applications of sensors. As can be imagined, lengthy leadwires function as antennae in that any magnetic fields passing through the leadwires will induce a voltage, creating noise and adding to the measurement uncertainty and ambiguity, and overall signal distortion and degradation. The ability of differential amplifiers to subtract the voltages available between the two signal leads means any signals common to both input terminals will be effectively eliminated by being subtracted out of the output. Since induced noise will be common to both inputs, if the noise is of a similar phase it will be subtracted and minimized, if not eliminated. In other words, any voltages common to both inputs will be rejected. Hence the term Common Mode Rejection.

CMR is a characteristic associated with operational amplifiers and is included on op amp spec sheets as common mode rejection ratio, or CMRR, measured in decibels (dB). CMRR is the ratio of differential gain, Ad, in reference to (divided by) common mode gain, Acm. Common mode gain is found by applying a common signal to both inputs of a differential amplifier. Ideally the output would be zero volts but unfortunately this is rarely the case. Typically, an operational amplifier designed and assembled for a differential gain of 100 will exhibit a common mode gain of around .001. Note that the common mode gain is always less than one. In this example the CMRR in dB would equal,

$$CMRR_{db} = 20 \log_{10} (Ad/Acm)$$
$$CMRR_{db} = 20 \log_{10} (100/.001)$$
$$CMRR_{db} = 20 \log_{10} (10^5)$$
$$CMRR_{db} = 20 (5)$$
$$CMRR_{db} = 100 \text{ db}$$

The common mode gain is found experimentally by measuring the output with a common AC voltage applied to both inputs. The common mode gain is the ratio of the applied input, say 10vpp, divided into the measured output voltage of 10 millivolts pp.

The values used in the previous example are realistic and representative. Differential gains are usually greater than one since the differential signals are millivolt level, and common mode gains are always less than one since the inputs are subtracted and minimized in the output. The more precision the operational amplifier (those with extremely large Avol, $\geq$ 1,000,000, are considered precision op amps) the smaller the common mode gain and even less of the common mode input will appear in the output.

Overall, common mode rejection ratio is not just a measure of how well a differential amplifier can eliminate common mode noise, but is more appropriately an assessment of an amplifiers ability to amplify a signal while minimizing noise. As an example and using the above quantities, assume a .1 volt differential signal (+. 05 volts on the non-inverting input and -. 05 volts on the inverting input) is applied to a differential amplifier with a gain of 100. Also assume a .1 volt peak to peak AC noise signal is also available to both inputs. (The measurement format of the AC noise is insignificant as long as the same units are used to measure the input and output). In essence, a similar amount of noise and signal are available at the amplifier's input. However using the differential gain of 100 and the common mode gain of .001, the amplifier output will contain 10 Volts of signal and .0001 volts of noise. The output will represent a signal to noise ratio of 100,000 to 1, or 100 decibels. In this manner CMRR is best observed as a measure of amplification in conjunction with noise elimination.

The op amp specification for CMR is typically between 60 and 80 dB (1000:1 to 10,000:1). The largest determining factor in a configured amplifier (with resistors applied) for CMR is the matching of the resistors. Even 5% tolerance resistors can degrade CMR performance significantly. Given that most differential amps will be used with precision sensors, 1% resistors should be considered to optimize CMR.

Differential Amplifier Configurations
The evolution of differential amplifiers has conformed to the changes in electronic technology. "Diff. Amps," also occasionally referred to as "DA's," have been assembled using vacuum tubes, transistors, operational amplifiers and also appear self-contained in single 8 and 14-pin integrated circuit dual in-line packages (DIP's). Presently a variety of differential amplifier configurations are available to the designer. The choice of circuit configuration will depend upon the demands of the application, the type of sensor output available, and the demands of the amplifier output load (display, PC or other data acquisition apparatus). Specifically, the sensor requirements for input impedance, gain, differential inputs, and common mode rejection (CMR) will determine the most appropriate amplifier configuration. The individual circuits and associated specifications will be developed in the following chapter material.

Currently four or five styles of differential amplifiers are popular. The first and least expensive is assembled using two identical transistors. Finding identical transistors can be somewhat difficult however. Anyone that has worked with transistors is familiar with the shortcomings, among the most obvious is the inability to match the intrinsic gain, Beta, between two transistors of identical part numbers. To get around this problem semiconductor manufacturers have matched and packaged matching transistors in 14 pin dual in-line packages. Most transistor differential amp IC's contain multiple sets of differential amps. Although limited in gain, Zin and CMR, the transistor differential amplifier is the least expensive approach to differential amplification.

The following figure shows a transistor differential amplifier. This type of differential amplifier is available in integrated circuit packages containing two diff-amps per package. All of the components within the dashed line are included in the IC package.

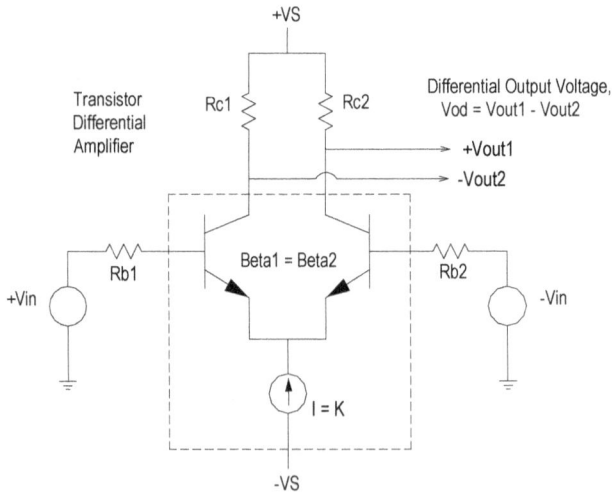

Figure 13 - A transistor differential amplifier with inputs, outputs and required components indicated. The components inside the dashed lines are commercially available in integrated circuit form with matched transistors.

In operation, the transistor differential amplifier functions as two individual common emitter amplifiers sharing a common emitter resistor, or in this case a common constant current source. In the corresponding transistor differential amplifier diagrams, the voltage sources on the transistor bases appear as separate voltage sources. In reality both originate from the same source but are different output polarities of the same source since differential signal sources contain two opposing (positive and negative) output connections. It should also be noted that the designated input polarities might not represent the absolute voltage polarity, but the direction of the input voltage change. In the case of a bridge output voltage appearing at the inputs of the differential amplifier, each input is equal (typically one-half of the applied supply voltage) at the initial, or null condition. As the input to the sensor changes, the individual outputs from the bridge also change but in different directions.

From the following circuit diagram, each input voltage establishes an input current, Ib, created by the difference in potential between the input voltage and the negative supply voltage, -Vs. As a point of interest, the negative supply voltage would establish an input current (base current) even if the input were grounded or set to zero volts. Since ground appears as a more positive potential when compared to the negative supply voltage, a closed circuit path between the negative supply and ground will result in electrons moving from the negative supply to the ground connection. Since the electrons will move to ground from the emitter through the base, the current flowing to ground will be base current in the micro to milli-amp range. Because of the ability of base current to flow with zero volts on the base the transistor differential amplifier is responsive to both negative and positive input voltages, where a positive going input increases the base current and a negative going input voltage reduces the base current. Typically with differential input voltages both are occurring simultaneously.

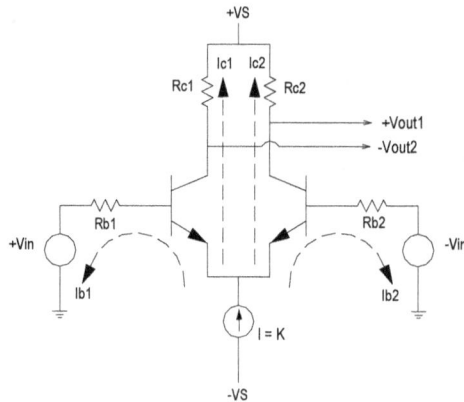

Figure 14 – A Transistor differential amplifier with base and collector currents labelled.

Once the base currents are established, a corresponding collector current, Ic, will flow in each transistor. The specific amount of Ic will be contingent upon and the intrinsic gain, Beta, of each transistor, the amount of Ib, which is contingent upon the amount of Vin applied at the individual inputs. As the collector currents pass through the individual collector resistors, Rc, a voltage is established across each resistance. The collector output voltages at each transistor (Vout1, Vout2) will be the difference between the positive supply voltage, +Vs, and the voltages appearing across the individual collector resistors.

In addition, a single emitter resistor or potentiometer could replace the constant current source in the diagram and the amplifier would operate with only a minor degradation in performance. The advantage to the current source is best observed by assuming a single input to the differential amplifier is changing while the other input remains constant. The current generator will create a constant value of current to share between the two transistor collector paths. The sum of each collector current (assume the base currents negligible) at any time under any input condition will equal the current generator output. If a single input is applied and made variable, the collector current in both transistors will change, and the output from each transistor will also change. In this manner an output change is generated at each transistor even though only a single input is changing. If the inputs are equal (a common-mode voltage) the two collector currents will be equal and the individual output voltages will be equal. The difference between the output voltages (the output differential voltage) will then be zero. Any difference between the two input voltages will result in different currents, different collector voltages and a non-zero output voltage. The effect of the constant current source enhances the transistor differential amplifier gain, CMR and overall performance. For this reason current generators are included within the integrated circuit version of transistor differential amplifiers.

The transistor differential amplifier provides another opportunity to investigate differential outputs. Although rare, amplifiers occasionally provide outputs of a differential nature. The two output leads function in a manner consistent with the differential outputs of a bridge circuit, where the difference in voltage between the two output leads determines the actual amplifier output. The difference between the two outputs is the amplified difference between the two inputs. Should the difference between the two inputs be zero, the difference between the two outputs should also be zero. A balancing potentiometer is occasionally installed to provide a means of balancing the output under a condition of input null, or zero difference.

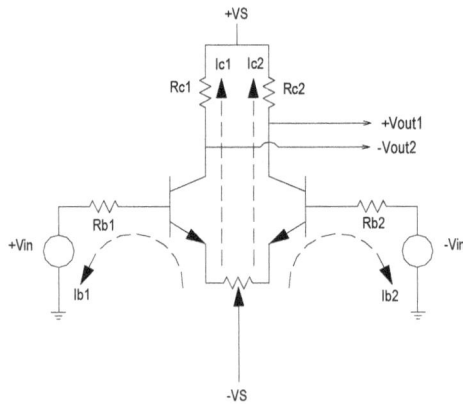

Figure 15 – A variation of the transistor differential amplifier includes a potentiometer to balance the differential output under a condition of common differential inputs. To null-balance the differential outputs, both inputs are normally grounded and the potentiometer is adjusted for a differential output of zero volts.

All of the previous sounds very complicated but in fact, the operation of the transistor differential amplifier is quite understandable if the manner with which transistors operate is understood. Even better, the transistor differential amplifier is rarely used, and if none of the previous explanation was understood, don't be concerned. Limited gain (about 200 max.) and limited input impedance (100K Ohms) in conjunction with thermal sensitivity, drift, differential outputs, common mode rejection limitations and overall ease of design make the operational amplifier form of the differential amp far more desirable.

The Op Amp Diff Amp

Currently the most popular approach to differential amplification is the operational amplifier configured as a differential amp. Four external resistors "program" the op amp to perform in a differential amplification capacity. When observed, the circuit appears very similar to either of the biased amplifiers of previous chapters and analyzing the differential amplifier circuit is performed in an identical manner to the analysis of the biased amplifiers.

From the following circuit, assume Vin1 is on the non-inverting input and Vin2 is connected to the inverting input. The output voltage can be derived to equal Vin1 multiplied by the Beta (reduction factor of the resistive network at the input) of the voltage divider, multiplied by the non-inverting gain. Sum this quantity with the product of the inverting input (Vin2) times the inverting gain and the result is an amplified differential input with a single-ended output voltage referenced to ground. In simplified form the output equation becomes,

$$V_{out} = V_{in1} \text{ (Beta) } A_{Vin1} + V_{in2} \text{ } (A_{Vin2})$$

Indicating that the output voltage is the result of two inputs multiplied by their respective gains (A_{Vin1} and A_{Vin2}), and summed. The Beta term is included in the first term since a voltage divider network is between V_{in1} and the op amp non-inverting input. The Beta network located before the direct acting amplifier reduces Vin1 and may seem unnecessary, but is performed to match the individual gains "seen" by each input. Further substitution into the previous equation yields,

$$V_{out} = [V_{in1} \{Rf' \div (Rin' + Rf')\} \text{ } (1 + Rf/Rin)] + [V_{in2} \text{ } (-Rf/Rin)]$$

The appropriate input voltage and resistance values are substituted into the equation to determine the output voltage value.

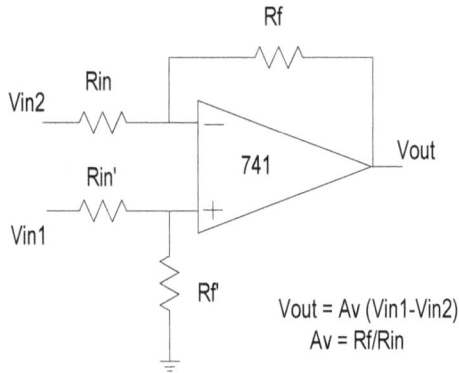

Figure 16 – Appearing as a biased inverting or a biased non-inverting amplifier, the differential amplifier works in an identical manner to the combined biased amplifiers. When the input resistors are equal, and the other resistors are equal, the gain and output equations are as indicated.

As an example, assume the following values of resistances are placed around the amplifier.

$$Rin = Rin' = 100K$$
$$Rf = Rf' = 1Megohm$$

Using the analytical convention provided earlier, an output equation can be written;

$$Vout =$$
[Vin1 * (Beta of resistive divider) * (Non-inverting Gain)] + [Vin2 * (inverting Gain)]

Or,

$$V_{out} = V_{in1} (Beta) A_{Vin1} + V_{in2} (A_{Vin2})$$

And,

$$V_{out} = [V_{in1} \{Rf' \div (Rin' + Rf')\} (1 + Rf/Rin)] + [V_{in2} (-Rf/Rin)]$$

The amplifier diagram with component values becomes;

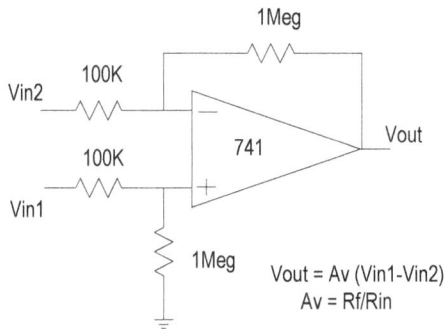

Figure 17 – Differential Amplifier with component values indicated. The differential gain of this amplifier is 10, the Zin is approximately 100K Ohms and the CMR is 60-80 dB depending upon resistor tolerances.

Continuing with the circuit analyses by subbing the values into the acquired equation,

$$Vout =$$
$$[Vin1 * (100K/1.1M) * (1 + 1M/100K)] + [Vin2 * (-1M/100K)] =$$
$$[Vin1 * (.90909) * (11)] + [Vin2 * (-10)] =$$
$$[Vin1 * (10)] + [Vin2 * (-10)] =$$
$$10\ Vin1 - 10\ Vin2 =$$

Therefore, factoring yields a Vout equal to the following,

$$Vout = 10\ (Vin1 - Vin2)$$

Or from the equation in the circuit diagram,

$$Ad = Rf\ /\ Rin$$
$$Ad = 1M\ /\ 100K = 10$$

And Vout from the circuit diagram equation is verified as,

$$Vout = Ad (Vin1 - Vin2)$$
$$Vout = 10 (Vin1 - Vin2)$$

The differential gain, Ad, and the output voltage can be computed as previously indicated only if Rin and Rin' are equal values of resistance, and Rf and Rf' are equal. From the circuit diagram of figure 17, if 2V were applied at V_{in1} and 1V were applied at V_{in2}, V_{out} would equal 10 times the difference, or 10V.

$$Vout = 10 (2V - 1V)$$
$$Vout = +10V$$

Diff Amp Circuit Analysis

As mentioned previously an identical approach to the biased non-inverting or biased inverting amplifier analysis is applied to the differential amplifier circuit analysis and yields the correct results.

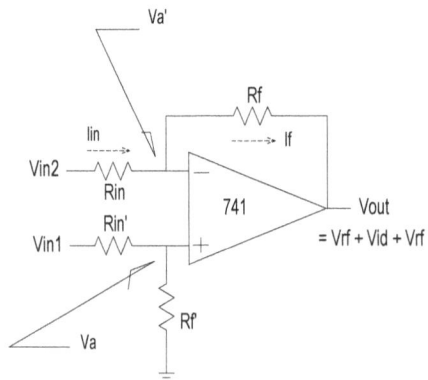

Figure 18 – Circuit analysis of the op amp differential amplifier is identical to the previous biased op amp circuits.

If Vin1 and Vin2 were known a value of output could be determined from the resulting equation. Given the following diagram, assume differential voltages are placed at both inputs and are labeled Vin1 and Vin2. Vin1 is reduced by the reduction ratio of the voltage divider and appears as Va at the non-inverting input. Since the input differential, Vid, is approximately zero volts, a voltage equal to Va will appear at the inverting input terminal. The voltage appearing at the inverting input is labeled as Va' because it is equal to yet different than Va. The voltage difference between Va' and Vin2 will create a current flow through the input resistance, Rin. Since the input current and feedback currents are equal in all negative feedback op amps, the current created at Rin also passes through Rf. As the input current passes through Rf a voltage is established. The output voltage will be the sum of all voltages from the op amp output to ground using any available path. The following diagram illustrates.

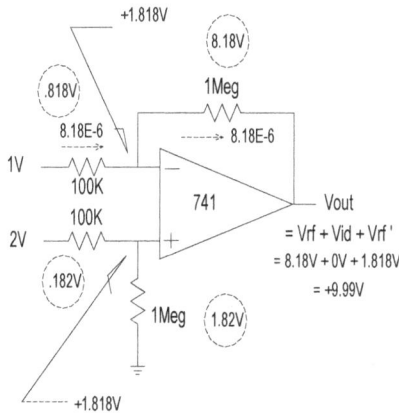

Figure 19 – Differential amplifier circuit analysis is performed in a similar manner to biased amplifier analysis. The voltage on the non-inverting is determined first, using this voltage and Vin2 the voltage across Rin is determined, the current through Rin and Rf is determined, and the voltage across Rf is found using the current through Rin and Rf. Once the voltage across all resistors is known, or once the voltage across all components in any path from the output to ground is determined, Vout is found by algebraically summing all the voltages from the output to ground. The circled voltages are the voltage drops across components. The voltages at the arrowheads are measured with reference to ground.

299

Diff Amp Circuit Analysis with Sensor Input

Assume an operational amplifier differential amplifier is connected to a two active element RTD temperature sensor and associated full-bridge circuit. The assumed temperature span is 0 to 100 Centigrade and the RTD are both Platinum exhibiting .285 Ohms per degree Centigrade of resistance change (gain) and 100 Ohms at zero degrees Centigrade. Since both RTD's are measuring the same temperature the sensors will be located in opposite legs of the bridge circuit. The RTD bridge appears following.

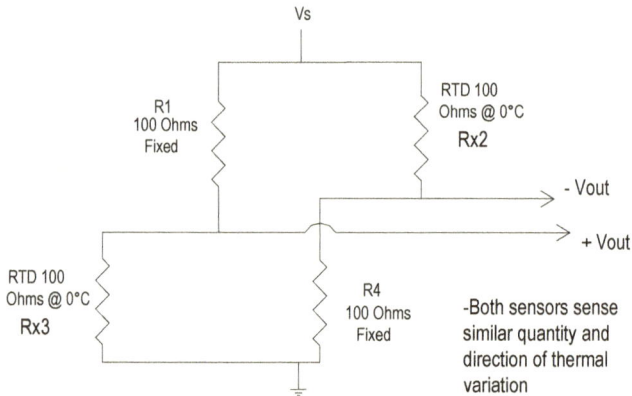

Figure 20 – Dual RTD Bridge. The outputs are connected to the inputs of the previous differential amplifier example. As the temperature applied to the RTD's increase the RTD resistance increases, causing the output voltages to move in the directions indicated by the output voltage polarities.

Since the RTD exhibits a positive resistance coefficient, the RTD resistance will increase with an increase in temperature. As the RTD's increase in resistance, the voltage across each RTD in each leg of the bridge circuit will also increase since each leg is a two-component series circuit. Therefore as Rx3 increases in resistance the voltage across Rx3 increases. Since the +Vout is taken across Rx3, The +Vout terminal voltage will increase with an increase in temperature.

Since the –Vout terminal voltage is across the series-dropping resistor R4 in the Rx2 leg of the bridge, as the applied temperature increases and the Rx2 RTD increases in resistance and voltage, the voltage across R4 and at the –Vout terminal will decrease.

At the minimum applied temperature, all of the bridge components will exhibit the same resistance and the voltage available at each output will be equal at one-half the bridge supply. The following diagram illustrates the null-balance condition of the bridge.

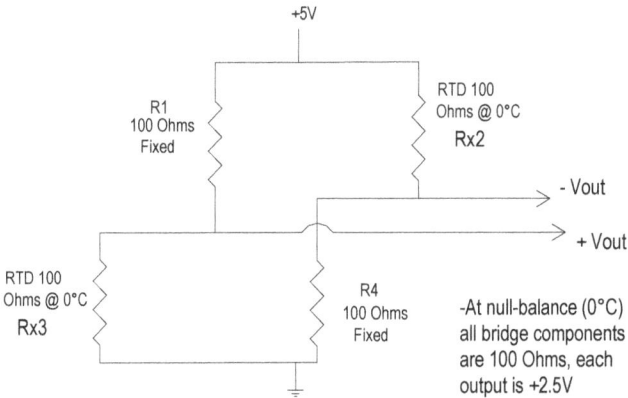

Figure 21 – The Null-balanced Bridge, all resistances are equal at the minimum applied temperature. This condition holds for all bridge circuits at the initial condition of minimum applied physical variable.

Assume the bridge circuit is connected to the corresponding differential amplifier. The differential amplifier has been assembled to develop a differential gain, Ad, of 10 and has included input resistances to properly impedance match the bridge to the amplifier. The complete bridge and amplifier circuit is shown in the following diagram.

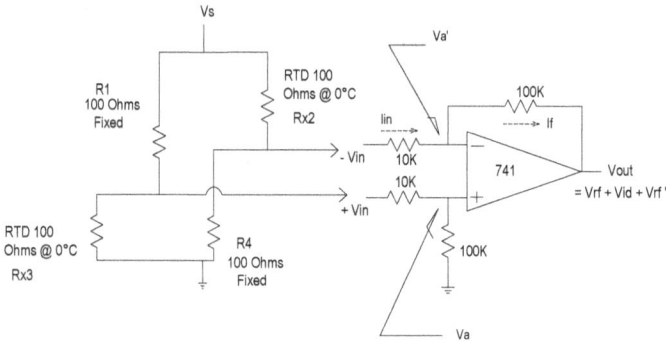

Figure 22 – A dual RTD bridge and differential amplifier.

To analyze the combined circuit the RTD resistances at the minimum and maximum temperatures must be determined. Once the temperature extremes are known the individual bridge output voltages can be determined. If the bridge outputs are known the differential amplifier circuit can be analyzed and the output can be determined. It should be mentioned here that easier techniques for determining the output voltage from the amplifier are available but the focus of this section of differential amplifiers is to demonstrate the circuit analysis of a differential amplifier and the bridge is used as a signal source. Again, the following analysis will be performed twice, once at each temperature extreme.

RTD Resistance at 0° Centigrade,

RTD resistance at 0° Centigrade = .385 Ohms/C° * (Temperature) + 100 Ohms.

RTD resistance at 0° Centigrade = .385 Ohms/C° * (0°C) + 100 Ohms.

RTD resistance at 0° Centigrade = 0 Ohms + 100 Ohms.

RTD resistance at 0° Centigrade = 100 Ohms.

This computation verifies the RTD offset value of 100 Ohms at zero degrees Centigrade. When the RTD resistance values are included in the bridge, the circuit appears as figure 21. From the circuit diagram the individual output voltages can be determined. It should be remembered the individual output polarities represent the output direction when the applied temperature is increased and not the output polarity with respect to ground. With this in mind the bridge outputs are computed using the voltage divider equation as follows.

$$-Vout = Vs * [R4/(Rx2+R4)]$$
$$-Vout = 5V * [100/(100+100)]$$
$$-Vout = 5V * [100/200]$$
$$-Vout = +2.5V \text{ (negative going)}$$

$$+Vout = Vs * [Rx3/(Rx3+R1)]$$
$$+Vout = 5V * [100/(100+100)]$$
$$+Vout = 5V * [100/200]$$
$$+Vout = +2.5V \text{ (positive going)}$$

From the previous computations, the differential output voltage (the difference between each output) is zeros volts meaning the bridge is null-balanced at 0° Centigrade. Knowing each voltage available at the inputs allows computation of the voltages distributed throughout the op amp circuit and a determination of the amplifier output voltage. The voltage computations follow.

Knowing the voltage available at +Vin, the voltage at the non-inverting input can be determined. Since +Vin = 2.5V, the non-inverting voltage, or Va, equals;

$$Va = +2.5V * [100K(100K+10K)]$$
$$Va = +2.5V * [.90909]$$
$$Va = +2.2727V$$

It's beneficial to recognize that the computed Va is also the voltage across R_F' and the voltage across R_{IN}' is the difference between Va and the applied 2.5V, or

.227V. These values are labeled in the following diagram. Since the input differential voltage, Vid, between the op amp input terminals is nearly zero volts, the voltage at the inverting terminal is approximately equal to Va. This is also indicated in the following diagram.

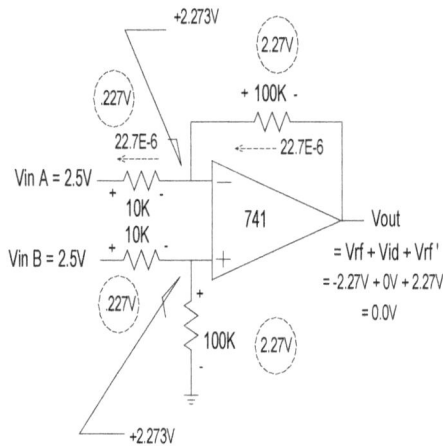

Figure 23 – The differential amplifier with a null-balanced bridge on the input. Note the canceling effect of the voltages distributed around the amplifier circuit. An amplifier output of zero volts is the result of the voltages and their corresponding polarities summing to zero between the amplifier output and any path to ground. Resistor tolerance can significantly alter the actual output voltage.

Knowing the voltage at the inverting terminal and the voltage at the inverting signal input (Vin A), the input current through Rin can be determined. Finding the difference in voltage between Vin A and Va', and dividing by the value of Rin yields the input and feedback current values. The current direction and polarity should also be noted on the circuit diagram.

$$Iin = If = (Vin\ A - 2.2727V) \div 10K$$
$$Iin = If = (2.5V - 2.2727V) \div 10K$$
$$Iin = If = (.227V) \div 10K$$
$$Iin = If = 22.7 \times 10^{-6}\ Amps,\ or\ 22.7\mu A$$

304

From the current computation the feedback resistance voltage can be determined.

$$V_{Rf} = I_F * Rf$$
$$V_{Rf} = 22.7 \times 10^{-6} \text{ Amps} * 100K$$
$$V_{Rf} = 2.27V$$

All the voltages distributed around the circuit are now known. Summing the voltages in any path to ground will result in determining Vout. The diagram of figure 24 shows a path from Vout to ground through the feedback resistance, the iput terminals and the Rf resistance at the non-inverting input. Algebraically summing the voltages and using this path would yield the following result.

$$Vout = V_{Rf} + V_{id} + V_{Rf}$$
$$Vout = -2.27V + 0V + 2.27V$$
$$Vout = 0V$$

Initially an output of zero volts may be a surprise but given the circuit is a differential amplifier and amplifies only the difference between the two inputs, when the inputs are the same the output will be zero volts. Note also that an output of exactly zero volts is contingent upon the voltages being distributed around the circuit in such a manner as to cancel when summed. This works perfectly on paper but in reality resistors have a tolerance that keeps them from being exactly the rated value. Resistor tolerance and the subsequent mismatch of components can result in unacceptable, non-zero results in the output. To minimize the possibility of this occurring, consult the section on CMR optimization in the chapter on op amp specifications and testing and _DON'T_ attempt to measure and precisely match all resistors used in the differential amplifier circuit. An easier, potentiometric method of compensation exists.

The previous computations allowed an analysis of the temperature sensor, bridge circuit and differential amplifier at zero degrees Centigrade. Using the same approach to determine the RTD resistances at 100 degrees Centigrade yields;

RTD Resistance at 100° Centigrade,
RTD resistance at 100° Centigrade = .385 Ohms/C° * (Temperature) + 100 Ohms
RTD resistance at 100° Centigrade = .385 Ohms/C° * (100°C) + 100 Ohms
RTD resistance at 100° Centigrade = 38.5 Ohms + 100 Ohms
RTD resistance at 100° Centigrade = 138.5 Ohms

The circuit appears as follows at 100 degrees Centigrade. Knowing the resistance values throughout the bridge at each temperature extreme, the individual differential output voltages can be determined using the voltage divider equation.

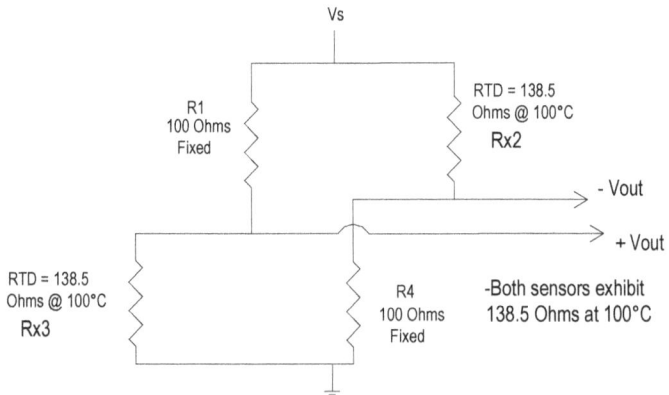

Figure 24 - A two RTD Bridge at 100 degrees Centigrade.

Finding the bridge output voltages at 100 degrees, and again using the voltage divider technique the individual output voltages can be determined. It should again be mentioned the individual output polarities represent the output direction when the applied temperature is increased and not the output polarity with respect to ground. With this in mind the bridge outputs are computed using the voltage divider equation as follows.

$$Vin\ A = -Vout = Vs * [R4/(Rx2+R4)]$$
$$Vin\ A = -Vout = 5V * [100/(138.5+100)]$$
$$Vin\ A = -Vout = 5V * [100/238.5]$$
$$Vin\ A = -Vout = +2.096V\ (negative\ going\ bridge\ output)$$

$$Vin\ B = +Vout = Vs * [Rx3/(Rx3+R1)]$$
$$Vin\ B = +Vout = 5V * [138.5/(138.5+100)]$$
$$Vin\ B = +Vout = 5V * [138.5/238.5]$$
$$Vin\ B = +Vout = +2.904V\ (positive\ going\ output)$$

From the previous computations, the differential output voltage (the difference between each output) is the difference between the two computed bridge output values. The resulting bridge output at 100° C Vin B =. Becomes;

$$Differential\ Bridge\ Vout\ at\ 100°\ C. = Vin\ B - Vin\ A$$
$$Differential\ Bridge\ Vout\ at\ 100°\ C. = +2.904V - +2.096V$$
$$Differential\ Bridge\ Vout\ at\ 100°\ C. = .8076V$$

In practice, knowing the differential output voltage from the bridge and recognizing the differential amplifier gain of 10, it can be accurately assumed the amplifier output should be approximately 8.1 volts. However, given this is an exercise in amplifier circuit analysis the assumption will be proven using the technique applied in the previous example at a temperature of zero degrees.

Before proceeding, a final note about bridge output polarity. The previous equation used to determine bridge output voltage subtracted Vin A from Vin B, and implies Vin A must always be subtracted from Vin B. This is not the case. Since neither output contains a ground or other reference, polarity becomes a non-issue. For example, if the bridge were connected to an amplifier, display or other load and a negative output voltage or indication were observed, the problem could be corrected by reversing the two output leads from the bridge. In a similar manner the precise location of either output voltage in the differential output equation becomes a non-issue. If the output computes negative, it is treated as being a "polarity-free," absolute value quantity. In this manner, differential voltage equations in this text are assembled to yield a non-polarized value.

Knowing the bridge output at 100° C. and continuing with the amplifier analysis, the voltage across the resistive divider components at the non-inverting input is determined.

$$Va = +2.904V * [100K(100K+10K)]$$
$$Va = +2.904V * [.90909]$$
$$Va = +2.64V$$

Again, this is also the voltage across Rf, and the voltage across Rin' is the difference between Rf and the applied bridge output voltage of 2.904V, or .264V. The input and feedback currents can be determined next.

$$Iin = If = (Vin\ A - 2.64V) \div 10K$$
$$Iin = If = (2.096V - 2.64V) \div 10K$$
$$Iin = If = (-.544V) \div 10K$$
$$Iin = If = -54.4x10^{-6}\ Amps,\ or\ 55.4\mu A$$

Again, the negative sign is irrelevant and indicates the current will be moving in opposition to the current direction at zero degrees. From the current computation the feedback resistance voltage can be determined.

$$V_{Rf} = I_F * Rf$$
$$V_{Rf} = 54.4 \times 10^{-6} \text{ Amps} * 100K$$
$$V_{Rf} = 5.44V$$

The circuit diagram with computed values labeled follows. Finding Vout requires summing all voltages and their corresponding polarities between the output and ground.

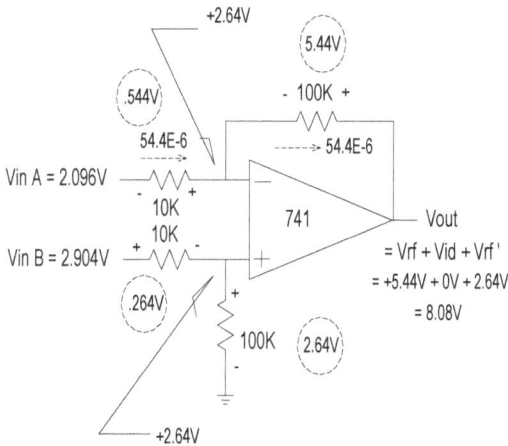

+2.64V

5.44V

.544V   - 100K +

54.4E-6   54.4E-6

Vin A = 2.096V
10K
10K   741   Vout
Vin B = 2.904V   = Vrf + Vid + Vrf '
                                        = +5.44V + 0V + 2.64V
.264V   = 8.08V

100K   2.64V

+2.64V

Figure 25 – Circuit analysis of a differential amplifier at 100°C.

$$Vout = V_{Rf} + V_{id} + V_{Rf}$$
$$Vout = +5.44V + 0V + 2.64V$$
$$Vout = +8.08V$$

As predicted earlier, the output computed to approximately 8.1 volts from the circuit analysis. In practice, the expedient method of circuit recognition, bridge output voltage measurement and approximation of the amplifier output by rapid mental computation is commonly employed, and circuit analysis is rarely performed. Circuit analysis is usually executed during an initial performance check when the amplifier or measurement system is not working as expected and a specific discrepancy must be located. The real benefit to knowing differential amplifier circuit analysis is in the facilitation of amplifier design.

Differential Amplifier Design

Designing the Differential Amplifier is considerably less tedious than the circuit analysis. In fact, design of any operational amplifier linear application is less demanding than the circuit analysis if the fundamentals are understood. In order to effectively design differential amplifiers for the purpose of sensor signal conditioning, the concepts of impedance matching and differential signal sources should be understood. Differential amplifiers are used for two primary purposes, conversion of differential to single-ended signals and amplification of a small signal voltage (usually milli to microvolts) into a larger signal that ca be easily utilized by displays, A/D's and so forth. Given the circuit configuration of the differential amplifier, offset biasing is not a consideration when designing differential amplifiers. Output offset shifting may be required for signal compatibility reasons however, but will be performed in subsequent stages utilizing previously introduced biased inverting and non-inverting amplifiers.

To begin the study of differential amplifier design, the signal source must be considered and understood to the greatest extent. Since a differential amplifier is being considered as the first stage of signal conditioning, the signal source must be of a differential nature such as a bridge. If possible, each resistance in the bridge should be known so that the output resistance of the source can be approximated if not provided in the specifications. The input resistances of Rin and Rin' should be at least ten times the provided or approximated value of the bridge output resistance. Again, if the output resistance is not provided a means of approximating the output resistance (such as Thevenizing) should be used.

Once the input resistances on the amplifier are determined, the feedback resistances are determined from the desired amplifier gain. The desired amplifier gain is contingent upon additional conditioning amplifiers, the sensor output and the desired voltage at the load. An example follows.

Differential Amplifier Design Example

Assume a 100-psi bridge type pressure sensor specifies a sensitivity of .1mV/V/psi, a 5-volt supply, an output offset of −20 millivolts and although the output impedance (resistance) is not specified, bridge resistances of 1.2K Ohms in each leg are shown in a schematic diagram. In addition, the recording instrument to be used as the ultimate load requires a 0V to 10-Volt input span. Determine the appropriate signal conditioning amplifier(s) to adapt the sensor output to the recorder.

A good starting place in the design of the pressure sensor signal-conditioning amplifier is to determine the functional block diagram of the complete amplifier. Although the specific amplifier design including component values cannot be performed in this step, the number of amplifier stages, the approximate gain of each stage, output offset and power supply requirements can be determined. To determine the block diagram characteristics, the sensor output voltage span and recorder input span need to be determined. The recorder input span was given initially but the sensor output span needs to be computed from the given sensitivity and offset quantities.

Since,

$$\text{Sensitivity} = \Delta \text{Vout} / \text{Vs} / \Delta \text{Pressure}$$

And since,

$$\text{Gain} = \text{Sensitivity} * \text{Supply Voltage}$$

Given,

$$\text{Vs} = 5\text{V}$$
$$\text{Sensitivity} = .1\text{mV/V/psi}$$

$$\text{Gain, Kx} = \text{Sensitivity} * \text{Vs}$$
$$\text{Kx} = .1\text{mV/V/psi} * 5\text{V}$$

$$\text{Sensor Gain, Kx} = .5\text{mV/psi}$$

Since,

$$\text{Sensor Gain, Kx} = \Delta\text{Vout}/\Delta\text{Pin}$$
$$\Delta\text{Vout} = \text{Gain} * \Delta\text{Pin}$$

Given,

$$\Delta\text{Pressure} = 100\text{psi}$$
$$\Delta\text{Vout} = .5\text{mV/psi} * 100\text{psi}$$
$$\Delta\text{Vout} = 50\text{mV}$$

The computed output voltage span provides the amount of output change given an input pressure change of 100psi. To determine the precise output voltages at 0psi, 100psi or any other pressure between the extremes, the linear (Y=MX+B) model is required.

Since,

$$Y = MX + B$$

In a pressure sensor context,

$$\text{Vout} = (\text{Gain} * \text{Pin}) \pm \text{Offset}$$

Since the sensor offset was given as −20mV, Vout at 0psi is,

$$\text{Vout} = (.5\text{mV/psi} * 0\text{psi}) + (-20\text{mV})$$
$$\text{Vout at 0psi} = -20\text{mV}$$

Vout at 100psi is,

$$\text{Vout} = (.5\text{mV/psi} * 100\text{psi}) + (-20\text{mV})$$
$$\text{Vout at 100psi} = +30\text{mV}$$

Knowing the sensor output span, and knowing the recorder input span, an overall gain and offset can be approximated. Since the overall amplifier characteristics are being determined as we progress, the computations are termed approximations. It may be determined after investigation that portions of the design requires subtle alterations. In which case a temporary return to an appropriate location within the design will allow variation to the initial assumptions.

Once the sensor output and the load voltage spans are determined, overall amplifier gain and offset values can be approximated.

Since the desired amplifier output is,

$$\Delta Vout = 0\text{-}10V$$

And the estimated amplifier input is,

$$\Delta Vin = 50mV$$

The overall gain (A_{Vtotal}) becomes,

$$A_{Vtotal} = 10V / .05V$$
$$A_{Vtotal} = 200$$

The required output offset (B) becomes,

$$B = Vout - (A_{Vtotal} * Vin)$$
$$B = 10V - (200 * .03V)$$
$$B = +4V$$

At this stage the designer can decide how to share the required circuit gains, offsets and other functions should they be required. As presented later in the text all the necessary functions can be performed in a single stage of instrumentation amplifier, or with multiple stages of operational amplifier assembled as a differential amplifier and used in conjunction with a biased non-inverting amplifier. At this point in the text the choice of circuits is obvious. Knowing nothing of Instrumentation Amplifiers, a differential amplifier and subsequent offset-biased amplifier will be used. The block stages will appear as follows. Suggested gains and functions are included. Notice the overall total voltage gain is distributed among both amplifier stages. The individual amplifier gains can be made to equal other values as long as the total combined gain is 200. The output-offset or biasing function has been designated at the second stage since the differential amplifier in the first stage is incapable of performing an output offset. With both inputs required to amplify the differential output voltage from the signal source, biasing must be performed in the second amplifier stage.

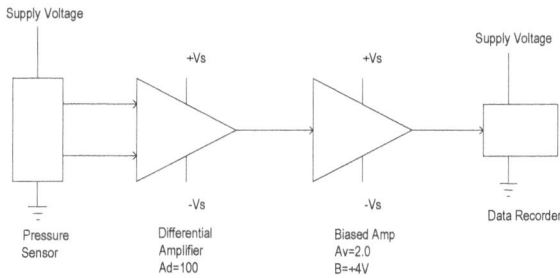

Figure 26 – The functional block diagram of the current example. Possible amplifier gains and offsets are also included.

Again, the distribution of the amplifier gains is only suggested. Alternative approaches might set the differential gain (Ad) at 10 and the biased non-inverting amplifier gain at 20. Another alternative might set the differential gain to forty and the biased amplifier gain to 5. Or possibly even reverse the sensors' outputs and design the second stage as a biased reverse-acting amplifier. Any combination is possible as long as the objective of an overall gain of 200 and an output offset of +4V is met. For this example, assume a differential gain of 100 and a biased non-inverting gain of 2, with a +4 volt output offset is desired. Determining the overall sensor and amplifier gain and offset characteristics is the usual first step in signal conditioning amplifier design.

As a second step, acquire the sensor or other signal source schematic and determine the output resistance of the differential signal source. Once the output resistance is known, or the signal source schematic is acquired the remainder of the amplifier design is a routine two-step mechanical procedure. If a schematic is not available, measure the resistances between each of the connection pins and draw a working diagram. Although this approach will not yield an accurate schematic in terms of internal components, values and connections, it will provide an approximation of the sensors' electrical characteristics as they appear to the subsequent amplifier stage. In this example and in most sensor applications the individual internal bridge resistances are provided. A diagram follows.

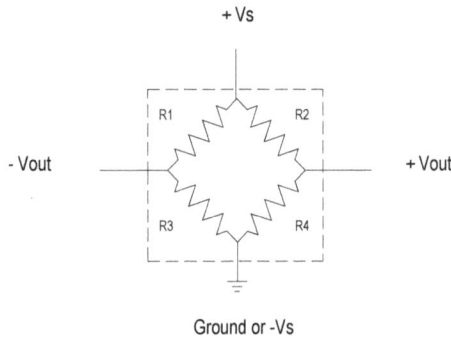

Figure 27 – A full bridge, each resistor is 1.2K Ohms.

In determining the bridges' output resistance, two approaches can be taken. The first involves a Thevenin analysis of the entire bridge. In performing the Thevenin analysis, short the supply connections (with the supply disconnected) and compute (or measure) the resistance between the output leads. This inevitably results in a Thevenin resistance value equal to the value of any individual bridge leg resistance. Therefore in the current example, the bridge output resistance as seen by a single load placed across the bridge output, equals 1.2K. Experience will dictate that any bridge containing equal value resistors in each leg will exhibit an output or Thevenin resistance equal to the value of any single bridge leg resistance. As mentioned, this is the resistance exhibited by the bridge to a single load placed across the bridge, like a meter. This may not be appropriate for a differential amplifier, which appears to the bridge as two separate loads. The differential amplifier and four element bridge circuit is shown in the following diagram.

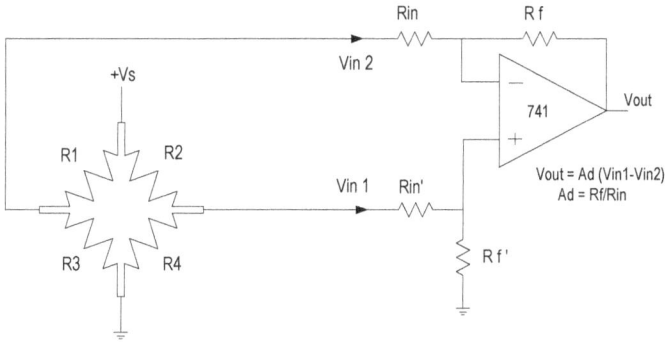

Figure 28 – A four-element full bridge circuit connected to an operational amplifier configured as a differential amplifier. The "Ad" quantity in the output equation represents the differential gain of the amplifier.

The second and most appropriate technique of determining the signal sources' characteristics involves investigating each of the differential signal sources' outputs. Each of the bridges' outputs connects to the amplifier input resistances of Rin and Rin'. Although the input impedance quantities associated with the amplifier inputs involves both Rin and Rf at each input, as a worst case design scenario it is best to assume the input impedance at each input equals only the value of Rin and Rin'. By assuming the input impedance equals the respective input resistance value, the amount of transferred signal from the source to the amplifier will always be greater than assumed.

In the current example the bridge circuit source contains two outputs. Each output appears as a 600-Ohm source resistance to each of the amplifier inputs (1.2K||1.2K) since the Thevenin equivalent of each output of the bridge "sees" two paths to ground when looking back into the bridge from each output.

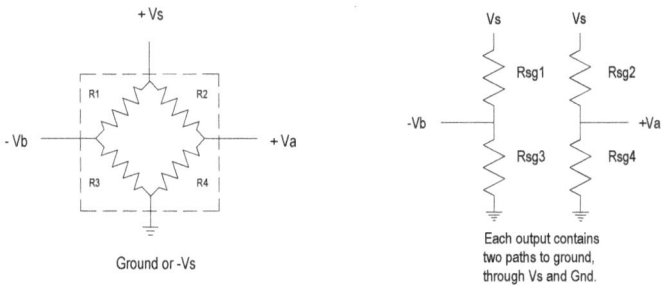

Figure 29 – The common bridge circuit can be redrawn as two series voltage dividers, each with its own output voltage and source resistance. The source resistance each side of the bridge exhibits is equal to the resistance seen from the output to ground through each resistance.

Given the bridge appears as two separate signal sources, each with its own output resistance and output voltage, another electronic symbol is occasionally used to represent bridge circuits, common-mode voltages, differential voltages and differential signal sources in general. The following diagram presents an alternative differential signal source schematic symbol. The symbol includes any voltage common to both outputs (Vcm), the differential signal voltage (Vdiff), the individual output resistance values and the equivalent bridge circuit resistance combinations, and the individual output labels (Va, Vb). The symbol is normally used as an abbreviation to drawing the complete signal source circuit diagram.

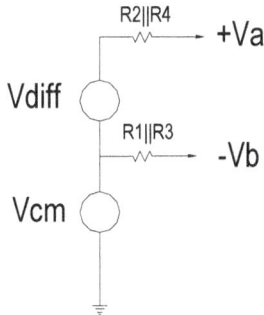

Figure 30 – The schematic symbol occasionally used to represent a differential signal source.

Since the desired amplifier input resistance is to be at least ten times the output resistance of the signal source, and since each of the signal sources output leads are 600 Ohms, the input resistances of the differential amplifier (Rin and Rin') will be selected at 6K Ohms or greater. Again, the actual input resistance to each differential amplifier input is approximately Rin plus Rf (and Rin' plus Rf'), and will result in an actual input resistance greater than 6K.

The preceding steps appear lengthy due to the detailed support explanations provided. In practice the differential amplifier can be designed in a matter of seconds, usually without putting a pencil to paper. The remainder of this example requires determining Rf and Rf' for the differential amplifier, and determining the component values for use in the second amplifier stage. In finding the feedback resistor values for the differential amplifier, the equation for differential gain must be algebraically rearranged.

Since,

$$Ad = Rf/Rin,$$

If,

$$Rf = Rf'$$

And,

$$Rin = Rin'$$

Then,

$$Rf = Rf' = Ad * Rin = Ad * Rin'$$
$$Rf = 100 * 6K (6.2K)$$
$$Rf = Rf' = 600K \text{ Ohms } (620K)$$

The values in parentheses represent the nearest standard value resistors. The sensor and amplifiers will appear as follows. Verification of the biased non-inverting amplifier design is left to the student. It is performed with a minimum of explanation including only the required steps following the diagram.

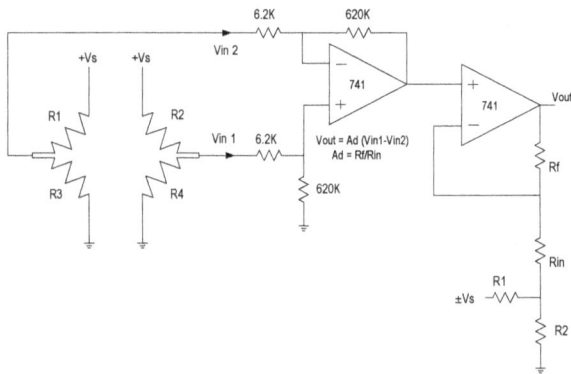

Figure 31 – The bridge circuit sensor, differential amplifier and biased non-inverting amplifier.

320

Since the biased non-inverting gain is 2.0, and assuming Rin is 100K,

Finding Rf,

$$Av = 1 + Rf/Rin$$
$$Rf = Rin (Av-1)$$
$$Rf = 100K (2-1)$$
$$Rf = 100K$$

Since,

$$B = +4V$$

And,

$$B = V_{bias} * A_{Vbias}$$
$$V_{bias} = B/A_{Vbias}$$

Since,

$$A_{Vbias} = -Rf/Rin = -100K/100K = -1$$

$\therefore V_{bias}$ equals,

$$V_{bias} = +4V/-1$$
$$V_{bias} = -4V$$

Sizing bias voltage resistors to the desired voltage drops, and using available Proto-Board supplies,

$$R1 = 11K$$
$$R2 = 4K$$
$$Vs = -15V$$

Since 11K and 4K resistors are not usually available, doubling or halving each results in the following values. Either of the following will suffice.

$$R1 = 22K$$
$$R2 = 8K (8.2K)$$
$$Vs = -15V$$

Or,

$$R1 = 5.5K\ (5.6K)$$
$$R2 = 2K$$
$$Vs = -15V$$

The completed circuit appears following.

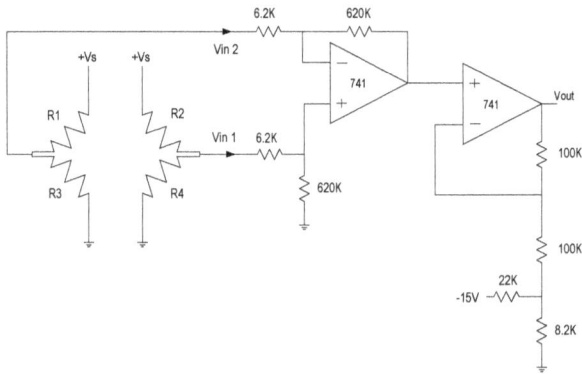

Figure 32 – The completed amplifier design with amplifier component values.

Differential Amplifier Deficiencies

The completed amplifier circuit is relatively inexpensive (about $1.00 with resistors at Radio Shack) and performs well but has some shortcomings. The inputs on the differential amplifier are not exactly balanced (unequal input impedances seriously degrades the CMR), nor are the input impedances exceptionally high. To establish a respectable Zin for most applications, the input resistors should be 10Meg or larger. The feedback resistor determines the op amp gain but the feedback resistors are limited to values under 10 Meg to avoid significant offset and parasitic current related errors. The size of the input resistors in conjunction with the limitation placed upon the feedback resistors limits the gain of the op amp differential amp to a maximum of about 100. Given that most signals derived from differential sources are in the microvolts to millivolt range, gains of over 100 are common to differential amplifiers. The additional gain requirement is not a large obstacle and is easily resolved with additional op amp stages assembled for gains of 10 or more. Since gains are multiplied when amplifiers are cascaded, the total gain of two amps with gains of say, 10 and 100 becomes 1000. However other differential amplifier circuits are available if two operational amplifiers are to be utilized.

Two Op Amp Differential Amplifier

A variation of Differential Amplifier is constructed from two op amps and appears as two cascade non-inverting amplifiers.

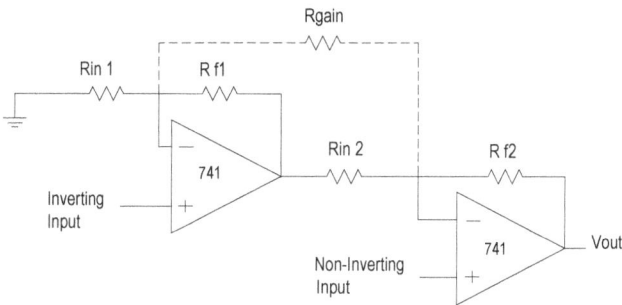

Figure 33 – Two Op Amps configured to improve performance over a single op amp differential amplifier.

The dual op amp Differential Amplifier represents a substantial performance improvement over a single op amp, differential amplifier. The amplifier's Zin is improved by applying the differential input signals to the non-inverting terminals of both op amps. As mentioned previously, whenever an input is applied directly to the non-inverting terminal, a high (1Meg or greater) Zin will be exhibited due to the inherently high input resistance of the op amp enhanced by the negative feedback characteristics of the amplifier. Zin characteristics of the inverting input are generally considered to be equal to the input resistance value, depending upon the configuration of the amplifier. As a result, the input impedance of the non-inverting input is preferred whenever available. Note in the following chapter instrumentation amplifier configurations, the inputs are always applied to the non-inverting inputs.

The amplification characteristics are best observed through an equation derivation of the output voltage. Given the following diagram, note that Vout1 is the result of Vin1 being applied to a non-inverting input at the first stage amplifier. Momentarily assume this gain is Av1. The output of the first stage is applied to the inverting input of the second stage and will see a corresponding inverting gain (-Av2) at the second stage where Vin2 sees a non-inverting gain (+Av2).

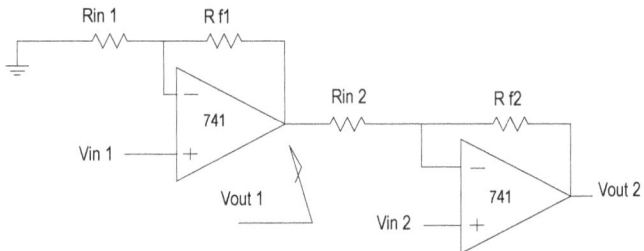

Figure 34 – Although each input sees a non-inverting polarity, Vout 1 of the first stage sees an inverting input of the second op amp stage, effectively subtracting or differentially amplifying the two inputs.

The output equation can be written as follows.
$$Vout2 = Vin2 \ (+Av2) + Vout1 \ (-Av2)$$

Since,
$$Vout1 = Vin1 \ (+Av1)$$
$$Vout2 = Vin2 \ (+Av2) + Vin1 \ (+Av1)(-Av2)$$

Including resistor designations,
$$Vout2 = Vin2 \ (1+Rf2/Rin2) + Vin1 \ (1+Rf1/Rin1)(-Rf2/Rin2)$$

At this point the resulting output equation is about impossible to observe. However if the correct values of resistors are applied to the circuit the equation reduces to appear simpler.

Assume;

$$Rin1 = Rf2 = 10K$$
$$Rf1 = Rin2 = 1K$$

The circuit becomes;

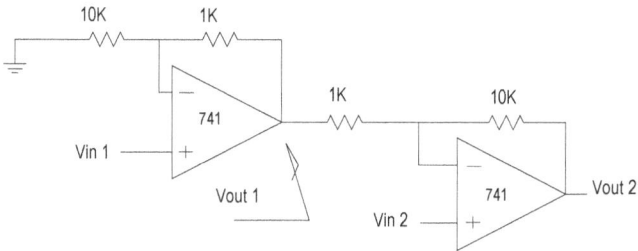

Figure 35 – The two-op amp Differential Amp is recognized by its' configuration symmetry, and the similarity of assigned resistor values.

And the output equation becomes;

$$Vout2 = Vin2 (1+10K/1K) + Vin1 (1+1K/10K)(-10K/1K)$$

$$Vout2 = Vin2 (11) + Vin1 (1.1)(-10)$$
$$Vout2 = 11Vin2 - 11Vin1$$
$$Vout2 = +11 [Vin2 - Vin1]$$

Note several things in the resulting Vout2 equation. First, the gain seen by each input is numerically equivalent, but opposite in polarity. Second, note the differential amplification appearance of the output equation by the subtraction function. Third, note the polarities (inverting and non-inverting) associated with each of the inputs. And finally, note the resistor configuration required to establish the output equation.

Differential Amplifier Homework Problems

#1
A full bridge contains a single Platinum RTD and three 100-Ohm resistors. Determine the bridge output span between 0 and 100 degrees centigrade. Show a clearly labeled circuit diagram and all computations.

#2
Design an amplifier to convert the output span of the previous problem into an output span of 0 to 1 Volts. Include a circuit diagram.

#3
Design an amplifier to condition the output of problem #1 into a single-ended .32 to 2.12 volt span.

#4
Analyze a single op amp differential amplifier designed for a gain of unity with 5 volts applied to each input. Find the voltage across each resistance and the resulting output voltage.

Chapter 8 – Instrument and Isolation Amplifiers

This chapter advances applications of operational amplifiers and linear integrated circuits. Multiple op amps are presented in the configuration of Instrumentation Amplifiers. Emphasis is placed upon using discrete op amps as Instrument Amps and the corresponding circuit considerations and analysis techniques. Instrument Amp integrated circuits are introduced and typical applications are designed and analyzed. Isolation Amplifiers are introduced in a conceptual manner and typical applications are provided.

Objectives:
Upon completion of this chapter, you should be able to:
- Explain the purpose and benefits of Instrument Amplifiers
- Explain the purpose and benefits of Isolation Amplifiers
- Design instrument amplifiers for sensor applications
- Address instrument amplifier calibration concerns
- Optimize CMR with instrument amplifier circuits
- Recognize potential applications of instrument and isolation amplifiers
- Interpret instrument amplifier specifications

Introduction

Operational amplifiers assembled and applied as differential amplifiers perform adequately as low-level signal conditioners, yet each exhibits an undesired effect. The single op amp differential amplifier has unbalanced input impedances that result in one input being amplified slightly more than the other input, a condition that results in degraded rejection of common-mode noise. The two-op amp differential amplifier is a popular amplifier but requires an additional amplifier if biasing is desired. And offset biasing is always desired in low-level signal conditioning to aid in setting the amplifier output to the correct value, or calibration.

The Classic Three Op Amp Instrumentation Amplifier

Instrumentation Amplifiers, or Instrument Amps, are a class of amplifiers intended for use with low-level sensors and are commonly characterized by high gain, high input impedance, and high common mode rejection. Other features common to Instrument Amps are a single gain determining resistance and an applied offset bias voltage equal to the desired amount of output offset (an Offset-Bias gain of unity). Available in integrated circuit or discrete op amp configurations, the instrumentation grade of sensor amplifier can cost from under $1.00 to $20 and higher in quantity. To fully understand applications of the various forms of instrument Amp, the op amp version will be investigated in detail.

The culmination of operational amplifier low-level, differential sensor signal conditioning is a three operational amplifier configuration referred to as the "Classic Three Op Amp Instrumentation Amplifier." The circuit is composed of two non-inverting amplifiers in a cascode circuit configuration, followed by a single op amp, differential amplifier stage. The combination of the three op amps provides a high gain, high input impedance and high common-mode rejection (CMR) amplifier that works well in all but the most extreme of applications. The majority of Instrument Amplifiers exhibit signal gains of up to a thousand, single-ended and differential input impedances of a thousand Megohms (10^9), common-mode rejection of 120 decibels, and a single resistance determines the gain of the entire amplifier. Output biasing and compensation provisions are also available in all but the least expensive models. An Instrument Amp constructed from op amps appears in the following diagram.

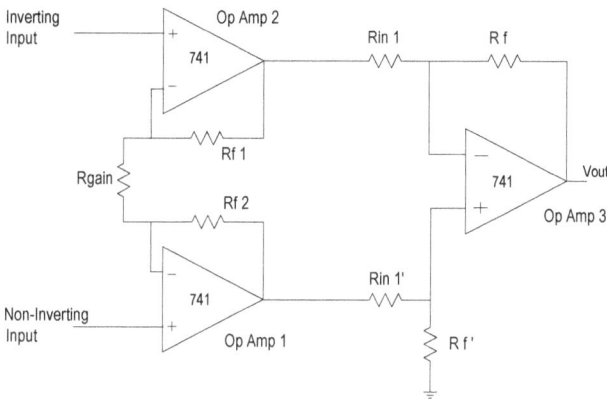

Figure 1 – The Classic Three Op Amp Instrumentation Amplifier. Note the cascode non-inverting amplifiers in the first stage sharing a common "Rin" as Rgain, and the differential amplifier in the output stage.

From the diagram it can be seen the amplifier has inverting and non-inverting differential inputs and a common input resistance, Rgain that will be used to determine the gain of the entire amplifier. Upon inspection the first stage appears as a modified non-inverting amplifier. The corresponding gain equation also shares a non-inverting similarity (1+2Rf/Rgain) as will be proven later in the chapter. The output stage, a common, op amp differential amplifier as introduced in the previous chapter, performs a differential to single-ended signal conversion, and can be used for additional gain, output biasing and output leadwire compensation if required. Note the supply connections to the individual op amps have been omitted from the diagram for organizational reasons. It is suggested the circuit configuration be studied for familiarity. Once familiar, the analysis and design becomes straightforward.

Triple Op Amp Instrumentation Amplifier - Circuit Analysis

A combination of circuit analysis techniques associated with the non-inverting and differential amplifiers are used to determine the circuits operation and corresponding voltage gain and Vout equations. Begin the analysis by assuming, as with all negative feedback op amp circuits, Vid = 0V. Given this and the following diagram, any voltage placed upon Vin "A" will appear at Node "A," and any voltage placed upon Vin "B" will appear at Node "B." Each voltage now appears as Vin A' and Vin B' above and below Rgain. If the input voltages are equal no difference of potential will appear across Rgain and no current will flow through Rgain, an input null-balance condition. However if the two inputs were different voltages, a difference of potential would exist across Rgain and a resulting current would flow through Rgain as indicated by the arrows in the following diagram. Given the direction of the arrows, Vin "B" must be greater (more positive) than Vin "A."

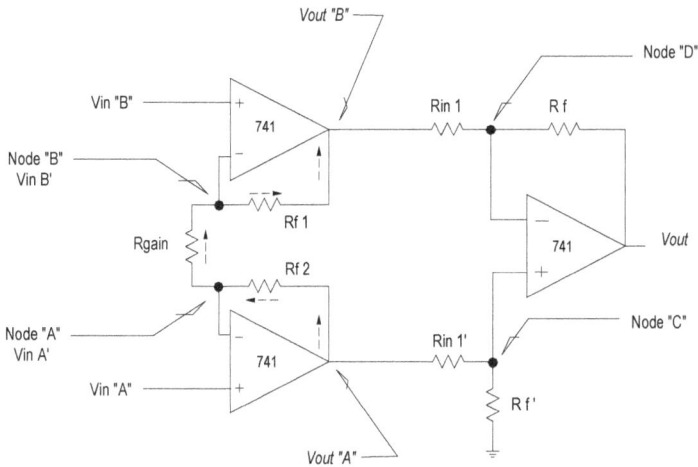

Figure 2 – The input and feedback currents in the first stage are the result of any difference in voltage across Rgain.

Since Rgain is in the location of Rin for both of the input stage amplifiers, the current passing through Rgain is the input current, Iin, for both non-inverting op amps. The current passing through each feedback resistor is the corresponding If, created by the input current through Rgain since input and feedback currents are equal in all negative feedback op amp circuits.

The currents passing through all three resistors (Rf1, Rgain, Rf2) of the input stage establish voltages across each resistance. The combined sum of all three voltages is applied to the output stage differential amplifier as shown in the following diagram. The resulting output voltage from the first stage is a differential voltage equal to the sum of the three resistor voltages, or the difference between Vout "A" and Vout "B" in the circuit diagram. The second stage differential amplifier subtracts, amplifies and offsets (if applied) the resulting first stage output voltage.

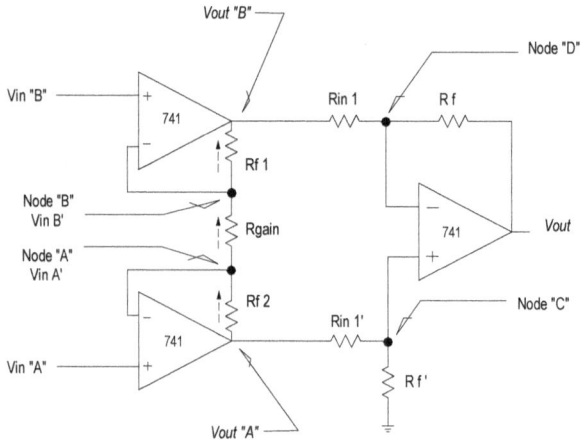

Figure 3 – The combined resistor voltages of the first stage are applied as a differential voltage to the second stage differential amplifier.

In deriving the output voltage and overall circuit gain equations, recall from earlier in the text the output of the second stage differential amplifier is;

$$V_{out} = R_F / R_{in} (V_{out} \text{ "A"} - V_{out} \text{ "B"})$$

When,

$$R_{in} = R_{in}'$$
$$R_f = R_f'$$

Since,

$$(V_{out} \text{ "A"} - V_{out} \text{ "B"}) = (V_{Rf1} + V_{Rgain} + V_{Rf2})$$

Vout becomes,

$$V_{out} = R_F / R_{in} (V_{Rf1} + V_{Rgain} + V_{Rf2})$$

Since,

$$(V_{Rf1} + V_{Rgain} + V_{Rf2}) = [(If1*Rf1) + (Iin*Rgain) + (If2*Rf2)]$$

332

Vout becomes,

$$V_{out} = (R_F / R_{in}) [(If1*Rf1) + (Iin*Rgain) + (If2*Rf2)]$$

Since Iin and If are equal in negative feedback op amps, substituting Iin for If yields,

$$V_{out} = (R_F / R_{in}) [(Iin*Rf1) + (Iin*Rgain) + (Iin*Rf2)]$$

Factoring Iin yields,

$$V_{out} = (R_F / R_{in}) [Iin * (Rf1 + Rgain + Rf2)]$$

Since,

$$Iin = (Vin "A" - Vin "B") / Rgain$$

Substituting for Iin yields,

$$V_{out} = (R_F / R_{in}) \{[(Vin "A" - Vin "B") / Rgain] * [(Rf1 + Rgain + Rf2)]\}$$

Rearranging,

$$V_{out} = (Vin "A" - Vin "B") (R_F / R_{in}) [(Rf1 + Rgain + Rf2)/Rgain]$$

Since Rf1 always equals Rf2,

$$V_{out} = (Vin "A" - Vin "B") (R_F / R_{in}) [(2Rf + Rgain)/Rgain]$$

Rearranging,

$$V_{out} = (Vin "A" - Vin "B") (R_F / R_{in})[(1 + 2Rf/Rgain)]$$

Since Gain equals Vout ÷ Vin, and Vin equals (Vin "A" − Vin "B"), the overall and total amplifier gain becomes,

$$Av = (R_F / R_{in})[(1 + 2Rf/Rgain)]$$

The final equations for Vout and Av are representative of all Instrumentation Amplifiers. However the second stage is only occasionally used for additional gain and normally has the Rf / Rin ratio set to unity. Given this, the most common form of Instrumentation Amplifier gain appears as;

$$Av = 1+(2Rf \div Rgain)$$

And Vout becomes,

$$V_{out} = (Vin \text{ "A"} - Vin \text{ "B"}) * (1+2Rf/Rgain)$$

Note the gain determining resistance appears in the denominator of both equations. As such, the amplifier gain increases with a decrease in Rgain.

Instrument Amplifier performance is exceptional in most every aspect except frequency response. Given the amplifier may be used at high gains, high frequency response is sacrificed due to the op amps' limited gain*bandwidth specification. Extended range frequency response is possible when high frequency op amps are used in place of the general performance 741's shown in the previous diagram. Limited frequency response is usually not a problem in experimental measurements since the inputs are acquired from sensors measuring physical phenomena that change relatively slowly. Very rarely will a physical signal applied to a sensor be expected to vary in the kilohertz range and never above. The most recent generation of Instrumentation Amplifiers utilize Operational Transconductance Amplifiers internally and exhibit greatly enhanced frequency response characteristics over the previous generation of Instrumentation Amplifier.

Instrument Amp Features

For any amplifier to qualify as an instrumentation grade of amplifier, an amplifier must conform to certain common specifications. The input impedance must equal or exceed 10^9 Ohms, the common mode rejection capability must be greater or equal to 100dB, the gain must be determined by a single resistor, and the output offset value (B) must equal the input bias voltage (Vbias). In other words, A_{Vbias} must equal +1. Instrumentation Amplifiers' of all manufacturers and part numbers exhibit high input impedance since the inputs are applied directly to the op amp inputs. High gain is accomplished by sharing a common input resistance (Rgain) in the first stage. From the gain equation, as Rgain is *decreased*, the amplifier gain increases. Since the second stage gain is normally held to unity, the overall gain equation for all Instrumentation Amplifiers appears similar to the single op amp non-inverting amplifier gain equation. The second stage differential amplifier performs a number of beneficial functions, to convert the signal format from differential to single-ended, to add additional gain (if needed) and to provide output offset, CMR (noise elimination) and amplifier "sensing" provisions. The following diagram demonstrates these features.

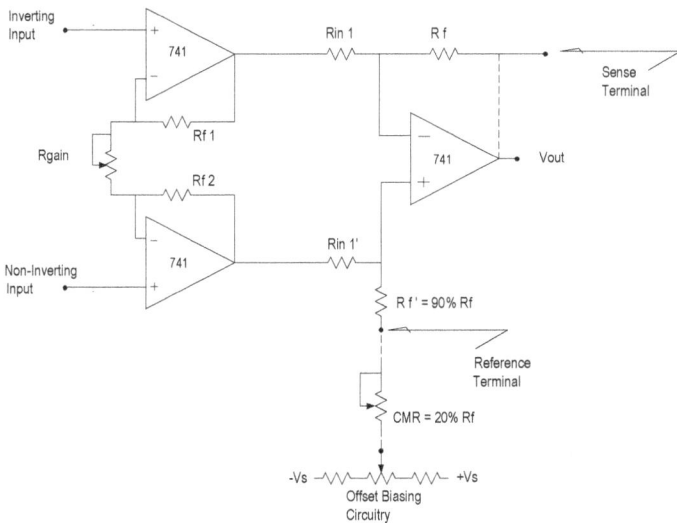

Figure 4 – Instrumentation amplifier configuration with offset, CMR optimization, sensing and gain features indicated.

Removing and replacing the ground connection (formerly at the bottom of the diagram, see previous diagrams) with a potentiometric circuit will apply an output-offset voltage to the Instrumentation amplifier output voltage span. To facilitate ease of use, the gain seen from the reference terminal to the output equals unity. This is due to the divider created between the "Reference Terminal" and the output of the first stage op amp, and the complementary gain observed at the non-inverting input of the output stage. The "Gain to Output" of the reference terminal is explained in greater detail later in the text.

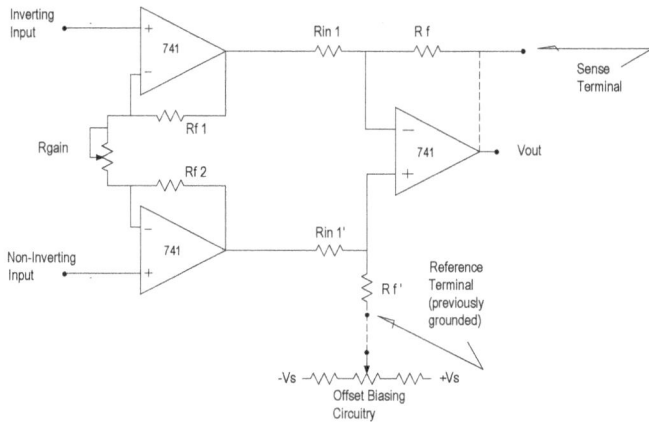

Figure 5 – Replacing the ground connection at the Reference terminal with a variable DC voltage provides a convenient means of biasing an Instrumentation Amplifier. The gain seen from the Reference terminal to the output is unity.

The variable CMR resistance balances the common-mode voltages around the last stage amplifier circuit to cause an algebraic sum of zero at the output. The summing of the common-mode voltages is identical to the manner described with the single stage, op amp differential amplifier. When additional resistances are applied to allow variable gain, offset and sensing functions, CMR balancing becomes necessary. To set the CMR optimization circuitry, a common mode input voltage is applied to both inputs simultaneously. The common-mode voltage is usually AC at a frequency equal to, or close to 60 Hertz since 60 Hz noise is the most prevalent noise source (120 Hz is easier to view on an oscilloscope). The output is observed as the CMR resistance is varied. The resistance is varied to achieve minimum AC voltage in the output. In order for the CMR optimization circuit to function correctly the CMR potentiometer should be a value to allow variation above and below the value of Rf in the feedback path of the last stage op amp. As an example, if Rf equals 10K, the combination of Ra and the CMR resistance should vary from 9K to 11K ohms. In this example Ra would equal 9K (9.1K) and the CMR potentiometer would equal 2K. Ideally, the variable CMR resistance should be close to centered when the circuit is optimized for CMR.

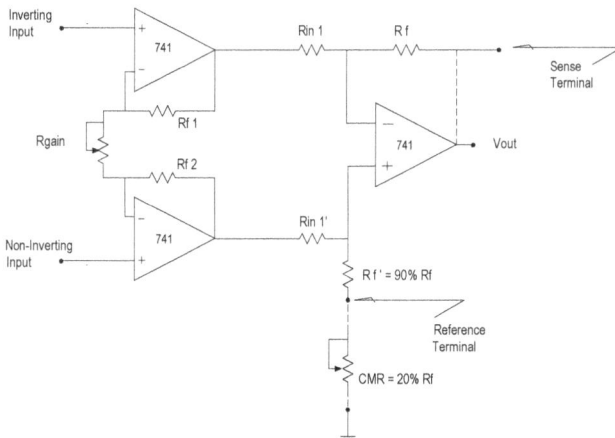

Figure 6 – The addition of a CMR resistance allows common-mode voltages on the inputs to distribute among all circuit resistances in such a manner as to cancel in the output. By varying the CMR resistance, the common-mode voltages distributed throughout the circuit will be altered and "matched" to create a minimal effect in the output.

The "Sense Terminal" designation provides two possible features. If additional gain in the output stage is desired, additional feedback resistance can be introduced between the Sense terminal and the output. If this is performed, the additional resistance should also be included in the "Reference" leg at Ra. The second and primary intended feature of the Sense terminal is to compensate for leadwire length between the output and the next stage. Often the distance between the output and next stage can be significant when the amplifier is in the "field" with the sensor and the data acquisition system, PC or display is at a remote location. Since the leadwire length will introduce resistance between the output and the next device, the voltage received at the device will be less than at the output location. To compensate for the signal loss, an additional leadwire is connected between the sense terminal and the load, placing the additional leadwire resistance within the feedback of the instrumentation amplifier output stage, slightly increasing the last stage gain and providing the desired output voltage at the load location. The following diagram illustrates.

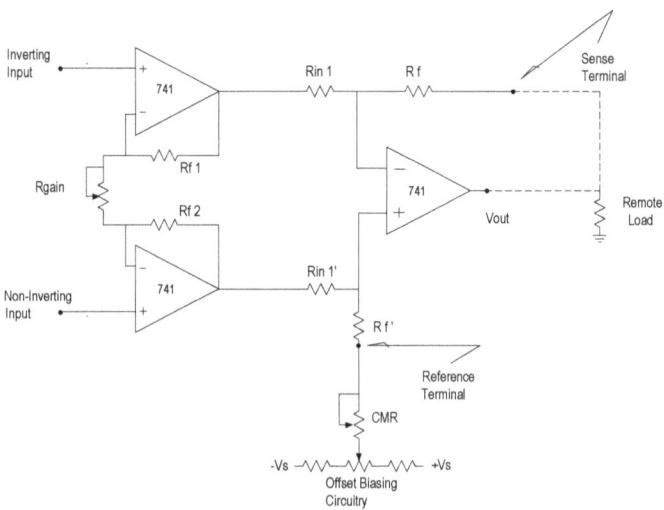

Figure 7 – The Sense terminal can be used to increase the gain of the output stage and to compensate for leadwire resistance between the output and a remotely located load. In either case, an increase in the feedback resistance in the Sense leg should be balanced by an increase in resistance in the Reference leg or by including a CMR optimization variable resistance.

Circuit Analysis Example

Assume an Instrument Amp constructed of three op amps is to amplify the output of a strain gauge bridge circuit. For convenience, the potentiometers associated with bridge span and zero, amplifier gain and offset have been omitted. The variable resistance for CMR has been retained for demonstration purposes. The bridge is composed of two active strain gauges and two fixed 120-Ohm resistors. Assume as a result of an applied tensile strain, the gauges each increase in resistance by .1 Ohms, to 120.1 Ohms (a larger than typical quantity). The circuit including amplifier resistance values follows.

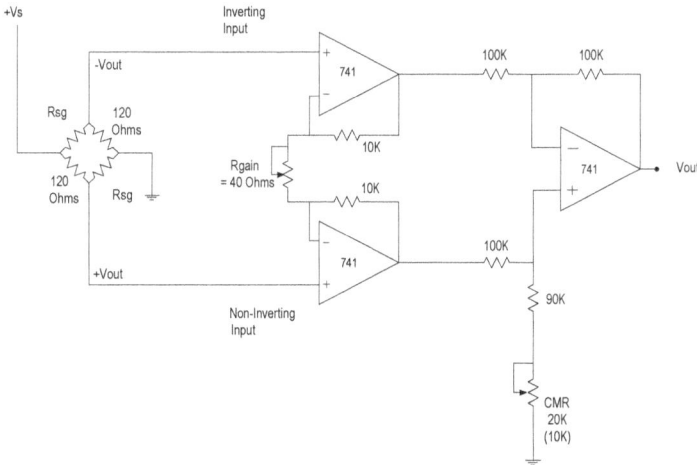

Figure 8 – Instrumentation amplifier and strain gauge bridge.

In analyzing the amplifier and bridge circuit, a logical first step would be to determine the bridge output voltage. The differential output from the bridge is the difference between the individual outputs at the positive and negative output locations. Using the voltage divider equation to determine each output yields;

Finding the positive Vout,

$$+Vout = 5V [Rsg/(Rsg+120)]$$
$$+Vout = 5V [120.1/(120.1+120)$$
$$+Vout = 5V [.500208247]$$
$$+Vout = 2.501041233V$$

Finding the negative Vout,

$$-Vout = 5V [120/(120+Rsg)]$$
$$-Vout = 5V [120/(120+120.1)]$$
$$-Vout = 5V [.499791753]$$
$$-Vout = 2.498958767V$$

Finding the differential Vout from the bridge,

$$Vout = +Vout - (-Vout)$$
$$Vout = 2.501041233V - 2.498958767V$$
$$Vout = .002082466V \text{ or } 2.08 \text{ millivolts}$$

Knowing the differential input voltage to the Instrument Amp, and knowing the component values of the amplifier, the amplifier gain can be determined. Although the goal is not specifically to determine the output voltage but rather to provide a circuit analysis example, the output voltage can be determined if the amplifier gain is known. Finding the amplifier output voltage will assist by providing a reference value to be checked after the analysis is performed. Finding the amplifier gain is determined by using the equation derived previously and subbing the corresponding resistance values. From the gain equation,

$$Av = (R_F / R_{in})[(1+2Rf/Rgain)]$$

Where R_F and Rin are values of the second stage differential amplifier, and Rgain and 2Rf are acquired from the first stage amplifier. Substituting values yields;

$$Av = (100K / 100K)[(1+(2*10K)/40)]$$
$$Av = (1) [1+(20K/40)]$$
$$Av = (1) [1+500]$$
$$Av = 501$$

Therefore, the bridge output voltage is computed as;
$$Vout = Av * Vin$$

Or,

$$Vout = 501 * .002082466V$$
$$Vout = 1.043315466V$$

Or, the Instrument Amplifier output should equal approximately one volt. This value will be used for comparison after the circuit analysis is performed.

Continuing with the analysis, the individual voltages are applied to their respective inputs. Note the positive bridge out is applied to the non-inverting input and the negative bridge output is applied to the inverting input. In practice, an accidental reversing of the bridge outputs would result in a negative amplifier output. In this case, if the strain gauges are "Loaded" further by the application of additional force, the amplifier output would move further in the negative direction. A sure sign the amplifier inputs need to be switched.

Applying the input voltages results in the following diagram. As can be seen, the individual input voltages appear at their respective amplifier inputs and at each end of the Rgain resistance. Since the input differential voltage (Vid) between the op amp input terminals equals zero in negative feedback applications, each input voltage is felt at Rgain as well. The difference between the individual input voltages results in a differential voltage of .00204V across Rgain, and creates a current of 52 microamps (.000052A) through Rgain. The current is labeled as Iin due to Rgain's location as the common input resistance to each of the op amps.

The resulting input current through Rgain flows through the individual feedback resistors of the first stage op amps and establishes individual voltage drops of .52V across each feedback resistor. The end result is the appearance of three series voltages between the two op amp outputs. The algebraic sum of the voltages will be conditioned by the second stage differential op amp circuit. The input stage voltages and currents appear in the following diagram.

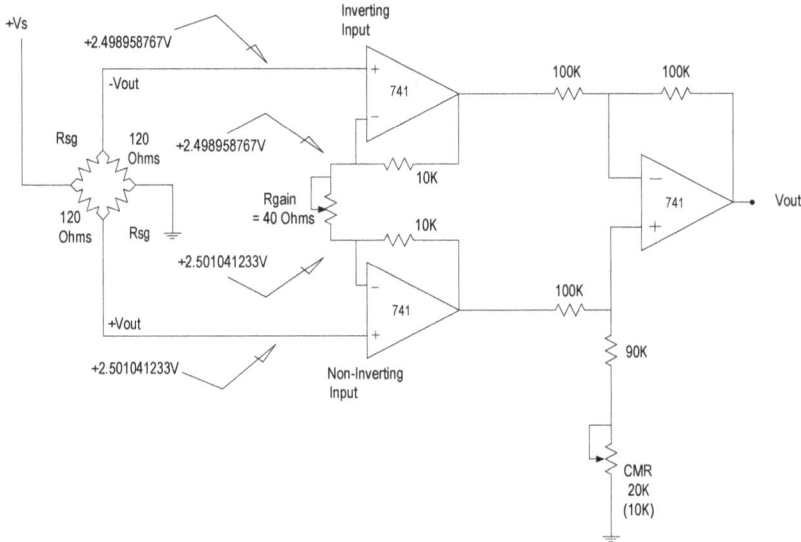

Figure 9 – Once the individual input voltages are determined, the voltages appear at each end of Rgain creating a differential voltage across Rgain, and a corresponding current through Rgain.

The second stage differential amplifier can be analyzed by recognizing the input voltage is the difference between the two outputs of the previous stage op amps. The output of the first stage is composed of the voltage across the three resistors between the op amp outputs. Once identified, the input differential voltage from stage one can be seen to be amplified by the second stage gain of unity since the stage gain is determined by the ratio of Rf/Rin. A rather lengthy alternative analytical technique also exists following.

As seen in the following diagram the voltage at the output of the first stage is a differential voltage, since no voltage in the output portion of the circuit is connected or referenced directly to ground. The voltages available at the first stage inputs however are referenced to ground. Knowing this and the voltage drops across all resistances in the first stage, the ground referenced or single-ended output voltages available at each op amp output can be determined. The following diagram illustrates the procedure for determining the ground-referenced outputs.

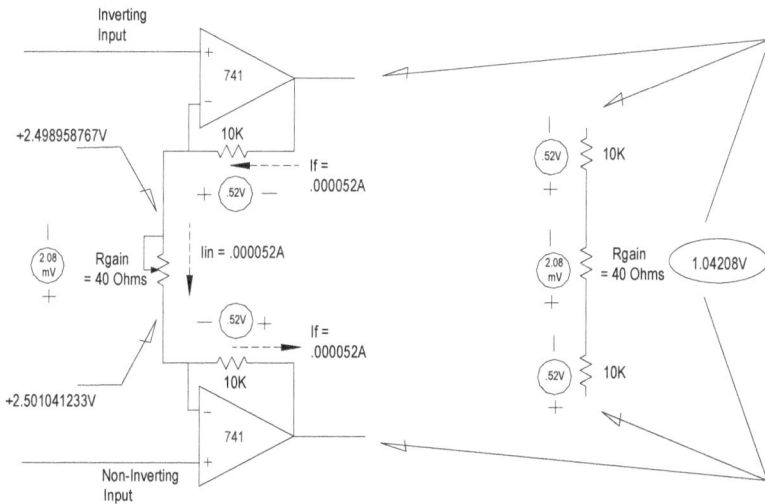

Figure 10 – The first stage of the Instrument Amplifier. The resulting current through Rgain and each feedback resistance establishes voltages across all three resistances seen in the diagram at the right. The sum of the voltages will be applied to the second stage differential amplifier.

The voltages indicated at each extreme of Rgain represent the ground-referenced voltages at each input. Although not depicted in the diagram, the voltage at Vin "A" is derived from a resistor connected directly to ground in the bridge circuit. Therefore the input to Vin "A" is a single-ended, or ground referenced voltage. The voltage at the Vin "B" input is also ground referenced but is slightly more difficult to recognize unless one realizes that voltage supplies are also a direct connection to ground due to the low internal resistance of common voltage supplies. Knowing this, Kirchoff Voltage Loop equations can be written to determine the ground-referenced voltages at any point in the amplifier circuit.

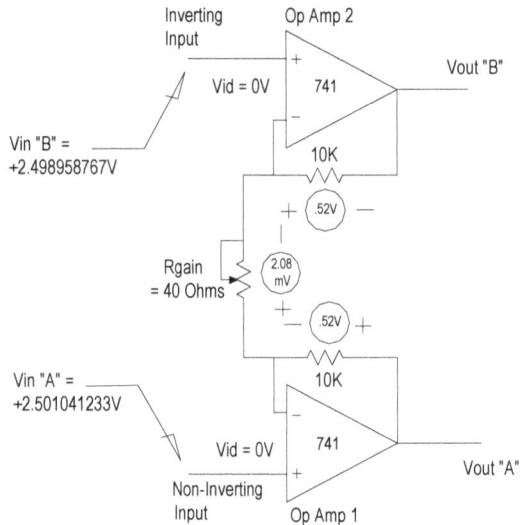

Figure 11 – The individual inputs are computed as ground referenced voltages from the bridge. When summed with Vid and the individual feedback resistor voltages, the individual ground referenced output voltages at op amps 1 and 2 can be determined.

A Kirchoff Voltage Loop (KVL) can be written from each op amp output, back to the individual ground referenced input voltages as follows.

$$V_{out} \text{ "A"} = V_{Rf} + V_{id} + V_{inA}$$

344

$$V_{out} \text{ “A”} = +.\, 52V + 0V + 2.501041233V$$
$$V_{out} \text{ “A”} = 3.021041233V$$

$$V_{out} \text{ “B”} = V_{Rf} + Vid + Vin_B$$
$$V_{out} \text{ “B”} = -.\, 52V + 0V + 2.498958767V$$
$$V_{out} \text{ “B”} = 1.978958767V$$

Note the output differential between the op amp outputs is verified at 1.04208V. Knowing these individual ground referenced voltages, a component level analysis of the second stage differential amplifier can be determined. Each of the ground referenced output voltages is applied at the individual inputs of the second stage, and a routine op amp circuit analysis is performed.

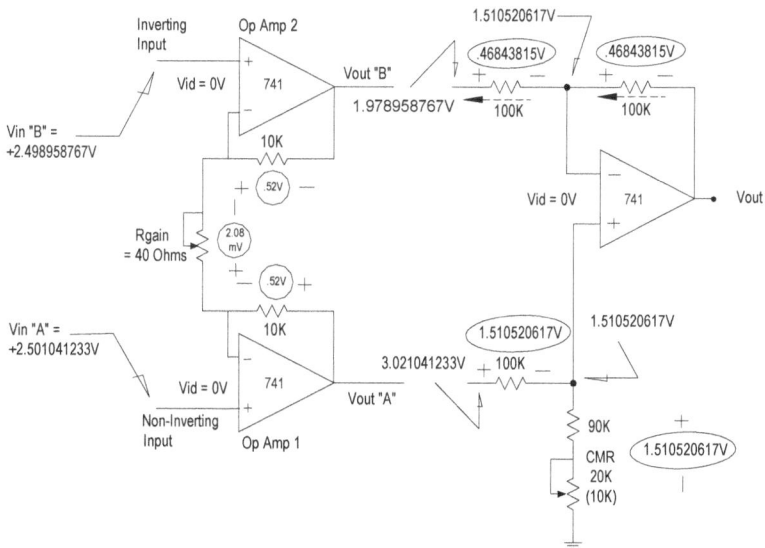

Figure 12 – It's not as complicated as it appears in this diagram. The outputs of each first stage op amp are applied to the corresponding inputs at the second stage differential amplifier. As with conventional op amp circuits, the voltage on the non-inverting input is determined and placed on the inverting input. The input and

feedback currents are determined, the voltages are computed from the input and fed back currents and a KVL is written to find the output voltage.

In analyzing the second stage, Vout "A" is applied to the resistive divider at the non-inverting input. Although the resistance to ground is composed of a CMR variable resistance and a fixed 90K resistor, assume the combination equals 100K. (As indicated on the CMR resistance, the total value applied to the circuit after CMR optimization will be close to the value of the op amp feedback resistance of 100K. The fixed 90K and 10K of CMR resistance will combine to equal the feedback value of 100K. In practice, this value deviates so slightly from the ideal that variable resistances on the order of 100 Ohms are commonly used in conjunction with fixed resistances in kilohms.) Continuing with the analysis, the resistive leg at the non-inverting input appears as a perfect voltage divider circuit reducing the output of Vout "A" by one-half to 1.510520617V. Since Vid equals zero, the same 1.510520617V, or very close to it will appear at the inverting op amp terminal. The difference between the 1.510520617V on the inverting terminal and the applied Vout "B" of 1.978958767V will appear across Rin. The difference of .46843815V across Rin will create a current flow, Iin.

$$Iin = V_{Rin} / Rin$$
$$Iin = .46843815V / 100K$$
$$Iin = 4.684 \text{ microamps } (4.684 \ \mu A)$$

Since the input resistance current (Iin) equals the feedback resistance current (If), the same 4.684 microamps of current will be passing through the feedback resistance. The resulting voltage across the feedback resistance becomes;

$$I_F = Iin = 4.684 \mu A$$
$$V_{Rf} = If * Rf$$
$$V_{Rf} = 4.684 \mu A * 100K$$
$$V_{Rf} = .4684V$$

From the voltages across Rin the polarities and current directions can be established and are included in the previous diagram. Knowing all of the voltages and polarities, a voltage loop equation can be written to determine the output voltage of the second stage.

$$\text{Vout} = \text{VRf} + \text{VRin} + \text{Vout "B"}$$
$$\text{Vout} = (-.4684\text{V}) + (-.4684) + 1.978958767\text{V}$$
$$\text{Vout} = +1.042082467\text{V}$$

In writing the voltage loop equation to determine Vout, the most convenient path through Rin was chosen. However, any path to ground would have resulted in the same output voltage quantity which turns out to be the same value as our initial estimate created by using the bridge output and the instrument amplifier gain equation. An additional suggestion would be to round the voltages to four places to the right of the decimal, or five significant digits. When working with millivolt and microvolt level signals, rounding can create substantial errors. A single or two decimal place quantity won't be adequate, but carrying-out quantities to the extent provided in this example is unnecessary and were performed here to amuse the skeptics and minimize the error.

Instrument Amp Design

Fortunately, Instrument Amp circuit design is far less demanding than the analysis. The analysis is necessary if one wishes to adequately prepare the circuit for an application since circuit modification is almost always necessary and effective circuit troubleshooting absolutely requires at least an awareness of the analytic quantities.

When designing the three-op amp instrument amp, the desired output span, available input span from the sensor, percent error associated with the input span, source impedance, load impedance and supply voltages should be known up front. The required gain and offset will be determined from the input and output signal spans, with the input percentage of error providing an indication of required gain and offset variation. Impedance matching probably won't be a problem since both inputs are applied directly to the non-inverting terminals. The resulting input impedance is at least 1 Meg and more realistically closer to hundreds of Megohms. Output impedance will be similar to a single op amp with values fewer than 100 Ohms common. The supply voltages are a concern if the amplifier is to be used in a remote location where batteries will be providing the available power. In which case the individual op amps used to assemble the instrument amp should be rated for the battery voltages. Op amps requiring only a few volts of power are rather common and should be considered in battery powered instrument amplifier applications.

As an application example, assume an SX30DN pressure sensor is used to monitor air compressor output pressure between zero and thirty pounds per square inch. The compressor is mobile and mounted on a utility vehicle, the sensor and amplifier power supply will be generated from a 12V unipolar (single polarity) battery/alternator power source. Develop an amplifier for the application.

In providing a design solution, two approaches are possible depending upon the specific value of output span desired. If the output is to represent the applied pressure in psi, such as zero to three volts represents zero to thirty psi of pressure, the power supplies will need to be ground referenced. In which case, the available 12V supply will need to be "divided" into two five or six volt supplies and a common ground with resistors, or 5V regulators of a Zener or solid-state variety (if the solid-state can use an input supply of six volts). The following diagram illustrates the power supply arrangement.

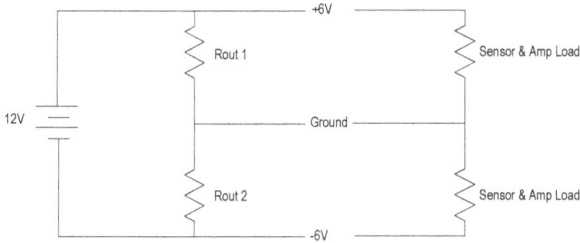

Figure 13 – The resistors on the right represent the sensor and amplifier loads to the power supply. The resistors on the left are required to divide the twelve-volt source into two bipolar six-volt sources with a common ground

Using the resistive approach to dividing the 12V supply is probably the most straightforward but also the most wasteful in power. Voltage regulation is also poor. Other alternatives, such as the Zener Diode voltage divider or solid-state regulator or commercially available switching power supplies perform the bipolar supply conversion with better results.

Continuing with the design, assume plus and minus 5 volt supplies are created from the single 12 volt supply, and the amplifier is expected to generate a 0V to 3V output span (in reference to the bipolar supply ground) over the 0-30 psi input pressure span to the sensor. Further assume the sensor will be connected to the +5V supply and from the pressure sensor specifications, will output –20 to +90 millivolts. The resulting circuit configuration appears as follows.

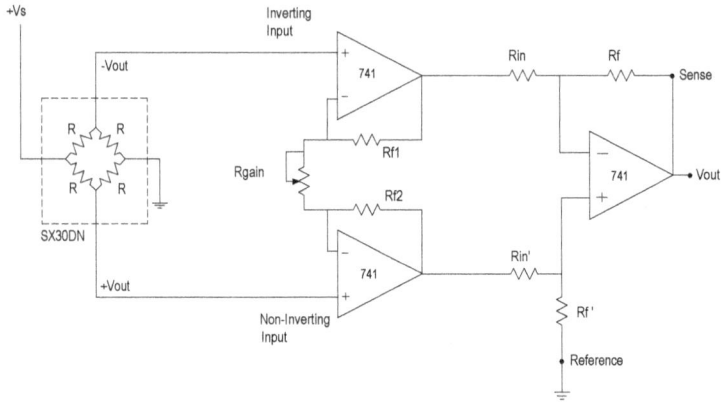

Figure 14 – The SX30DN pressure sensor and op amp instrument amplifier.

The instrument amp has been provided without any component values so that a complete design can be undertaken. Note the pressure sensor is provided as a bridge without component values since the output voltage span is provided in the sensor specifications (see Appendix).

Given the desired amplifier output span of 0V to 3.0V and the amplifier input span of -20 to +90 millivolts, the ideal gain and offset can be determined.

$$\text{Amp Gain, Av} = \Delta\text{Vout} \div \Delta\text{Vin} = 3V \div 110mV$$
$$Av = 27.273$$
$$B = \text{Vout} - (Av * \text{Vin})$$
$$B = 3V - (27.273 * .09V)$$
$$B = \text{Output Offset} = .545V$$

As mentioned previously these computations represent the ideal quantities for gain and output offset. In other words, if the sensor outputs precisely −20 millivolts to +90 millivolts over a zero to thirty psi input pressure span, the computed gain and offset will result in a 0V to 3.0V output voltage span. Assume for the moment, the sensor outputs the expected voltage span.

350

Knowing the desired amplifier gain and offset, the individual resistor values can be determined. First assume the gain of the differential amplifier in the second stage of the instrument amplifier is set to unity. Unity gain results in all four resistors around the second stage op amp being of equal value, say 10K ohms. Secondly, the voltage gain of the first stage requires assuming values for the Rf resistors in the feedback paths of the individual op amps since,

$$Av = 1 + (2Rf/Rg)$$
$$Rg = 2Rf /(Av-1)$$

Assume the feedback resistors are each 100K. Given the desired voltage gain of 27.273, the value of Rg becomes,

$$Rg = 7612.3778 \text{ Ohms}$$

And the value of Vbias applied to the amplifier equals the value of output offset computed earlier since A_{Vbias} equals unity. Therefore,

$$Vbias = +. 545V$$

Knowing these values, the amplifier diagram appears as the following.

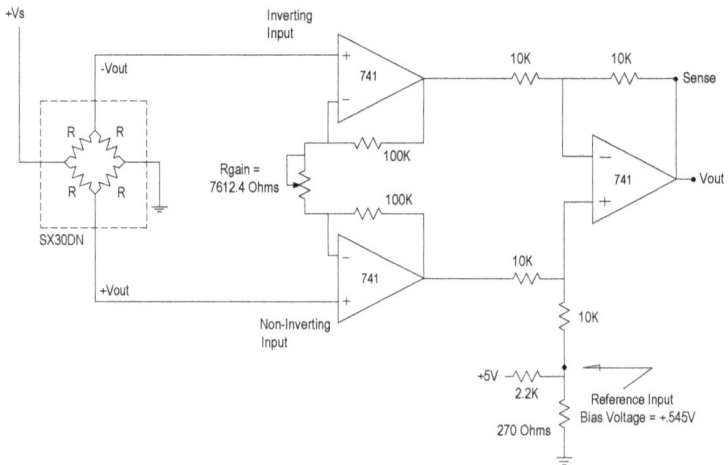

Figure 15 – The SX30DN and instrument amplifier with gain resistance and output offset bias voltage resistors labelled.

It would be easy to overlook the bias voltage applied to the reference input in the previous diagram. More specifically notice the voltage quantity is the same value that was computed from the input/output plot. Normally with an op amp, the amount of gain seen from the bias input (A_{Vbias}) would be considered before determining the amount of required bias voltage. As mentioned previously in the chapter, when all resistances in the second stage are equal, or when Rin = Rin' and Rf = Rf' the gain seen from the reference terminal to the output is unity. Therefore the amount of bias voltage applied to the reference input is the amount of offset computed from the I/O plot. In determining the resistance values for the output-offset bias voltage circuit, the same technique was used as introduced previously in the text. A voltage divider circuit was drawn off the reference input and the amount of desired bias voltage was placed across the divider output resistance. The divider supply was determined by the polarity and amount of needed Vbias, assuming Proto-Board supplies of +5V, +15V and –15V are available. The difference between the supply and the output voltage was placed across the dropping resistance as shown in the following diagram.

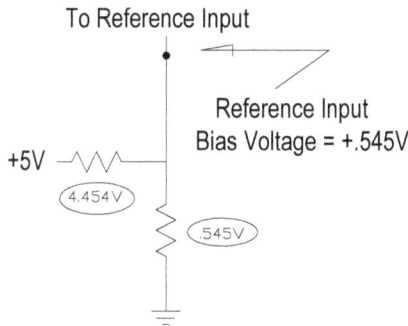

To Reference Input

Reference Input
Bias Voltage = +.545V

+5V ─/\/\─

4.454V

.545V

Figure 16 – To determine the required resistance values for a voltage divider circuit, assign the desired voltages and size the resistors accordingly.

Unfortunately, resistor values corresponding to the indicated voltage values (4.454K and .545K) do not exist, but dividing each resistance in half yields commonly available values of .27K and 2.2K. When 270 Ohms and 2.2K are added to the circuit, the voltage at the reference terminal computes to .546V, not exact but well within a 5% tolerance.

Instrument Amp Gain Variation

The specifications for the SX30DN pressure sensor indicate minimum, typical and maximum values of sensitivity and related operational specifications. The implication is that the sensor may exhibit considerable variation and may require amplifier modifications to accommodate the variation. Subsequently, the gain and offset bias indicated on the previous diagram need to be made variable. The amount of variation is contingent upon a number of factors including expected sensor variation. Other considerations include sensor and amplifier variations created by changes in the ambient (surrounding environment), leadwire lengths between the sensor and/or amplifier circuitry and the supplies, wishful thinking on the part of the sensor manufacturers and about every other conceivable influence upon the amplifier output. But the primary determining factor contributing to the variation of amplifier gain and offset will be the variation in the sensor characteristics. Because of this, the amount of gain and offset change will be dictated by the sensor variation with an additional amount added as a "fudge factor," to compensate for the unforeseen influences and to accommodate those sensors which may fall outside of the sensor specifications. The fudge factor has other benefits in determining the resistor values applied to the circuit, more on this later.

From the SX30 specifications, the typical amount of sensor sensitivity is indicated as .75mV/V/psi. Note the units are consistent with all bridge sensors where the applied supply determines the amount of output voltage span. To convert from the provided sensitivity information into units of sensor gain, the sensitivity is multiplied by the intended supply voltage of 5 volts, and results in the following.

$$\text{Resistance Bridge Gain = Sensitivity * Supply Voltage}$$
$$\text{Gain} = .75 \text{ mV/V/psi * 5V}$$
$$\text{Gain } 3.75 \text{ mV/psi}$$

Note the indicated variation in the sensor sensitivity of .5 mV/V/psi to 1.0 mV/V/psi translates into a percentage variation of $\pm 33\%$. The also translates into a sensor gain variation of $\pm 33\%$ and with the additional fudge factor percentage allows computation of the amplifier gain variation. In application, the "fudge" related unpredictables would probably alter the amplifier output within only a few percent, but to facilitate the circuit design (and make life in general a little easier) assume the variation span for the sensor gain is expanded from $\pm 33\%$, to $\pm 50\%$. With an expected sensor gain variation span of $\pm 50\%$, the required amplifier gain becomes the inverse of the expected variation extremes. Computing the amplifier gains at sensor gains above and below the ideal sensor gain yields,

Sensor gain of 50% of ideal ("typical" in spec's) equals,

50% of 3.75mV/psi = 1.875 mV/psi

Sensor output span becomes,

1.875 mV/psi * 30psi = 56.25 mV

Amplifier gain = ΔVout / ΔVin = 3V / 56.25 mV

Amplifier gain = 53.33

The amplifier gain is about twice the ideal value. Computing the amplifier gain at the other extreme yields,

Sensor gain of 150% of ideal ("typical" in spec's) equals,

150% of 3.75mV/psi = 5.625 mV/psi

Sensor output span becomes,

5.625 mV/psi * 30psi = 168.75 mV

Amplifier gain = ΔVout / ΔVin = 3V / 168.75 mV

Amplifier gain = 17.8

The required amplifier gain is about two-thirds (not quite half) of the ideal gain. The reason for the complementary amplifier values follows. Assume the product (cascade amplifier gains multiply) of the sensor and the amplifier gain equals unity (1). Therefore,

$A_{sensor} * A_{amplifier} = 1$

At the 50% sensor gain extreme,

$50\% \ A_{sensor} * A_{amplifier} = 1$

$A_{amplifier} = 1/(50\% \ A_{sensor}) = 2$, or twice the ideal gain

$A_{sensor} * A_{amplifier} = 1$

At the 150% sensor gain extreme,

$150\% \ A_{sensor} * A_{amplifier} = 1$

$A_{amplifier} = 1/(150\% \ A_{sensor}) = .667$, or two-thirds of the ideal gain

The amplifier gain variation should swing between *about* 17.8 and *about* 53.3. Since the required gain variation was increased by the fudge factor, the actual gain extremes can be close but are not required to be exactly the values indicated. In this example the gain extremes can be 15 to 55, or 20 to 50. Each would work well given the expected sensor gain variation from the specifications. As will be seen, the fudge factor variation concept can be applied to the gain resistance determination as well.

To determine the gain resistance variation, apply the computed gain extremes to the gain resistance equation.

Rg at Av minimum of 17.8,

$$Av = 1+(2Rf/Rg)$$
$$Rg = 2Rf/(Av-1)$$
$$Rg = 200K/(16.8)$$
$$Rg = 11904.8 \text{ Ohms}$$

Rg at Av minimum of 53.3,

$$Av = 1+(2Rf/Rg)$$
$$Rg = 2Rf/(Av-1)$$
$$Rg = 200K/(52.3)$$
$$Rg = 3824.1 \text{ Ohms}$$

The gain resistance variation should swing between *about* 3.9K and *about* 12K. Again, since the required gain variation was increased by the fudge factor, the actual gain determining resistance extremes can be close but are not required to be exactly the values indicated. From the computed resistance values a 3.9K fixed resistance and an 8.1K variable resistance would be required. Since an 8.1K variable resistance is an unavailable value, a 10K variable resistance and 2K fixed, or 10K variable and 3.9K fixed could be used depending upon which gain extreme was determined essential. In most variable gain applications the ideal gain is located near the center of the adjustment span. The modified circuit including the gain variation is following.

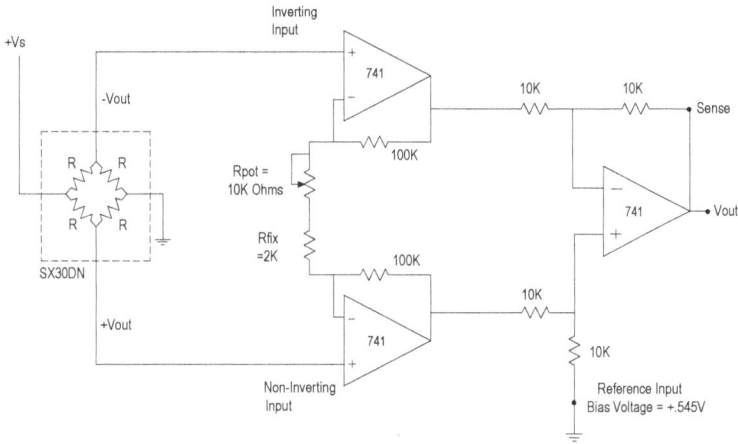

Figure 17 – Sx30DN pressure sensor and instrument amplifier with variable.

Offset Bias Variation

Determining the amount of bias voltage variation for a given amount of desired output offset variation is straightforward with an instrument amplifier. To determine the amount of required output offset variation the sensor specifications are again consulted. From the "zero pressure offset" specification the amount of variation is observed as −35mV to 0.0mV with −20mV being the typical sensor output voltage offset. From these values, it can be determined that the percentage of expected sensor offset will vary between −75% and plus 100% at the extremes. With a fudge factor, an adjustment range of ±150% should yield an all-encompassing variation span.

As mentioned earlier in the chapter, the reference terminal off the second stage non-inverting input also serves as the location to apply an output-offset bias voltage if desired. This is an ideal location to apply a bias voltage since the voltage gain observed between the reference input and the output (A_{Vbias}) is unity. The reason for the unity gain is due to the voltage divider resistor configuration between the op amp output at the input stage, and the second stage input. Recall the op amp output stage appears as a small resistance ([100 Ohms) to ground. The following diagram demonstrates.

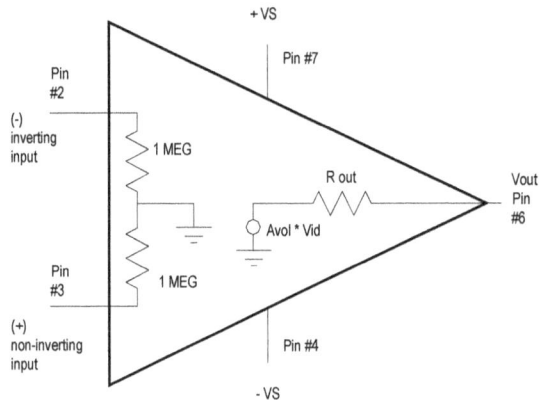

Figure 18 – The open loop op amp. Note the small amount of resistance between the output and ground. The Avol*Vid symbol contributes no resistance to the circuit and represents only the internally generated output voltage. The typical resistance between an op amp output and ground is less than 100 Ohms.

In effect, the amount of resistance introduced to the outside circuitry by the op amps' internal circuitry becomes negligible. Looking from the reference input in the following diagram, a resistive divider composed of two 10K resistors to ground (through the output of op amp 2) forms a voltage divider with a reduction factor (Beta) of .5. Therefore, any voltage applied to the reference input is reduced to one-half. The reduced voltage then appears at the input to op amp 3 where a complementary gain of 1+Rf/Rin amplifies the reduced voltage by 2. The result of reducing the voltage by a half and subsequently amplifying by two is an

overall non-inverted gain of unity. Likewise, the reference input exhibits unity
gain to the output.

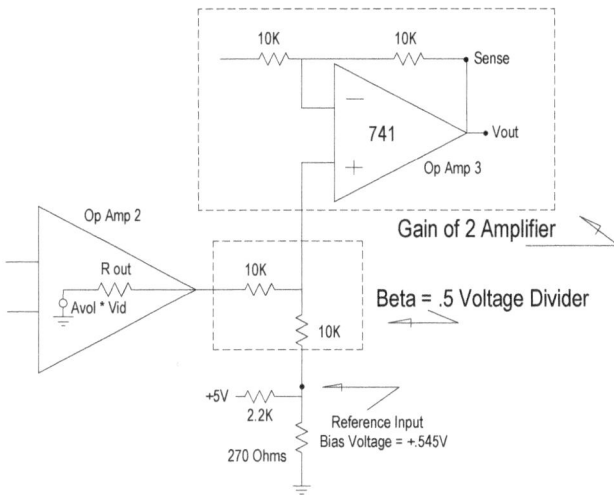

**Figure 19 – The voltage gain from the reference terminal to the output is unity due
to the voltage divider Beta (,5) and the complementary output amplifier gain (2).**

To compensate for the offset variation inherent in the sensors, the offset bias
voltage applied to the reference input must be made variable. Given the
percentage of variation computed earlier as −75% to +100%, and with the
included fudge factor results in a percentage of variation of ±150%. Therefore,
the ideal offset bias voltage of .545V is to be made variable by ±150%. The
resulting bias voltage variation becomes -. 2725V at −150%, and .8175V at
+150%. The bias voltage circuit needs to be modified to accommodate these
values.

To determine the required circuit for the variable offset bias, it should first be
recognized that positive and negative supply voltages are required because of
the bipolar nature of the desired bias voltage. In addition, a variable resistance

will be required to provide the needed variable voltage of -. 27V to +. 82V and two dropping resistors are necessary to drop the difference between the supply voltages and the variable resistance output. Begin the circuit design by drawing the circuit and labeling the intended voltages at each end of the potentiometer as follows.

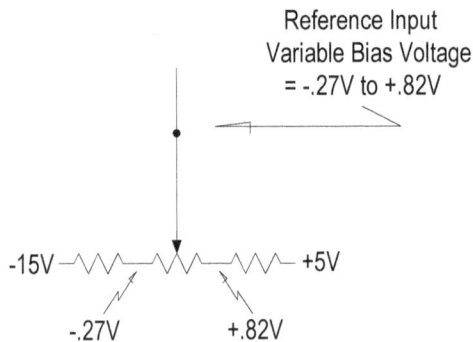

Figure 20 – Beginning steps to determining the resistance values for a variable bias voltage circuit. Label the supplies and assign the desired voltages to each side of the potentiometer.

Knowing the voltages wanted at each extreme of the potentiometer, the voltage drops across all resistors are determined and the individual resistance values can be assigned by making the resistances approximately equal to the voltages as follows. Obviously, the resulting voltages will not precisely equal the desired values but given the substantial fudge factor, close approximations should work well. After the available resistors have been applied, the resulting voltage at the wiper of the potentiometer computes to about 0V to +1V.

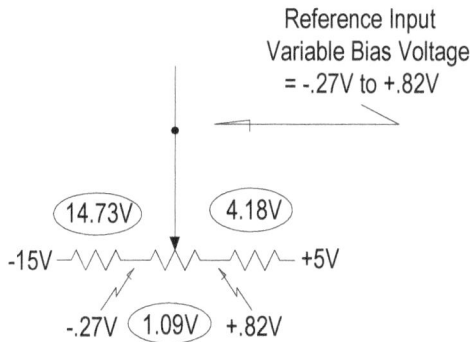

Figure 21 – Subtracting the voltages at each end of the resistors yields the drop across the resistors. The resistors are sized according to the desired voltage drop.

Note also, the assigned biasing resistance values in 1's of Kilohms may result in interaction with the second stage gain and CMR concerns, since the second stage amplifier was assembled with 10K resistors. The 10K's should be replaced with 100K's or the biasing circuit should be applied using a Unity Gain Follower op amp stage if the resulting interaction results in performance degradation. The completed amplifier with variable gain and offset appears as follows.

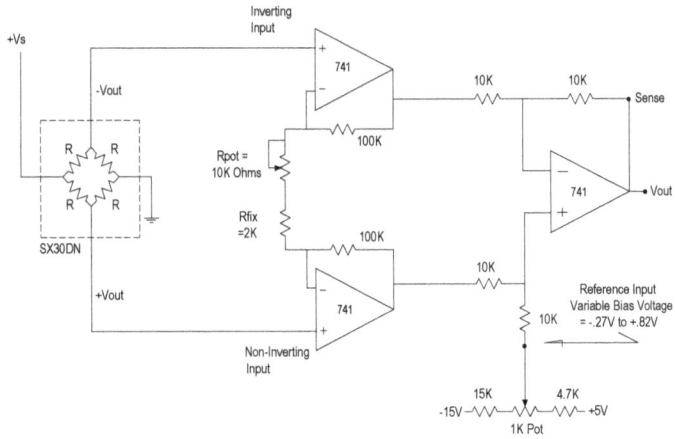

Figure 22 – SX30DN and Instrument Amp with variable gain and offset provisions.

A final note about gain and offset variation, determining the precise gain or offset value extremes or the precise resistor values to accomplish the desired extremes should not be the focus of one's efforts when designing variation into an amplifier. The real focus needs to be placed on circuit operation and understanding, so that if one finds they're in a situation where the adjustment extremes do not acquire the desired results, an appropriate modification can be made to the circuit from one's understanding. This applies to all technical systems and not just instrument amplifiers.

Integrated Circuit Instrumentation Amplifiers

Instrumentation Amplifiers are available in a variety of circuit options depending upon the amount of expense desired. All are self-contained in an integrated circuit package and require a minimum of external components. Some Instrument Amps require no external components, only connections to supply, input and output, and external jumpers to "program" the gain.

Figure 23 – Instrumentation Amplifiers are available in a variety of packages and specifications. The AD521 in the upper left is an Analog Devices equivalent of the Calex 178 hybrid Instrumentation Amplifier. Neither of the upper two is currently in production. The bottom two are *transconductance* Instrumentation Amplifiers but perform and are configured in an identical manner to the original, classic 3-op amp Instrumentation Amplifier.

The following text provides an overview to support the Calex model 178 and AD 621 Instrumentation Amplifier specification sheets following. This circuit is the preferred choice for sensor signal conditioning because of its high Zin (10^9), high CMR (120 db), and high gain (≥ 1000). For any amplifier to qualify as an Instrumentation Amplifier, an amplifier must conform to these specifications and two others. The gain must be determined by a single external resistor, and the output offset value (B) must equal the input bias voltage (Vbias). In other words, A_{Vbias} must equal +1.

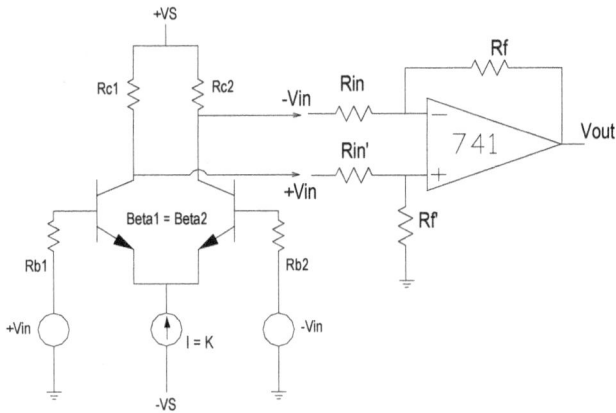

Figure 24 – An inexpensive approach to high gain differential amplification is the hybrid configuration of a transistor amplifier followed by an op amp differential amplifier.

The Classic 3 Op Amp Instrumentation Amplifier is so popular that for over twenty years it has been contained in a single IC package. Previously available as a $50 item, recent developments in IC technology has seen the IC Instrumentation Amplifier packaged in an 8-pin DIP and sold for less than $5 in quantity. The AD620 series from Analog Devices is representative of this new breed of Instrumentation Amplifier, and meets all of the basic qualifications for an instrumentation amplifier.

Instrumentation Amplifier Qualification Criteria:

-High Zin $\geq 10^9$

-High CMR $\geq$ 120 db

-High gain ≥ 1000

-A_{Vbias} = +1 [Output offset value (B) equals input bias voltage (Vbias)]

-Single gain determining resistor

When packaged as an integrated circuit, the Instrument Amp has a designated symbol as follows.

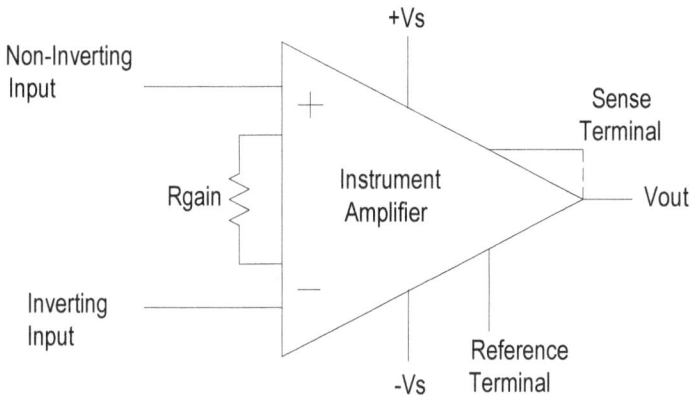

Figure 25 – The Instrumentation Amplifier symbol is easily distinguished from an operational amplifier symbol due to the gain determining resistance between the inputs. When not used for other purposes, the Sense terminal is connected to the output and the Reference terminal is grounded.

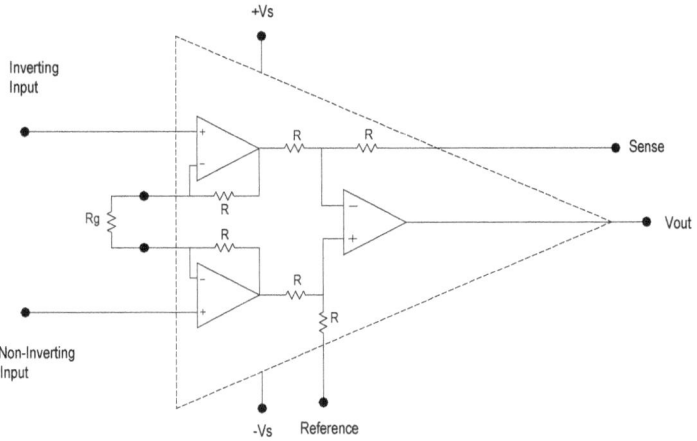

Figure 26 – The Instrumentation Amplifier symbol and internal equivalency.

As a visual aid, the included spec. sheets for the AD 621 hold numerous references to the "classic" 3 op amp instrumentation amplifier. On page 10 of the AD 621 spec. sheet the circuit diagram is obvious at the bottom of the page. This circuit is easily assembled using almost any general purpose op amp. The circuit on page 10 is assembled using 3 Op-07's, precision op amps from Precision Monolithics. Precision op amps exhibit open-loop voltage gains greater than or equal to a million (compared to 200,000 for the general purpose 741), and result in improved thermal stability, lower internally generated noise and more precise gain and output values.

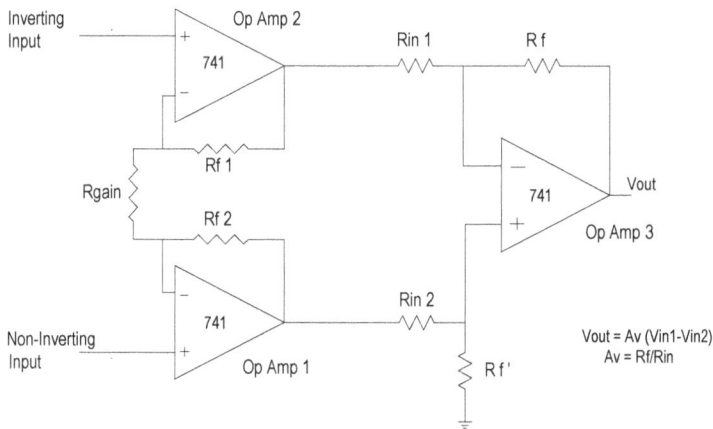

Figure 27 – The Classic 3-Op Amp, Instrumentation Amplifier

Returning to the In Amp characteristics, the most common gain determining equation for an integrated circuit In Amp's like the Calex 178, AD 621 and others is:

$Av = (1 + 2R_{F1} / R_{Gain})$

Where,

Av is the amplifier gain

R_{F1} is value of each feedback resistor in the first stage amplifiers

R_{Gain} is the value of the single gain determining resistor

Usually the desired gain is known (or can be found) and the equation is re-arranged to solve for R_{Gain}, this would appear as,

$$R_{Gain} = 2R_{F1} / (Av-1)$$

Obviously, this equation can be expressed a variety of ways. Since the R_{F1} resistors are contained within the IC and not accessible, a constant is usually expressed in the equation. Note the Calex 178, a value of 20K is given the gain equation, indicating each first stage Rf resistor equals 10K. Try this equation on the classic 3 op amp In Amp circuit of page 10 contained in the AD 621 spec. sheet (according to the document the gain should be 100).

Another commonality of instrumentation grade amplifiers is the availability of the 'sense' and 'reference' terminals. The 'sense' terminal is used to compensate for extended output leadwire lengths. As the output voltage is expected to travel longer distances, the leadwire resistance reduces the output voltage by "dropping" the voltage. An additional length of wire connected between the 'sense' terminal and the load can compensate for the leadwire by establishing the desired value for the output at the load instead of the amplifier output. If excessive leadwire runs are not used on the output, the sense terminal is shorted directly to the output terminal. Upon inspection the sense terminal is connected to the gain determining resistor in the feedback portion of the final amplifier stage of the In Amp., and ultimately should be connected to the amplifier output to avoid saturating the output.

The 'reference' terminal is generally connected directly to ground but can be used to bias or offset the amplifier's output, as well as optimize the amplifier CMR. Examples of both applications are common throughout the spec. sheets of all In Amp's. Briefly, applying the desired output offset voltage quantity to the reference terminal develops the output-offset function. Upon inspection the gain "seen" from the reference terminal to the amplifier output is exactly unity when the final stage is configured as an op amp differential amplifier (Rin = Rin' and Rf = Rf').

CMR optimisation is achieved by placing a low value (10 to 100 ohms) variable resistor between the reference terminal and ground. Connected as a rheostat, the variable resistance will cause the voltage distribution around the circuit to precisely cancel common mode voltages. This adjustment is usually fine-tuned with a common mode signal applied to both inputs and the sensor signal removed. Adjusting the rheostat will slightly vary the portion of the common mode input voltage dropped on the CMR rheostat which varies the voltage across the upper Rin resistor, which varies the input current and feedback current, which varies the feedback resistor and output voltages. In this manner a single resistance variation influences the voltages distributed around the entire amplifier. Since the voltages applied at the input are common, the distributed voltages across all the last stage resistors will be summed in a cancelling manner. This last statement can be easily verified with a circuit analysis example. If CMR optimisation is to be used, the sense circuit should contain a balancing resistance equal to the approximate rheostat value as an aid to improving CMR.

The Calex 178 spec. sheet diagram on page 10 demonstrates the sense and reference terminal applications, including CMR and offset functions.

In addition to the circuit and component connections, review the enclosed spec. sheets for typical applications involving thermocouples, bridges and other sensors.

Hybrid Amplifiers

The Calex model 178 included in the spec. sheet following the text, contains an op amp configured as a differential amplifier. The amplifier is contained in the final, output stage of the 178, and appears in the following diagram.

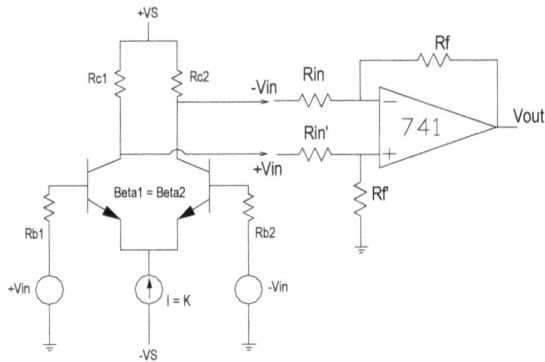

Figure 28 – A hybrid Instrumentation amplifier is composed of a transistor differential amplifier followed by an operational amplifier differential amplifier configuration. Less expensive than conventional instrumentation amplifier, the gain, CMR and input impedance specifications are acceptable for many sensor applications.

Isolation and Isolation Amplifiers

Several years ago I observed a student attempt to trigger an SCR (Silicon Controlled Rectifier, a solid-state power control component used to control the current through a 120 VAC load) from a common signal generator. Although correct in concept, the student had overlooked the fact that the signal generator output ground would be connected to the SCR ground that was connected directly to the 120 VAC source. The resulting cloud of smoke from the signal generator caught my attention. The student learned an invaluable lesson about isolation (I hope), a means of electrically removing two circuits that must share common circuit or signal considerations. A single 1:1 transformer placed between the low power DC signal generator output and the 120 VAC SCR circuit would have isolated the two circuits while providing the required trigger signal to the SCR gate. Occasionally it is desired to keep a sensor's circuitry electrically isolated from the signal conditioning circuit. Isolation is also performed for reasons of safety and noise immunity, and with some industrial and biomedical applications of signal conditioning is performed routinely.

370

A better example of isolation is the common remote control used with television. Assume you noticed an electrical shock as you adjusted the tuner or volume control located on your television at home. Such a shock might occur if the metal chassis or controls became connected with the AC or DC power that resides within the television. Does this mean that if you use the remote control you will also receive a shock? Of course not, the remote is isolated from the television. Specifically the remote control is optically isolated from the television, meaning the electronic signals are converted to light, transmitted to the TV, and converted back to electronic signals again.

Two types of isolating mediums have existed until recently in electronics, optical and inductive. Optical isolation is very common with computer communication circuits. Light emitting diodes act as transmitters in converting digital signal pulses into light pulses. Phototransistors and photo-diodes are used as receivers of the light pulses from the phototransistors. The transmitter and receiver pairs are usually encapsulated into a single device called an optical isolator. Transformers are used as inductive isolators. A single 1:1 transformer is capable of performing isolation by leaving the secondary "floating," where neither secondary lead is connected to the AC ground at the transformer primary. This type of isolation is very common in hospital applications. Should the patient come between with either of the secondary power leads and a metal bed frame, the ground connection will be made through the patient but the current will flow through the load between the secondaries and away from the patient. Obviously every effort is made to avoid having the patient come in contact with either secondary lead but, should the contact occur the patient will be safe. In a similar manner a sensor input can be isolated from the signal conditioning circuitry.

Sensors are often directly connected to metal housings, frames and chassis of experimental fixtures or manufacturing processes. Metal parts are often directly or indirectly grounded to AC or DC power sources. In this situation the possibility exists that the signal generated from the sensor will be applied to the signal conditioning amplifier in conjunction with any ground or power supply voltages. Running a ground connection between the sensor signal source and the signal conditioning circuitry also runs the risk of a "ground loop" fault. When the sensor is grounded at the process and the same ground wire is connected at the amplifier, a large current through the ground lead will flow if the ground at the amplifier and the ground at the sensor are at different potentials. Ground loop currents are common if the sensor and amplifier are connected to two different power sources. Unfortunately, the ground loop current flows through the signal leads and becomes amplified in conjunction with the signal resulting in a huge amount of AC noise at the signal conditioner output. To avoid this situation ground leads (often the wire mesh shield around and within the leadwire insulation) are connected at the source or at the load (amplifier) *but are not grounded at both ends.*

As an example of ground loop elimination, early in my career I was working as an engineer in a PBS affiliate television station in the mid-seventies. While troubleshooting a problem in the audio board I noticed the wire shielding coming from all of the studio microphones was cut at the audio board input location, but was physically connected within the connectors to all microphones in the studio. The microphone being a sensor and an audio board an accumulation of numerous high gain amplifiers, this system is no different that any sensor based data acquisition system used in experimental testing. In this example it is very unlikely that the studio and control room where the audio board was located were on different AC grounds, yet as a matter of practice the ground was connected at the source but not at the load. Again, this is done to eliminate even the remote possibility of ground loop noise being introduced to the microphone signal line.

For reasons of noise immunity and electrical power isolation some instrumentation amplifiers are internally isolated. This is accomplished by including a transformer (or transformers) or capacitors within the amplifier module. Isolation is a common option with signal conditioning modules, and although more expensive than non-isolated instrumentation amplifiers, isolated amplifiers are well worth the additional expense. Within the isolated instrumentation amplifier, often referred to as an isolation amplifier, is at least one transformer containing a primary and a secondary. The sensor signal is applied to the primary. Since the vast majority of sensor signals are DC and transformers require AC or pulsating DC to induce a voltage within the secondary, a "chopper" amplifier resides between the sensor signal and the primary of the isolation transformer. A chopper amp modulates or "chops" the sensor signal into a series of DC pulses that are applied to the primary. A primary magnetic field is established and induces a pulsating voltage into the isolation transformer secondary, which is demodulated back into a DC voltage by the usual AC to DC rectifier and filtering means. The DC is then amplified and offset with a conventional instrumentation amplifier.

Some isolation amplifiers isolate only the signal in the manner described. But one should recognize that the chopper amplifier on the transformer primary requires a power supply to operate just as the instrumentation amplifier on the transformer secondary requires a power supply. If the chopper and instrumentation amplifier both utilize power from the same power supply connection, a non-isolated path between the chopper and instrumentation amplifiers will exist. Without isolating the power supply connections between the two amplifiers the degree of isolation is compromised. In many light industrial applications isolation amplifiers of this type are perfectly acceptable. However in medical and other safety applications such as EEG's or EKG's complete isolation, including both the amplifier and power supply is required. When working with isolation amplifiers note the equivalent circuit diagram of the amplifier module and check for the degree of isolation. Two transformers, one for the amplifier and one for the supply provide complete input to output isolation.

Recent generations of isolation amplifiers are incorporating optical and capacitive technologies to couple signals using non-contact means. Capacitive coupling transfers the signal by connecting one plate with the input signal and the remaining plate with the output stage. The modulated AC signal passes across the plates by rapidly charging and discharging the capacitor. The signal is then demodulated and converted back to DC on the output side of the capacitor. For detailed information on differential, operational, isolation and instrumentation amplifiers consider looking up WWW.Burr-Brown.com, WWW.Analog.com or performing a subject search on any one of the amplifier types.

As a final example of an isolation application, while installing a number of sensors and a programmable logic controller (PLC) in a large rock crushing operation a single input channel on the PLC was observed to be drifting. Isolation was not immediately suspected as the solution of the problem since the leadwire run was short and the AC power at the sensor and at the PLC was from the same source. The drift problem appeared as an input that would only occasionally remain stable. Yet while all other sensor inputs would appear rock solid in their stability, a single input would wander about a plus and minus 10% to 20% around the correct value. After closer observation it was noticed that the drift was appearing to carry-over into the channels immediately adjacent to the problem channel. As the problem channel would change the channels on each side of the problem would also vary but to a much lesser degree. Out of frustration in not being able to specifically locate the source of the drift an isolation module, with a gain of one and zero offset was installed between the sensor and the PLC. The drift was eliminated. In looking back the symptom was very similar to the waveshape that might appear if the ground lead of an oscilloscope were removed while making a measurement, especially if a digital display were reading the voltage value of the measured waveshape on the scope. The random wandering of the voltage around the correct value was the symptom appearing at the PLC input.

Whenever the budget allows, isolation should be included in the sensor signal conditioning system.

Chapter 8 Problems

#1
Verify the available offset bias voltage at the wiper of the pot in figure #23 is approximately 0V to +1V.

#2
Design a classic 3 op amp instrumentation amplifier to condition the output of the LCA-20 load cell (spec sheet included in the Appendix, request if not available). Assume the output voltage span to be 0 to 2 Volts.

#3
Use an AD 620 and a type "T" thermocouple to design an amplifier that outputs .32 to 2.12 volts over a 0 to 100°C temperature range. Determine the thermocouple output from the millivolts vs. temperature table in the appendix.

#4
Design an amplifier using an AD 620 to condition the LCA-20 load cell of problem #2.

#5
Search the web for isolation amplifier specifications. Check to see if the isolation is partial (amplifier only) or complete (amplifier and power supply).

Chapter 9 - Stress, Strain and Strain Gauges

This chapter introduces the subjects of stress, strain, strain gauges and related measurement considerations. Ductile materials are investigated and stress, strain and related strain gauge computations are provided. Strain gauges of various types are examined, and resistance bridges, signal conditioning and temperature compensation techniques are introduced.

Objectives:

Upon completion of this chapter, you should be able to:

-   Explain the elasticity coefficient of ductile materials
-   Differentiate between the physical variables of stress and strain
-   Explain the purpose and operation of a strain gauge
-   Determine the change in resistance of a strain gauge
-   Determine the output voltage signal from a strain gauge bridge circuit
-   Determine the required signal conditioner gain and offset to calibrate a strain gauge-based measurement system
-   Interpret strain gauge specifications
-   Explain the function and details of gauge/gain factor
-   Suggest strain gauge bridge forms of signal conditioning
-   Suggest amplifier characteristics for strain gauge signal amplification
-   Perform strain gauge shunt calibration
-   Explain the purpose and operation of load cells
-   Interpret load cell specifications
-   Suggest amplifier characteristics for load cell signal amplification
-   Perform load cell shunt calibration

Introduction

Pressure, stress and strain are common measurements in the construction, manufacturing, experimental testing, research and product development industries. The strain gauge, used to measure mechanical deformation, is the primary sensing element of numerous instruments and is possibly the most universally applied sensor in the industry. Used to measure stress, strain, pressure, vibration and numerous related physical variables, the strain gauge is also an intrinsically simple sensor if one is familiar with the basic concepts of resistance in conductors.

Figure 1 – Over 60 strain gauges were applied in this test of an agricultural cart. Each cable represents a single input to a data acquisition system.

Physical Principles

The study of strain gauges begins with an examination of pressure, or *stress*. Elastic materials and their corresponding yield characteristics are then investigated in the subsequent pages. And finally, the strain gauge, it's associated computations and many of the required conditioning circuits are provided in the hardware portion towards the end of the chapter.

Stress

Pressure exhibited by, or within solids is often referred to as *Stress* and is symbolized with an upper case "S". Measured in the same dimensions of pressure as pounds per square inch (or Newtons/m^2), stress is also defined as a force exerted over an area. If one can imagine placing a cubed object of substantial weight over a support of fixed and known area, the resulting stress can easily be determined by dividing the material weight by the size of the supporting area. One can further imagine the supporting area yielding to the applied force of the object's weight. The change in physical dimension experienced by the supporting structure is the result of an applied stress, and is referred to a *Strain*.

To begin an explanation of stress, strain and strain gauges, an introduction to applications of both stress and strain is appropriate. As mentioned previously, stress is the application of a force that creates a resulting deformation or strain. As can be seen from the equation, stress is a pressure developed within solid materials,

$$Stress = Force / Area, psi$$

The applied force can assume the forms of tension (pulling), compression (squeezing), shear, bending or torque (twisting). For the sake of simplicity, and to assist in focusing upon the objective of strain gauges, only tension and compression forces will be considered in the chapter computations.

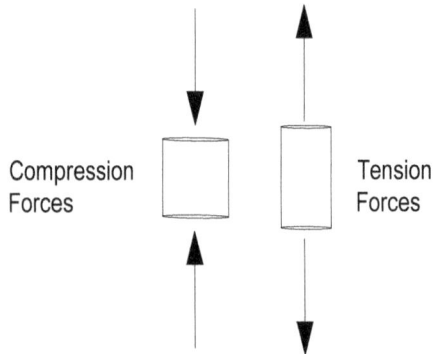

Figure 2 - Compression forces cause objects to become shorter and wider. Tension forces cause objects to become longer and narrower.

Ductile Materials

All materials will deform when placed under a load, or under an applied force. If the load is small and creates a stress that is within the material's "elastic limit," the material will return to its original dimensions when the load is removed. Ductile materials exhibit elastic characteristics if the applied force or "load" is relatively small. Metals, plastics, fabrics, wood, concrete and even glass are examples of ductile materials. All elastic materials conform to a predictable set of characteristics when destructively tested by being subjected to gradually increasing loads. The characteristics are obvious when graphed upon a stress versus strain characteristic plot.

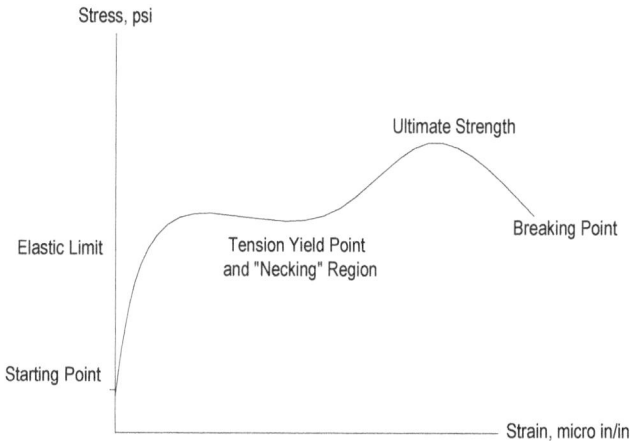

Stress, psi

Ultimate Strength

Breaking Point

Elastic Limit

Tension Yield Point
and "Necking" Region

Starting Point

Strain, micro in/in

Figure 3 – Ductile (elastic) materials will exhibit common characteristics as indicated. Note the applied input force is plotted along the Y-axis with the resulting elongation is plotted along the X-axis. This is the only case in this course where input is plotted along the Y-axis and output is plotted on the X-axis.

Strain

Longitudinal Strain (symbolized with a lower case Greek letter sigma, ε_L), is defined as a single dimension (distance) fractional deformation and is measured in fractional units of change in length divided by the original length, unit dimensions are inches/inch.

$$\text{Longitudinal Strain, } \varepsilon_L = \text{change in length / original length}$$

Or,

$$\varepsilon_L = \Delta \text{ length / Lorig, } (\mu \text{ in/in})$$

The fractional longitudinal strain units of inches per inch are more commonly expressed as micro-inches/inch and are often termed micro-strain. Strain is not just a single dimensional change however. As an object's length changes under an applied stress, its area must also deform in the opposite direction. For example, as a sample length increases with an applied tension (pulling) force, the cross-sectional area decreases. And if a sample length decreases as resulting from a compression force, the cross-sectional area increases. The terms transverse (across) strain and longitudinal (lengthways) strain are used to represent physical deformation in the two dimensions.

$$\text{Transverse Strain, } \varepsilon_T = \text{change in area / original area}$$

Or,

$$\varepsilon_T = \Delta \text{ area / Aorig, } (\mu \text{ in}^2/\text{in}^2)$$

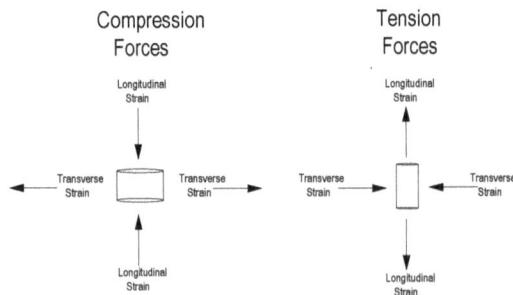

Compression Forces   Tension Forces

Longitudinal Strain   Longitudinal Strain

Transverse Strain   Transverse Strain   Transverse Strain   Transverse Strain

Longitudinal Strain   Longitudinal Strain

Figure 4 - Stress forces and the resulting strains.

Transverse strain, ε_T, is measured in units of micro-inches squared per inches squared (often called micro-strain transverse) since both the numerator and denominator contain area quantities. Transverse strain and longitudinal strain are related through a coefficient called the *Poisson Ratio*, symbolized as μ.

$$\mu = \text{Transverse Strain} / \text{Longitudinal Strain}$$

Or,

$$\mu = \varepsilon_T / \varepsilon_L$$

The previous description provides a good example of a basic property in instrumentation and experimental measurement. Stresses and strains are not always visually obvious in mechanical systems. When a large truck drives over a bridge the resulting stresses and strains are not obvious to the human eye. Yet, if one has ever stood upon a bridge when a large truck passes, one can surely feel the bridge yielding to the applied load. In instrumentation, it is often useful to remember that any applied force, independent of how large or small, will create a deformation, and the resulting deformation can therefore be used to determine the magnitude of the applied force. If one retains this fundamental principle, it will most certainly be observed repeatedly in practice.

Elasticity Modulus

As can be seen in the following plot, stress/strain graphs covering a span of zero to breaking stress exhibit five distinct points between the application of an initial force and the material breaking point. The starting point, elastic limit (also called the limit of proportionality), tension yield point, ultimate strength and breaking point (point of fracture) are encountered as one continues to apply force (stress) from zero through breaking of a ductile material sample. As long as the applied stress remains within a specific maximum limit, the sample being stretched or compressed will return to its original dimensions, acting much as a spring. Observing the stress versus strain characteristics of a ductile (elastic) material operating within its elastic limit exhibits a linear characteristic. The slope of a given materials stress/strain plot within the linear region is referred to as the Modulus of Elasticity (E) and is referred to as Young's Modulus.

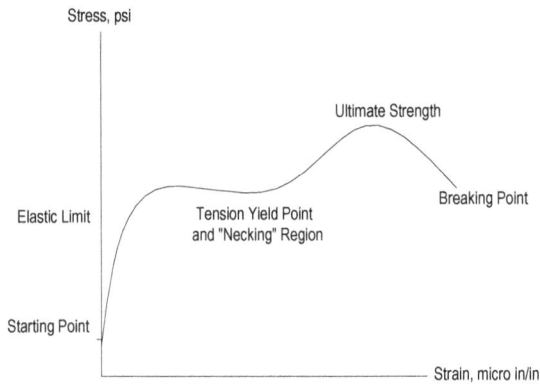

Figure 5 – A typical tensile (tension) characteristic for ductile (elastic) materials. Note the applied input force is plotted along the Y-axis with the resulting elongation is plotted along the X-axis.

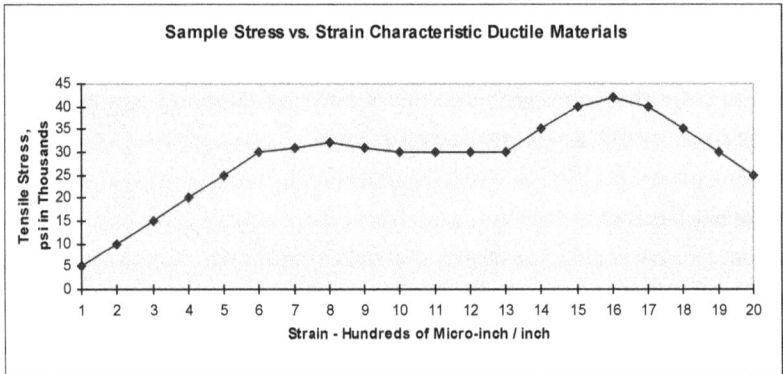

Figure 6 – Stress vs. Strain characteristic, data taken from a 1020 steel tensile test result.

E = Stress / Longitudinal Strain, within the elastic limit

Young's Modulus allows the determination of a materials strain if a given stress is known.

Longitudinal Strain, ε_L = Stress / E

Typical values of "E" are 31 million for 1020 steel and 10.7 million for T-3 aluminum. Should the elastic limit be exceeded however, the sample reaches the Tension Yield Point (TYP) and will appear permanently deformed with the material's elastic characteristics greatly reduced. Stressing a material beyond the elastic limit and into the Tension Yield Point results in "Necking." The material will continue to strain without any significant increase in stress, and the cross sectional area will become substantially and obviously narrower. Further increasing the applied stress beyond the necking region will exceed the material's Ultimate Strength (US) followed immediately by the sample's point of fracture, or breaking point. Commonly available materials handbooks contain values for Young's Modulus, the Tension Yield Point, the Ultimate Strength and other relevant information.

Figure 7 - A standard sample of 1020 steel with a strain gauge attached is ready for a tensile (tension) test.

The information provided in reference texts is usually supplied as fractional quantities to facilitate universal application to material samples of all shapes and sizes. However, standard samples are often used as well. A standard material sample for tensile testing is 2 inches long, ½ inch in diameter with threaded ends for gripping by a tensile testing machine. An initial force of 100 pounds, approximately 500 psi, is applied to assure the tensile testing machinery doesn't slip. Data is acquired in a manner similar to the following spreadsheet.

LOAD	STRESS	ELONG	STRAIN	MODULUS
(#)	(psi)	(in)	(in/in)	(psi)
0	0	0.0000	0.0000	
1000	4993	0.0002	0.0001	49926068
2000	9985	0.0004	0.0002	49926068
3000	14978	0.0007	0.0004	33284045
4000	19970	0.0010	0.0005	33284045
5000	24963	0.0014	0.0007	24963034
6000	29956	0.0017	0.0009	33284045
7000	34948	0.0020	0.0010	33284045
8000	39941	0.0023	0.0012	33284045
9000	44933	0.0026	0.0013	33284045
10000	49926	0.0029	0.0015	33284045
11000	54919	0.0032	0.0016	33284045
12000	59911	0.0036	0.0018	24963034
13000	64904	0.0040	0.0020	24963034
14000	69896	0.0044	0.0022	24963034
15000	74889	0.0053	0.0027	11094682
16000	79882	0.0068	0.0034	6656809
17000	84874	0.0103	0.0052	2852918
18000	89867	0.0214	0.0107	899569
18700	93362	0.0780	0.0390	123492
13700	68399	0.3350	0.1675	#VALUE!

E aver = 31619843

Figure 8 - Stress Vs. Strain characteristic for 1018 Cold Rolled Steel, lab data.

Example computations of the quantities included within the spreadsheet appear later in this chapter. Plotting the stress and strain data points yields an obvious linear region. The slope of the linear region is averaged to determine the

Modulus of Elasticity, or Young's Modulus. The Modulus was determined to be 31.6 X 10^6 psi/in/in.

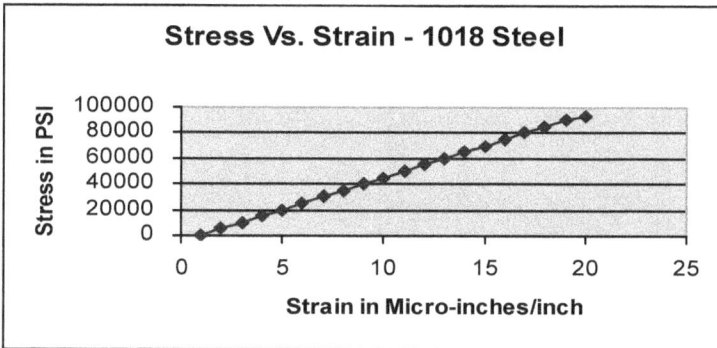

Figure 9 - Stress vs. Strain 1018 Steel

Alloy, Treatment and/or Temper	Tensile Strength, Normalized	Yield Strength
1020 Steel	64 ksi	50.3 ksi
1050 Steel	108.5 ksi	62 ksi
1080 Steel	146.5 ksi	85 ksi
4150 Steel	167.5 ksi	106.5 ksi
2024-T3 Aluminum	70 ksi	50 ksi
2219-T42 Aluminum	52 ksi	27 ksi
5005-H12 Aluminum	20 ksi	19 ksi
7175-T66 Aluminum	86 ksi	76 ksi

A table of mechanical properties for common carbon and alloy steels and wrought aluminum alloys. KSI equals thousands of psi.

Stress/Strain Hardware

Being a resistor, the strain gauge can be applied and effectively used in a multitude of circuit configurations. By far, the series voltage divider circuit arrangement dominates strain gauge signal conditioning configurations. However, a host of non-ideal temperature and leadwire considerations must be addressed if the strain gauge is to perform as desired. The following addresses some of the application concerns when using strain gauges.

387

Strain Gauges

The Strain Gauge is a resistive sensor used to measure stress, strain and related physical variables. Composed of a single piece of copper alloy wire, (an alloy is used to control temperature effects) the strain gauge is shaped for sensitivity to applied forces of various types. Chevron, rosette and a multitude of other patterns are available to respond to forces of torque, tension, compression and bending.

In application, the copper-alloy, foil-grid strain gauge is glued to the surface of the sample or device under test such that the physical dimensions of the strain gauge will change in an identical manner to the material sample surface undergoing the applied stress. In this manner, the strain of the sample becomes the strain of the gauge.

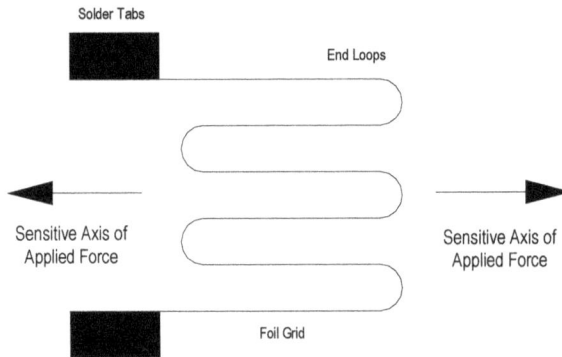

Figure 10 – A drawing of a strain gauge with a few locations indicated. Note the sensitive axis indicates the direction of the applied force. Forces can be applied as tensile (shown) or compressive with opposite results. Forces applied 90 degrees to the sensitive axis will result in zero or minimal change in resistance.

As the gauge's physical dimensions change, the wire's copper alloy grid dimensions also change. From fundamental electronics, the resistance of any wire equals the wire's resistivity coefficient ($\Re$ho, determined by the wire material type) multiplied by the wire length divided by the wire's area.

$$\text{Resistance} = \Re\text{ho} * \text{Length} / \text{Area}$$

If a strain gauge is glued to the surface of a metal column to be sensitive to longitudinal strain created by applied <u>tension</u> forces, the vertical stretch of the column will cause the strain gauge to stretch as well. The strain gauge, similar in appearance to the illustrations included in the following pages, increases its wire length and decreases its wire area resulting in an increase in the gauge's resistance according to the preceding equation.

Figure 11 – A single strain gauge applied to measure tension and compression forces. Note the three wires coming from the strain gauge. Three wires provide a means of temperature compensating the leadwires against changes in resistance.

Figure 12 – The previous strain gauge observed closely. Note the really bad soldering results but before you laugh, keep in mind this is a very small area for soldering.

A compressive applied force will result in a decrease in strain gauge resistance since all of the previous conditions have reversed. Compressive loads cause the length of the sample and strain gauge to decrease which according to the equation for resistance, will result in the gauge resistance decreasing. In either case, the change in strain gauge resistance is a linear representative of the applied stress and resulting sample strain.

Gauge Factor

The Gauge Factor (GF) or Gain Factor of a strain gauge relates the output (resistance) change to the input (strain) change of the gauge. Since gain is computed as output divided by input, the gain of a strain gauge is computed as its fractional change in output resistance divided by the fractional change in input strain. In equation form;

GF = fractional output resistance change / fractional input strain change

Or,

$$GF = (\Delta R / Rgauge) / (\Delta Length / Lorig)$$

And,

$$GF = \Delta R \,/\, Rgauge \;/\; \varepsilon_L$$

Where,

      GF = strain gauge Gain Factor

      ΔR = strain gauge resistance change

      Rgauge = original gauge resistance before stress

      ε_L = Longitudinal Strain developed within the sample under test

Rearranging to determine the amount of resistance change given longitudinal strain, gain factor and original gauge resistance yields;

$$\Delta R = Rgauge \;*\; \varepsilon_L \;*\; GF$$

Typical values of resistance change are under .1 Ohms for a 120-ohm and 350-Ohm strain gauges however, as a material sample approaches its limit of elasticity the strain gauge resistance change will commonly approach .5 Ohms. For non-destructive testing, a strain gauge resistance change of approximately .5 Ohms represents the maximum change it will encounter.

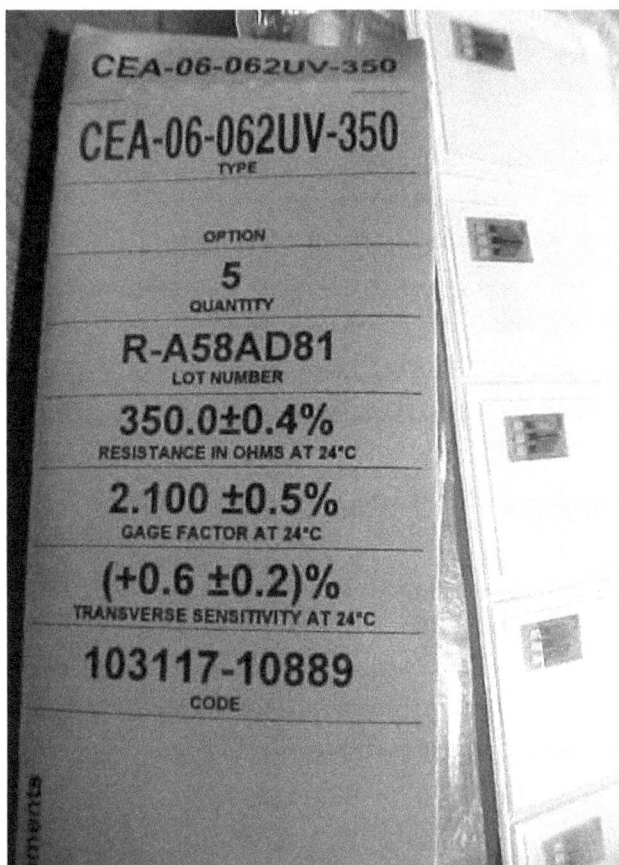

Figure 13 - A set of "Chevron" strain gauges used to measure torque and the enclosed specification sheet. Note the gauge resistance (350 ± .4%) and gauge factor (2.1 ± .5%) specifications.

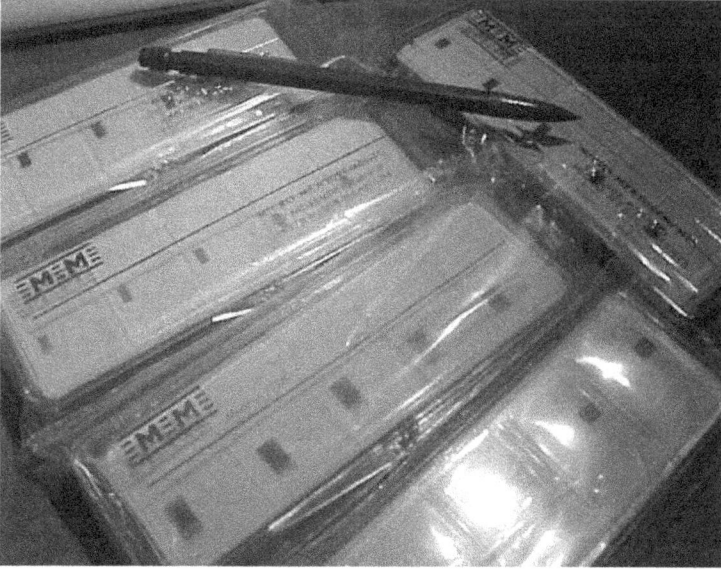

Figure 14 – Strain gauge sets of various types.

Bridge Conditioning

Strain gauge use is not limited to only stress and strain measurement. In fact the strain gauge is arguably the most popularly applied sensor, being used to also measure fluid (liquids and gases) pressure, weight, density, acceleration and an assortment of other physical variables. Simple in operation and application, the strain gauge appears universally within a variety of force and pressure measurement applications.

Acting as a variable resistor, the strain gauge must be incorporated into some form of *resistance bridge* circuit to convert the resistance change created by the applied stress and strain, into a voltage (millivoltage) signal for amplification. Two common conditioning circuits include the half-bridge and full, or Wheatstone bridge. Occasionally a one-milliamp constant current source is also used to generate an output voltage from the single strain gauge but this technique is somewhat rare.

In process automation and especially experimental testing applications, resistance bridge circuits can be found embedded as the active sensing element within a sensor, or in the "front-end" of a signal conditioner converting resistance changes into signal voltage for amplification. In bridge applications, the voltage difference across the bridge is considered the output signal.

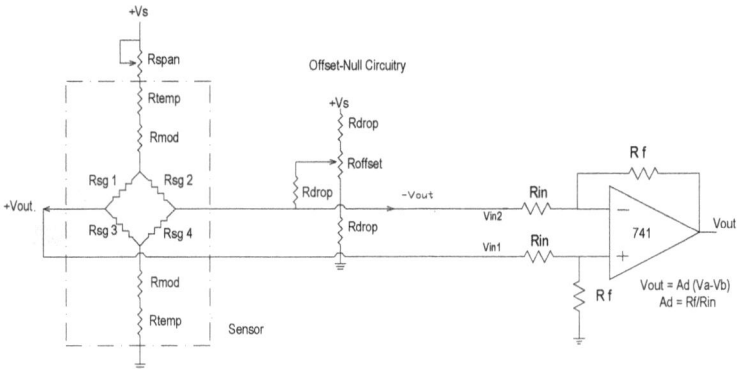

Figure 15 – A Resistive Bridge (diagonal components) and associated conditioning circuitry.

The bridge output is initially assumed to be "null-balanced" when zero physical force is applied. Null-balance occurs when each of the resistors in the bridge circuit are of equal value. Since the output voltage is taken by measuring the difference or subtracting the voltage available from each side of the bridge, if the resistor values are equal the voltages at each output are equal and the net output voltage will be zero, or *Nulled*.

The bridge gradually unbalances as the applied physical variable increases. As the bridge becomes unbalanced with the physical input changing, a *differential output voltage* is generated.

+ Vs

R1 R2

- Vout   + Vout

R3 R4

Ground or -Vs

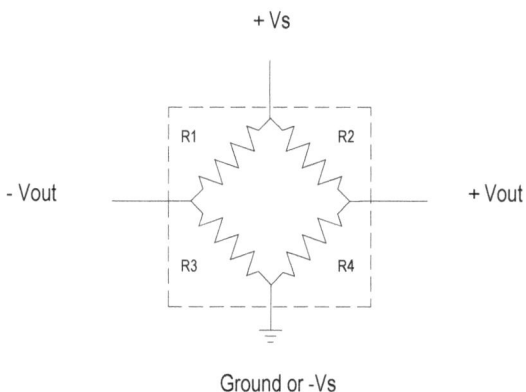

Figure 16 - A typical resistance bridge as used in sensor applications. Note the two outputs and corresponding polarities. In this example, the four resistor elements are included within a sensor housing indicated by the dashed line around the bridge. The diagram is representative of stress, strain and pressure sensors.

The differential output voltage is the result of having the voltage at each output (with respect to ground) moving in different directions when the physical variable is applied. This is caused by the manner with which the individual strain gauges are connected, and the physical variable applied internally to the individual strain gauges. Upon inspection of any four-arm bridge-type sensor, one will notice that each output has an associated polarity. The polarity indicates the direction that particular output will move when the input is applied and increased. Note this does not mean the polarity of the voltage at that output location, but the <u>direction</u> the output will change when the input variable is applied. Negative polarity indicates the output will be negative-going with an increase of the input variable, and positive implies the output will be positive-going with an increase in the applied physical variable.

The output polarities also imply something about the operational conditions of the individual resistors that compose the bridge. Typically, the resistors across each bridge output will change its resistance in a manner indicated by the output polarity.

If the polarity is negative, the resistor at the negative output (to ground or –Vs) will be decreasing in resistance value. At the positive output the resistor between the output and ground (or –Vs) will be increasing in resistance. The polarities associated with the resistor labels in the following diagram represent the direction the individual resistance changes when an input variable is applied and increased. As is the case with most housed, integral bridge sensors, all four internal bridge components are made sensitive to the input variable. In which case resistive sensors in opposite legs will vary their resistance in a corresponding manner, and resistors in adjacent legs will vary in an opposing manner.

The following diagram demonstrates the phase relationship among the four active, internal bridge components.

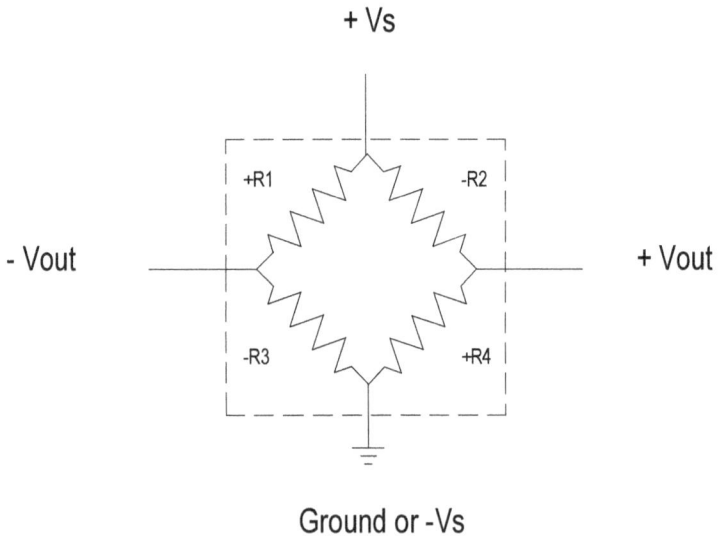

Figure 17 – A common bridge circuit representing the operation of many sensor circuits. The R1 through R4 resistor polarities represent the direction of the resistance change with an increase in the applied physical variable.

With all circuits employing differential outputs, neither output is connected directly to ground. Therefore, interpreting the output signal requires measuring (or amplifying) the *difference between the two voltages at the positive and negative output leads* independent of a reference to ground. Usually a ground connection is available but is not required to determine the output signal. Only a measurement of the voltage difference between the two outputs is required to determine the output voltage signal. Likewise, **when asked to measure the output from a bridge, the voltage measurement should be made <u>differentially</u> between the (positive and negative) outputs and not from either output to ground**. When used with strain gauges, the bridge output voltage is typically in the range of microvolts to millivolts and usually requires further amplification for the output voltage to be of any significant use.

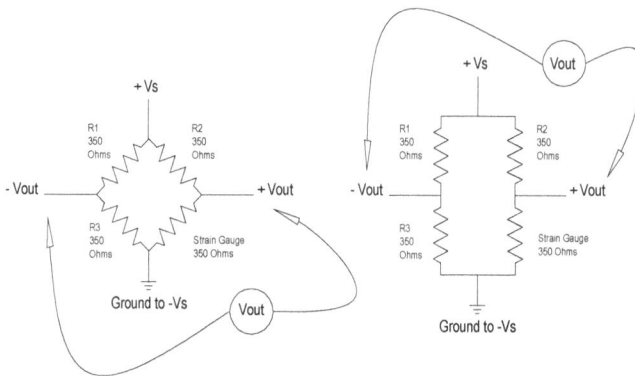

Figure 18 - Bridge circuits are composed of two side-by-side series circuits and may appear as either of these drawings indicates. The output signal voltage is measured between the differential (plus and minus) locations as indicated in both diagrams.

Resistive bridges may contain multiple (2, 4 or more) resistive sensing elements depending upon the desired sensitivity, application demands and the nature of the sensing elements. Bridges containing a single sensor are referred to as single active (sensor) element bridges or quarter active bridges, or simply quarter bridges.

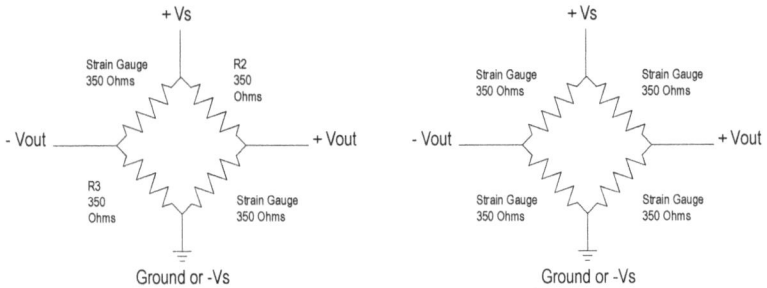

Figure 19 – A two-active element bridge (left) and a four active element or full-bridge (right).

Two and four active sensor bridges are commonly employed with strain gauges. The enclosed sensors discussed in the previous section were assumed to be full bridge circuits since all four of the resistive components were assumed to be sensing and changing with an applied input variable.

Figure 20 -A 3000 psig Viatran pressure sensor utilizes two strain gauges (barely visible in the upper right assembly) mounted on a metal diaphragm. The bridge completion resistors are visible on the adjoining circuit board.

Quarter bridges are rarely employed without additional bridge completion components since having only a single component in the circuit would result in an output voltage equal to the applied source voltage. When quarter bridges are used without additional bridge completion resistors, a *constant current generator* IC (integrated circuit) is placed in series with the single sensing element. The current generator develops a constant value (1, 5 or 10 milliamps is most common) of current, independent of imposed disturbance variables such as ambient temperature. The output voltage signal is taken directly across the sensor, as indicated in the following diagram.

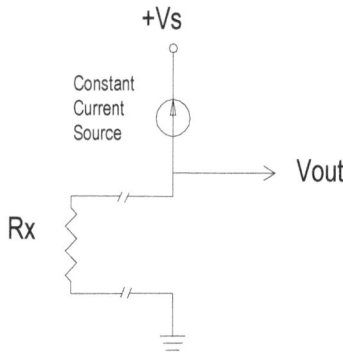

Figure 21 – A quarter bridge employs a single sensor Rx and a current generator. The output signal is taken across the sensor. Circuits of this type are common with RTD temperature sensors and single strain gauges. Temperature sensors for this purpose are available containing 4 leads, two for the current supply and two for the output to minimize the effects of leadwire runs.

Occasionally in bridge circuit diagrams the sensing element (shown as resistor Rx in the previous figure), is indicated as being removed from the bridge to place emphasis on the remote location of the sensor from the remainder of the bridge circuit. This is done to draw attention to the lead wire resistance. Bridges of this type often employ only a single resistive sensor such as a strain gauge at the point of measurement, and a single fixed resistance at a remote location (where the data acquisition system resides) to complete the bridge. It might be appropriate to again mention the original idea behind the bridge in sensor applications was to convert a resistance change into a voltage change. Therefore, bridge completion as referred to previous implies the insertion of an additional component(s) for the purpose of converting the resistance change into a voltage change.

Half-bridge circuits are common in experimental applications where a test fixture is assembled utilizing a single sensor to measure a specific characteristic associated with the device under test. The fixed resistor (R1 in the following figure) utilized in the half-bridge is selected to be a value equal to the sensors' resistance value at zero applied input.

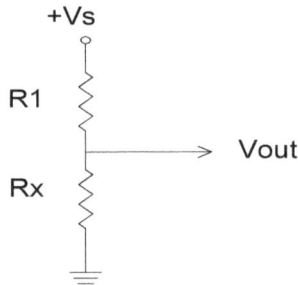

Figure 22–Schematic representation of a single active element half-bridge, or voltage divider circuit. The lower resistor labeled Rx is the single sensing element. The upper resistor is a precision, low temperature coefficient fixed value resistor. The circuit converts a change in resistance to a proportional change in voltage but exhibits an output-offset equal to half the supply voltage at the minimum input condition.

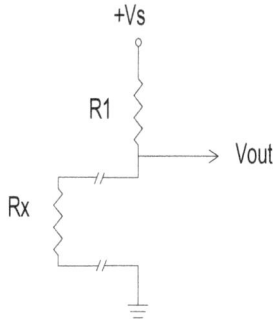

As an example, a single 120-Ohm strain gauge will employ a 120-Ohm fixed resistor as the bridge completion component and a single 350-Ohm strain gauge would require a 350-Ohm fixed resistor. Occasionally a variable resistor is used in the location of R1 to set the initial bridge output condition.

Figure 24 – An inexpensive two active element load cell (top) with two forms of bridge completion resistors (center and bottom). The resistor in the center is 499 Ohms at .01% tolerance; the two resistors at the bottom are low temperature coefficient, 01% tolerance, 120-Ohm bridge completion resistors intended for use with strain gauges.

It should be noticed that using half-bridges would result in a minimum output value of one-half the supply when the applied physical variable is zero. Since the value of both bridge components is equal initially, the resulting output will equal one-half the bridge supply voltage. If the applied supply voltage is 5 volts, the initial output from the bridge is 2.5 volts. This relatively large amount of initial voltage, or offset, (2.5V) can impose a severe limitation on the amplifier gain following the bridge. For this reason the initial bridge output is usually eliminated, or nulled-out, before amplification when using half-bridges.

The following section provides an example of strain gauge bridge circuit analysis.

Sample Computation

The following provides an example of strain gauge related computations. It is provided as reference for student computation and to demonstrate the proportional relationship between the related items of stress, strain and gauge resistance variation.

-Given the following information, determine the included values:

-Given:
      Area = 2 in^2
      Length = 10 inches
      F_{comp} = 5000#
      E = 31 E6 $_{PSI/in/in}$ (1020 steel)
      μ = .288 $_{in^2/in^2/in/in}$
      $R_{Strain\ Gauge}$ = 120 Ω
      $GF_{Strain\ Gauge}$ = 2.12
      Vs = 5 Volts

-Find:
A) Stress
B) Strain (ε_L)
C) ΔLength
D) Strain (ε_T)
E) ΔArea
F) ΔRsg
G) Bridge Vout

-Single active strain gauge element, full bridge configuration.

-Finding Stress:
      Stress = Force/Area = 5000# / 2 in^2 = _2500 psi_

-Finding Longitudinal Strain:

Since,

E = Stress / Strain

Strain = Stress / E = 2500 psi / $31*10^6$ PSI/in/in

Strain = 80.645×10^{-6} in/in or,

Strain = *80.645 μ Strain*

-Finding the change in length:
-Since,

$Strain_{Long}$ = ΔLength / Length

ΔLength = Strain * Length

ΔLength = 80.645×10^{-6} in/in * 10 in

ΔLength = *806.45 x 10⁻⁶ in.*

-Finding Transverse Strain:
Since,

μ = Transverse Strain / Longitudinal Strain

Or, $\mu = \varepsilon_T / \varepsilon_L$

$\therefore \varepsilon_T = \varepsilon_L * \mu$

Transverse Strain = .288 in^2/in^2/in/in * 80.645×10^{-6} in/in

Transverse Strain = *23.22 x 10⁻⁶ in²/ in²*

-Finding the change in area:
-Since,

$Strain_{Trans}$ = ΔArea / Area

ΔArea = $Strain_{Trans}$ * Area

ΔArea = 23.22×10^{-6} in²/ in² * 2 in²

ΔArea = *46.44 x 10⁻⁶ in².*

-Finding the change in resistance of the strain gauge:

-Since,

Gauge Factor, GF = ΔRsg / Rgauge / ε_{Long}

ΔRsg = GF * Rgauge * ε_{Long}

ΔRsg = 2.12 Ω/Ω/in/in * 120Ω * 80.645 x 10^{-6} in/in

ΔRsg = _0.020516088 Ohms_

-Finding the new resistance of the strain gauge:

New Rsg = Rsg $\pm$ ΔRsg

$\therefore$Rsg = 120 - 0.020516088 (compression decreases Rsg)

Rsg = _119.979483912 Ohms_

-Finding Vout from the bridge using the voltage divider equation:

Vout = [5V (119.979483912 /(120+ 119.979483912))] – 2.5 Volts

Vout = [5V * (0.499957254496)] – 2.5 Volts

Vout = 2.49978627248 Volts – 2.5 Volts

Vout = -0.0002137275202188 Volts

Or,   Vout = _-213.728 μ Volts_ (negative output designates compression)

It is suggested that the previous computations be performed in the sequence provided.

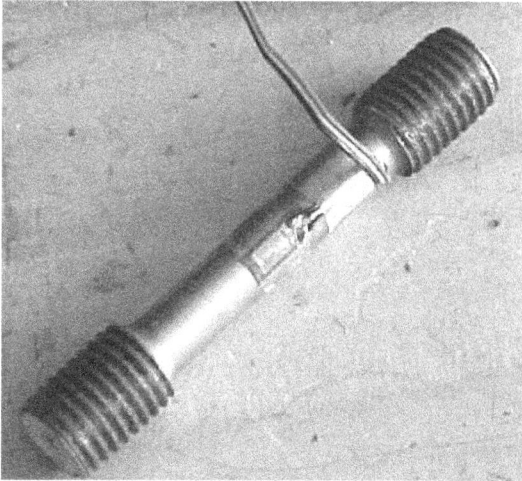

Figure 25 – A standard material sample with strain gauge applied. The previous example problem
was derived from applications utilizing samples similar to this one.

The preceding computations were performed around a single strain gauged sample. Two or four gauges mounted on the same sample and connected within the same bridge increase sensitivity by proportionally increasing the output, and assist in decreasing sensitivity to thermal errors.

Gauge Compensations

After performing the computations of stress, strain, ΔR and bridge output voltage for strain gauge circuits, one develops an appreciation of the low-level signals being developed. Since low-level bridge outputs are usually amplified many thousands of times before being recorded, it is desired to keep the signal as "noise-free" as possible. When working with "micro-measurements," subtle influences due to component noise, warm-up drift, magnetically induced noise, temperature effects and thermocouple connections need to be considered and minimized.

Among the most prevalent effects causing an output to shift in strain gauge circuits is temperature. Even the smallest thermal effect can cause a substantial change in the output of a bridge circuit and especially a high-gain amplifier's output. To minimize the effects of temperature upon the strain gauges and leadwire connections, circuit configurations and strain gauge placement and mounting techniques have been developed.

Frequently, all four bridge resistance components are mounted at the sample under test site, and are all strain gauges. In this case two of the strain gauges are *active* or sensitive to the applied stress, and two are passive or "dummy" gauges, and are insensitive to the applied stress. Dummy gauges are actual strain gauges mounted on the sample under test with the (two) active gauges, but are usually rotated 90 degrees from the active gauges to be insensitive to the applied stress and resulting strain. Since the (two) dummy gauges are wired in to the full-bridge circuit in the location of the fixed resistors, the dummy gauges perform the required voltage divider functions. Since the dummy gauges are located in the same temperature environment as the active gauge, any temperature-induced change of operating characteristics will appear at all gauges in the bridge. The thermal effects will be negated due to the differential nature of the bridge output, and since all of the four bridge components are changing their characteristics together. A result of this application, full bridges with all strain gauges mounted at the application are understood to be inherently temperature compensated.

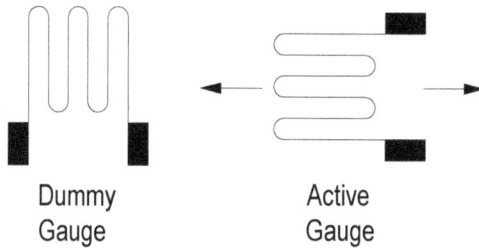

Figure 19 – Active and dummy strain gauge orientation. If applied as two gauges, both gauges would be applied in close proximity on the sample. Both gauges can also be purchased in this orientation as a *Poisson* gauge. Poisson connected gauges are used to correct for transverse strain errors but can also be used for thermal compensation.

However not all bridges contain two or four active strain gauges. Many contain only a single active strain gauge, and usually the remainder of the bridge circuit is located remote to the single strain gauge. For these circuits, the leadwire resistance must be compensated. The length of leadwire between the bridge circuit and the strain gauge may introduce considerable resistance into the bridge circuit. If the resistance of the leadwire were constant, the effect of the wire could be offset, or biased-out. Nevertheless, the effect of the wire resistance is rarely constant, since as the temperature of the wire varies so does the resistance. With the leadwire resistance capable of varying, it is conceivable that the bridge output could vary in relation to the leadwire temperature even if the applied stress or strain was to remain constant. This condition would yield a false stress or strain indication, or what is occasionally referred to as an _apparent_ microstrain.

Leadwire compensation is best understood if one is familiar with the differential manner by which the bridge circuit generates an output voltage. As an aid in understanding the following explanation, one should assume that for any single leg of a bridge circuit there exists a single resistance in an opposite location and two resistors in adjacent locations. Mentally visualize the bridge circuit and verify the preceding statement before proceeding. The bridge output voltage is equal to the difference between the two outputs, labeled $^-$V and $^+$V. Considering the output signal is the <u>difference</u> between each bridge output lead, any voltage common to each output will be subtracted. Any undesired effects introduced into adjacent legs of a bridge will cancel, having no contribution to the output signal. Leadwire compensation is accomplished by connecting one of the two leadwires to the strain gauge, in each of two adjacent legs of the bridge circuit. In this manner, the introduced leadwire resistances will establish equal voltages in adjacent legs of the bridge, canceling in the bridge output and effectively canceling the leadwire resistances. The following diagram demonstrates this condition.

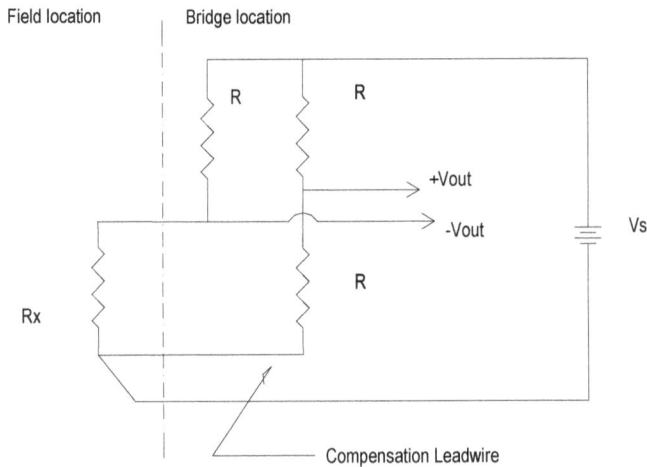

Figure 27 – Single resistance sensors commonly contained three wires connected to two terminals at the sensor site to provide thermal leadwire resistance compensation.

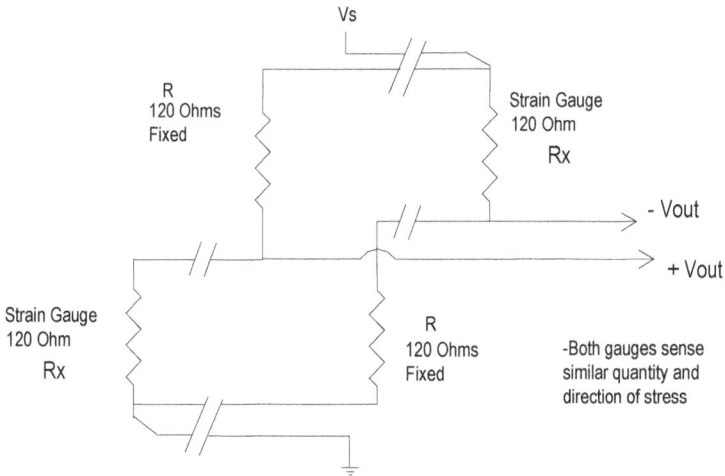

Vs

R
120 Ohms
Fixed

Strain Gauge
120 Ohm

Rx

- Vout

+ Vout

Strain Gauge
120 Ohm

Rx

R
120 Ohms
Fixed

-Both gauges sense
similar quantity and
direction of stress

Figure 28 – A bridge containing two remotely located and leadwire compensated sensors.

Bridge Gain and Offset

The Strain Gauge Bridge with 1, 2 or all 4 active gauges will often include provisions for gain and offset adjustment. Offset, zero, bias, or null adjustment is performed in one of two ways. The most direct method is to include a potentiometer within the bridge between any two of the bridge resistance components. With the potentiometer connected at one of the bridge outputs, the wiper will provide the output for the respective side of the bridge. As the wiper is adjusted, the voltage at that one side, and therefore the differential output voltage, will change slightly in a positive and negative direction. The potentiometer could also be connected between two components with the wiper to the supply or to ground. With the wiper connected to the supply or ground, the current in each leg of the bridge will vary in opposite directions as the wiper position is varied. In this method, the output at each side of the bridge will vary slightly positive and negative as the supply or ground potential is adjusted. Example diagrams of the techniques will provide additional insight. (Not included, 9/98)

Another common approach to bridge null adjustment is to include a voltage divider in parallel with either leg of the bridge circuit. This is accomplished with two fixed resistors and a variable resistance. The three are connected in series between the bridge supply and ground, with the potentiometer connected in between the two fixed resistors. This circuit provides a voltage that varies slightly above and below the voltage at one of the bridge outputs. The wiper of the potentiometer is then connected to the bridge output through a current limiting (usually around 100K resistance). Although sophisticated in appearance, the overall idea is to vary the voltage at either bridge output by only a few millivolts, enough to null any non-zero output voltage when the applied stress/strain is minimal. The following diagram illustrates the nulling circuit configuration.

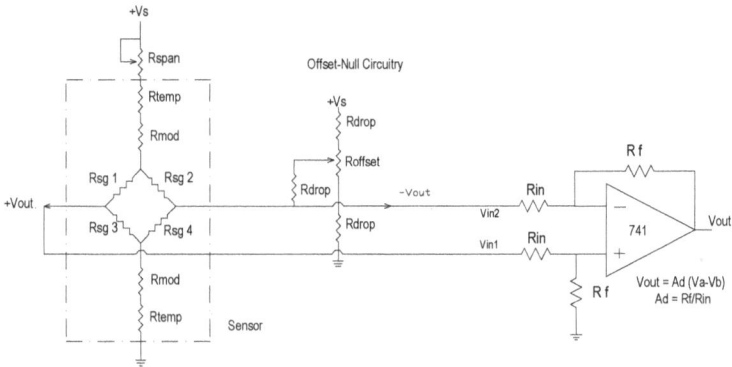

Figure 29 – Bridge circuits may also contain Span and Zero (gain and offset) adjustments as shown.

Bridge gain, or sensitivity, can be varied by including a correctly sized variable resistance, wired as a rheostat, in series between the bridge and the supply voltage. Given the output voltage as being the supply voltage multiplied by the ratio of the output resistance to the total resistance at each side of the bridge, Vout appears as being in direct proportion to the applied voltage source. Or,

410

$$\text{Vout} = \text{Left output voltage} - \text{Right output voltage}$$
$$\text{Vout} = \text{Vs } (R3/(R1+R3)) - \text{Vs } (R4/(R2+R4))$$
$$\therefore \text{Vout} = \text{Vs } [(R3/(R1+R3)) - (R4/(R2+R4))]$$

As can be seen from the equations, Vout varies in direct proportion to the supply voltage. For any given and constant value of bridge resistances, as the supply voltage increases, the output voltage span increases. As the supply voltage decreases, the output voltage span decreases. Consequently, any variation in the voltage supplied to the bridge produces a proportional change in the bridge output.

Bridges require compensating against two basic types of errors, gain errors and offset errors. Offset errors can usually be biased-out by introducing an external voltage to either output or by introducing a potentiometer into the bridge circuit. An initial adjustment of offset, bias or null before acquiring data from the bridge usually suffices if the components and their leadwires have been properly compensated. Introducing a temperature sensitive resistor into the supply voltage connection between the bridge and the supply minimizes gain errors.

As the ambient temperature of the bridge changes, the components within the bridge, including the strain gauges change their resistance values. The overall effect will appear as a change in the bridge's gain characteristics. In other words, if it is momentarily assumed the stress and strain are held constant and the temperature of the bridge varies, the output from the bridge would respond to the temperature change by varying in a proportional manner. If during the temperature variation the stress/strain of the gauged material were to change, the output would be greater or lesser than the same amount of stress/strain change at a normal operating temperature. This type of temperature sensitivity is often called *thermal drift*, and is measured in Ohms/C°, μin/in/C°, μstrain/C° or Volts/C° depending upon the specific reference of the drift.

To counter-act the effect of the resistance changes, a temperature sensor in series with the bridge supply can detect temperature variations and adjust the voltage supplied to the bridge to reduce or increase the bridge gain. In this way, the bridge sensitivity to an applied stress or strain remains constant, at least over a fixed (and usually fairly narrow) temperature span, typically 10 to 20 degrees Celsius.

For additional bridge circuit material, consult Chapter 2 in the ET 200 manuscript.

Scaling

A final computation can be made if a signal conditioning amplifier or other scaling device is used at the output of the bridge circuit. Since the signal at the bridge output is in the fractional millivolt or microvolt range, to properly display the measured units of stress or strain, mathematical quantities of gain (occasionally called scale factor) and offset will be required. If the desired value to be displayed is Stress, in psi, and assuming the bridge output is nulled at zero applied force, the scale factor would be 2500 psi divided by 213.728 microvolts. The resulting scale factor is 11,697,136.60385. This would be the value of a scale factor entered into a data-acquisition software program.

Scale Factor = Quantity to be displayed / Input voltage to the display amplifier

Figure 30 – A portion of an analog input signal conditioning rack. All modules contain gain and offset (span and zero) adjustments internally, as shown at the top of two modules.

In the case of an analog amplifier, an output voltage of .25 Volts might be used to represent the 2500psi of Stress developed within the strain-gauged sample. Again assuming the bridge output voltage is applied to the amplifier input, this would require an amplifier gain of 11,697.137, and would necessitate an instrumentation grade of amplifier.

Amplifier Gain, Av = Desired amplifier ΔVout span / Bridge ΔVout span

Note the amplifier input and output is indicated as being voltage spans. It is assumed the applied force at the sample will be varying between zero and 5000 pounds, the bridge output will be varying between zero and 213.728 microvolts, and the amplifier output will vary between zero and .25 Volts. It assumed that zero force results in zero bridge output, a null-balanced bridge. Although this is usually close to being the case, it should be assured by including a bridge null circuit if necessary. In any case, the bridge offset is often not known until the strain gauge bridge circuit is connected, and the bridge output is measured. With gains and scale factors as high as 10,000 to 10 million plus, even a slight bridge offset can yield a substantial initial error of the displayed value.

Figure 31 – Another older form of strain gauge signal scaling amplifier.

Stresscoat

At this point in the chapter, it should be obvious that strain gauges are applied where a mechanical designer needs to acquire more information about material or component deformation and strength. Upon closer investigation however it may be difficult to establish the specific locations to glue the strain gauge(s) and types of gauges to be laid. For this reason a product called Stresscoat has been developed to determine where the stresses appear and to determine the location to glue the strain gauges. Applied like paint, Stresscoat becomes brittle and exhibits cracks when subjected to stress. Once Stresscoat is applied and allowed to dry the device under test is run through a preliminary test. The resulting cracks allow engineers to determine where the stresses are formed and where to lay the strain gauges.

The following photos demonstrate.

Figure 32 - Stresscoat applied to the hitch of a cart. The cart will next be run through a test and the stresscoated area will be analyzed to determine where the gauges will be laid.

Figure 33 - Curious aliens are often found examining seed carts, not really of course. A technician applies Stresscoat to determine specific locations to lay strain gauges.

Figure 34 - Another example of a Stresscoat application, several strain gauges also appear in the photo.

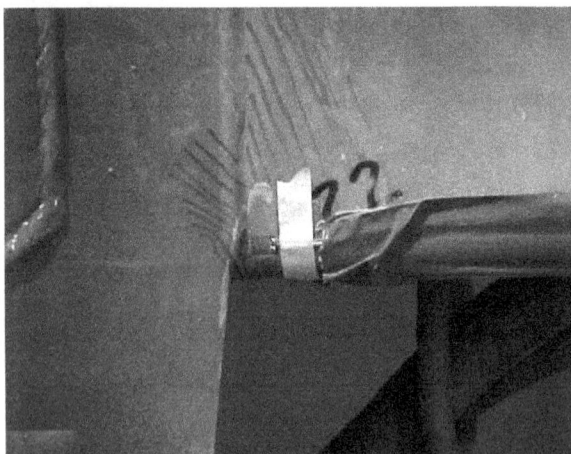

Figure 35 - The cracks resulting from the Stresscoat application has been highlighted with a red marker in this photo. The individual Stresscoat cracks are not obvious in the photo.

Figure 36 - Once the strain gauges are layed the leadwires are run to a junction box where the bridge completion resitors are located. The junction boxes are connected to a portable data acquisition unit.

Figure 37 - The portable data acquisition unit.

Load Cells

Among the more popular forms of strain gauge applications is the *load cell*. A load cell is a force sensor. More specifically, load cells are an accumulation of strain gauges connected to form an electrical (Wheatstone) bridge and mounted on a metal platform that is used for the purpose of determining mechanical load, such as weight or force. Load cells are commercially available in a variety of load ratings and configurations. "Pancake" styles are commonly used to measure forces in the 100's of pounds to hundreds of thousands of pounds. Platform load cells are typically found in grocery scales at the checkout counter and weight between ounces and a few pounds.

Operation

Load cells do *NOT* measure stress or strain. Although strain gauges are the primary sensing element within each load cell, the purpose of the unit is to output a signal that varies in relation to the applied *force* and not the applied stress or resulting strain. As will be demonstrated following, determining the output of a load cell is far less complicated than computing the output of a strain gauge bridge.

Figure 38 - Loads cells are available in a variety of sizes and styles depending upon the application. Pictured above are two "Pancake" styles of load cell (left), a load washer (center) and a platform load cell (upper right).

It is suggested that one become reacquainted with the operation of a resistive (Wheatstone) bridge before continuing as the majority of load cells contain a full bridge internally. Specifically, the concepts of differential output signals and *Ratiometricity* should be fully understood. All bridge circuits are ratiometric; the amount of supply voltage determines the amount of output voltage when the bridge is unbalanced. Since any bridge is merely an application of series resistance voltage divider circuits, it can be proven that a bridges' differential output voltage varies in direct proportion to the amount of supply voltage provided to the bridge.

The following demonstrates the relation between bridge supply voltage and output.

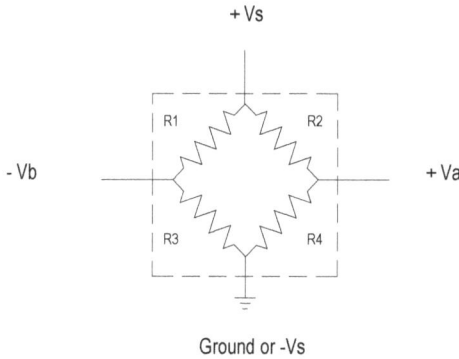

+ Vs

R1   R2

- Vb   + Va

R3   R4

Ground or -Vs

Figure 39 – Load cells are commonly composed of 4 active strain gauges connected in a full-bridge circuit configuration. When a force is applied to the load cell, the positive output (+Va) will increase towards +Vs and the negative output (-Vb) will decrease towards Ground or -Vs. The output signal is the difference between Va and Vb.

$$Vout = Va - Vb$$
$$Va = Vs\ [R4/(R2+R4)]$$
$$Vb = Vs\ [R3/(R1+R3)]$$
$$Vout = Vs\ \{[R4/(R2+R4)] - [R3/(R1+R3)]\}$$

As can be seen from the equation, the bridge output varies in a direct proportion to the applied supply voltage, Vs. In this way varying the supply voltage to the bridge varies the output voltage span and as such, is understood to control or determine the bridge "Gain." Gain is defined as being the distance between output voltage-span end points, and is occasionally referred to and labeled on schematics as bridge sensitivity or Span.

Compensation

As can be seen in the following load cell photo the resistive bridge circuit has been modified to include additional resistance between the supply connections and the strain gauge bridge. This is done for two possible reasons, to compensate the bridge output for changes in ambient temperature (temperature compensation) and secondly, to assure that all load cells of the same model numbers exhibit similar input/output characteristics (modulus determination). Strain gauge and load cell bridge circuits are often modified to compensate for thermal effects and to better match units of similar model numbers for replacement should one fail.

Figure 40 - This 20,000 pound Eaton 3169 load cell contains a bridge circuit diagram as an aid to making the proper electrical connections. Note the additional gain compensating resistors in series wth the supply voltage connections in the diagram.

Upon inspection of the engineering data sheet that is included with strain gauges one would notice the sensitivity of the strain gauge's gain (gauge) factor (GF) to temperature. A graph of the characteristic is contained in the following figure. The graph demonstrates the linear variation in sensitivity (Gauge Factor) as temperature changes. As temperature increases, the gauge factor, (or gain factor, GF) increases. This means the individual strain gauges that compose the load cell's bridge circuit become more sensitive as the temperature increases. Without some form of compensation the load cell output would tend to increase as a result of a temperature increase even though the applied load remained constant. Obviously, without thermal compensation the results of an experimental test would be disastrous if the product being tested is instrumented and calibrated indoor where the temperature is 70°F and is tested outside where the temperature is 30°F or worse. This is commonly the case with experimental testing of agricultural, military, automotive and construction products.

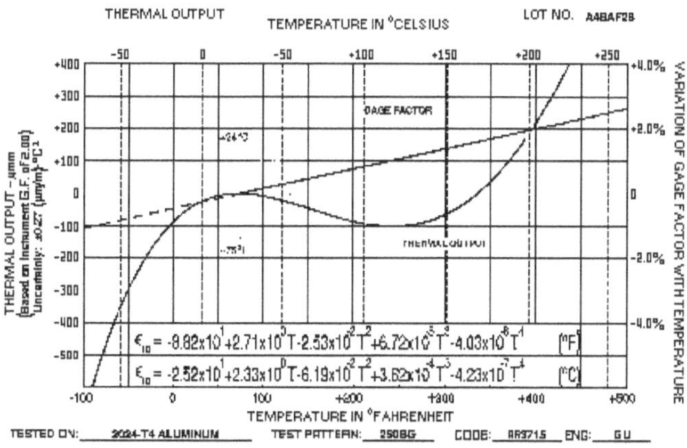

Figure 41 - Thermal effects on strain gauge factor (linear characteristic) and offset (sinusoidal appearing characteristic). Photo courtesy of Vishay Micro Measurements.

Full bridge strain gauge and load cell circuits are inherently temperature compensated against thermally induced *offset* drift. Offset refers to the bridge circuit output voltage at null-balance with no mechanical load applied. A temperature created shift of the strain gauges resistance at zero applied force (no-load) will cause all of the strain gauges in the bridge to change the same amount in the same direction, since the strain gauges are matched. Since the bridge is a pair of voltage divider circuits the effect will be offsetting and the output at no-load will continue to be null-balance, or zero volts at the output.

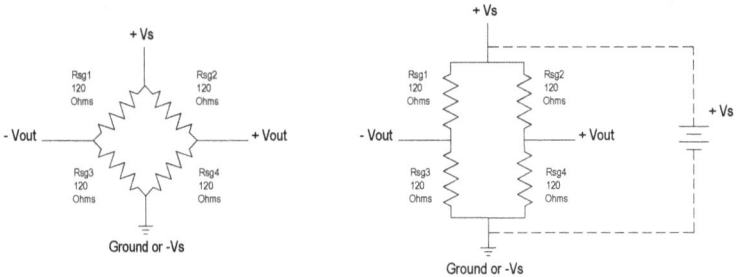

Figure 42 - All Wheatstone bridge circuits (left) can be analyzed and interpretted as series voltage divider circuits (right).

As a precaution against thermally induced *gain* (gauge factor) error, many load cell manufacturers include temperature compensation components within the load cell. Since the load cell contains strain gauges configured as a full (4-active element) bridge, and since the bridge output voltage is directly proportional to the amount of applied supply voltage (remember ratiometricity?), if the bridge sensitivity should increase with an increase in ambient temperature, the bridge output will increase with an increase in temperature – even if the applied load remains constant. To compensate for the increased sensitivity, temperature-sensing resistors (pure metal resistive temperature detectors, or RTD's) are placed in series with the bridge circuit.

The temperature compensating resistors increase in resistance as the ambient temperature increases. The increase in resistance results in less voltage being applied to the strain gauge bridge circuit. As the voltage applied to the bridge decreases the output voltage from the bridge would normally decrease but since the individual strain gauges have become more sensitive as a result of the temperature increase, the output voltage stabilizes. Overall, an increase in temperature causes the strain gauges to generate more output voltage from the bridge, but since the temperature sensing resistances in series with the bridge have reduced the supply voltage to the strain gauges the output signal voltage remains constant. However, temperature compensation is only valid over very defined limits of temperature extremes. Operation outside of the temperature extremes will result in erroneous data from the load cell.

Resistors are also placed in series with the load cell bridge to match the operating characteristics of similar models. If a load cell should go bad, and they do – mostly from abuse, to minimize the calibration procedure it is highly desirable to replace the bad unit with one of similar operating characteristics. Recognizing this, many load cell manufacturers place "Modulus" resistors in series with the strain gauge bridge. Modulus resistors reduce the sensitivity of the load cell in an attempt to provide all similar units with the same sensitivity to voltage supply and load. Modulus resistors are commonly found on platform load cells where the amount of calibration adjustment associated with the display amplifier may be limited. Some platform load cells contain both temperature compensation and modulus resistances in series with the supply voltage.

Figure 43 - Internal view of a platform load cell. Note the modulus resistor in the center of the photo.

One final consideration when applying load cells, the manner with which the load force is applied must also be considered critical. Off-center loading may result in an output signal that is less than expected and could result in damage to the load cell. Load cells will typically output a signal in proportion to the vertical component of the applied force, assuming the load cell is mounted in a position to accept vertical forces. If the force is applied at an angle, even a slight angle, the resulting output signal from the load cell will be misleading and will not represent the total applied force, only the sensitive axis component of the applied force.

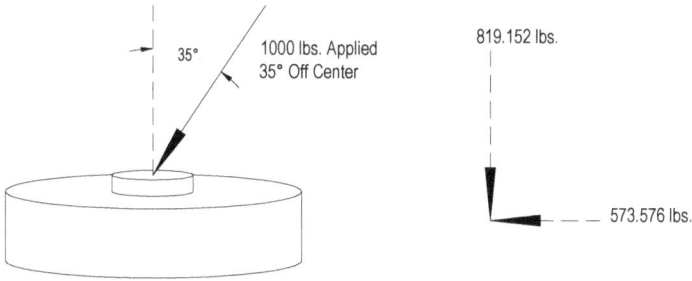

Figure 44 – Since load cells are only sensitive to forces applied upon a specific axis, a 1000 pound load applied 35 degrees from center will generate an output signal equivalent to 819.15 pounds.

Rated Output

Since the load cell contains multiple strain gauges configured into a full-bridge circuit configuration, one need not concern themselves with computations comparable to the stress/strain and strain gauges. To determine the output millivoltage signal of a load cell for a given amount of applied load (force or weight), only the *rated output* and supply voltage values are required.

As an electronic sensor of force, it would be expected that the load cell specifications would contain a statement of gain that involves volts per force. However, since the bridge circuit's output voltage is contingent upon the amount of supply voltage, the load cell's sensitivity coefficient will also contain a reference to the anticipated supply voltage to be used on the load cell bridge circuit. For this reason the "Gain" of a load cell is expressed in terms of overall sensitivity. The load cell's *Rated Output* is the output voltage generated at <u>full rated load</u> (the mechanical load rating of the load cell) divided by the supply voltage.

$$\text{Rated Output (RO)} = \Delta V_{out} \div \text{Supply Voltage}$$

Expressed as a load cell specification, rated output (RO) normally appears in specifications similar to the following. As an example, a load cell with a rated output of 2 millivolts per volt would appear as in the following.

$$RO = 2.0 \text{ mV/V}$$

In this example, the load cell will produce 2.0 millivolts <u>for each volt of applied supply voltage to the load cell, at the full load rating of the load cell</u>.

Figure 45 - A pair of 200-pound pancake load cells. The cover has been removed from the unit on the right to expose the strain gauges.

Example Problem:

Assume a load cell is specified at a rated load of 100 pounds, has a rated output of 2.0 mV/V and is to be used with a 5-Volt supply. Determine the output voltage at full rated load, 50 pounds, 20 pounds and 10 pounds.

Solution:

Given the supply is 5V and the RO is 2.0mV/V, the load cell will produce 10mV at 100 pounds.

Since,

$$RO = \Delta Vout \div Supply\ Voltage$$
$$RO = \Delta Vout\ /Supply\ Voltage,\ in\ mV/V$$

Rearranging to solve for the output voltage,
$$\Delta Vout = RO * Supply\ Voltage$$
$$\Delta Vout = (2\ mV/V) * (5V)$$
$$\Delta Vout = 10\ mV\ at\ full\ rated\ load\ (100\#)$$

The output voltage at applied loads other than the full rated load will be proportional to the output at the full rated load. Therefore if 50# is applied, the output will be one-half of the full rated load output, or 5 mV. An applied load of 20# equals,
$$\Delta Vout\ = 20\#/100\# * RO$$
$$\Delta Vout\ = 20\#/100\# * (2.0\ mV/V*5V)$$
$$\Delta Vout = 2\ mV$$

And an applied load of 10# equals,
$$\Delta Vout\ = 10\#/100\# * RO$$
$$\Delta Vout\ = 10\#/100\# * (2.0\ mV/V*5V)$$
$$\Delta Vout = 1\ mV$$

In short, the load cell is a linear device with an output of approximately zero millivolts (a null-balanced bridge) with no load applied. When a load is applied the output will vary in proportion to the rated output value of the load cell. To approximate the output, first determine the supply voltage and the RO of the load cell. Multiply the RO and supply voltage to determine ΔVout, or the output voltage change at full rated load. Then, as demonstrated previously, determine the output as a proportion of the full rated output by finding the proportion of the rated load applied. The equation follows.

$$\text{Vout} = [(\text{Applied load} \div \text{Full rated load}) * \text{RO}] \pm \text{Offset}[1]$$

[1]Note - this equation assumes the load cell output is close to zero millivolts at null-balance, with no load or mechanical force applied. Most of the time the null-balance output of the load cell will be very close to zero millivolts (if not, the load rating of the load cell may have been exceeded and the load cell should be considered defective). Otherwise the offset at no-load must be added to the computation.

Figure 46 - These signal conditioning modules are capable of accepting inputs signals from a variety of sensors including load cells and output both current (4-20 mA) and voltage (0-10V) signals. The calibration span and zero adjustments are obvious at the top of the modules.

Calibration

The process of assuring accurate information from any sensor-based measurement system is referred to as *calibration*. In most experimental measurement applications an amplifier or other form of signal conditioner follows the sensor and "drives" the display or data acquisition device.

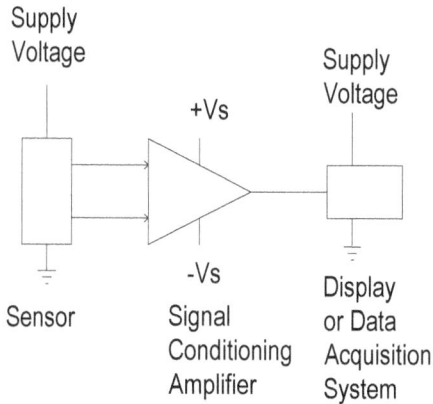

Figure 47 - The signal conditioning amplifier in this diagram is responsible for developing a voltage range that numerically represents the physical variable being measured. Adjustments of gain and offset are normally included to calibrate the output signal range.

Calibration allows the output of the amplifier or other signal conditioner to numerically represent the physical values being measured by the sensor. As an example, assume a load cell is measuring force between zero and 200 pounds. In this example the amplifier output would likely be set to 0-Volts at zero pounds and 2-Volts at 200 lbs. Signal conditioning circuits normally have adjustment provisions to precisely set the output to the desired values representing the variable being measured. Linear sensors are normally connected to signal conditioning amplifiers with gain and offset adjustments for calibration. The gain and offset adjustments are directly related to the mathematical functions of slope and intercept respectively. The gain and offset adjustments are commonly referred to as *span* and *zero* adjustments where gain determines the output signal span and zero determines the initial output offset or minimum output voltage. In the previous example the zero adjust would set the output to zero volts at no applied load and the span (gain) would set the output to 2-Volts with a 200 pound load applied.

Figure 48 - Each of the LED displays in this photo has a unique signal conditioning amplifier. The amplifier modules are located in the rack along the bottom of the photo.

Given that load cells are linear devices; calibration is normally performed at the extremes of the anticipated application forces. Load cells are rarely used throughout the entire range of full rated load however. More commonly, load cells are used between zero applied force and approximately 70% to 80% of full rated load. This is done as a precautionary measure to avoid inadvertently overloading the load cell as replacement costs normally run into the hundreds and in many cases, thousands of dollars.

Load cells rated at small load values, 10-100 pounds, can be calibrated by first applying zero force and then applying the maximum anticipated value of load. With zero force applied, the amplifier, transmitter or data acquisition unit Offset, or Zero adjustment is set for a minimum (usually zero) indication. When the maximum force is applied the amplifier or data acquisition Gain, or Span is set to indicate the value of maximum applied load. With smaller ratings of load cells an actual mass of known weight can be applied to simulate the maximum anticipated load during operation.

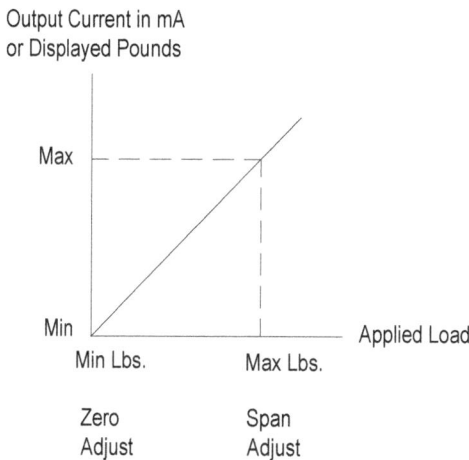

Figure 49 - Graphical representation of the load cell calibration process. Depending upon the application, calibration could be performed upon a process transmitter, digital data acquisition system, analog instrumentation amplifier or digital panel meter.

However, load cells rated into the thousands or tens of thousands of pounds must resort to an alternative calibration technique, as applying a 20,000-pound mass is not only difficult but can also be dangerous.

Shunt Calibration

Most load cells include a *shunt calibration resistor* when purchased. Shunt calibration allows the maximum load to be simulated rather than applied to the load cell. Again, the maximum applied load is anticipated to be 70% to 80% of the full rated load of the load cell and the shunt calibration resistor will simulate this amount when applied. The shunt calibration resistor is applied between two of the four load cell leadwires (see the following figure) with no mechanical load applied to the load cell.

When connected, the parallel combination of the shunt calibration resistor and the particular strain gauge in the load cell bridge will result in a total resistance slightly less then the strain gauge's no-load value. The parallel combination will create an equivalent voltage in the load cell output, simulating the load value specified with the shunt calibration resistor.

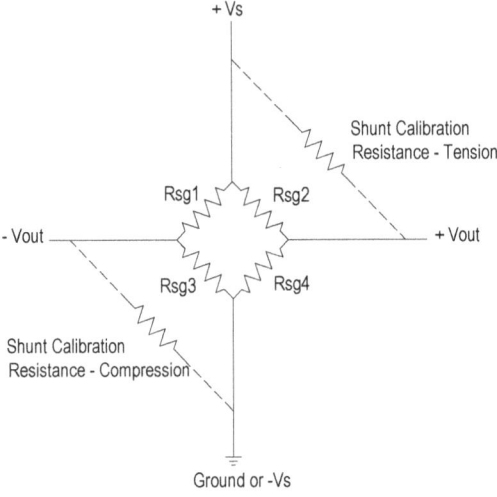

Figure 50 - Shunt calibration resistors are applied to simulate an applied physical load. Calibration resistors can simulate tension or compression loads.

As indicated in the previous figure, two connection possibilities are specified with the shunt calibration resistance, one for tension simulation and the other for compression simulation since most load cells can be used for both tension and compression loads.

Figure 51 - A microprocessor based data acquisition system. Up to 64 channels of sensors are connected to the four modules at the left. Calibration values and channel information is accessed via the keypad.

Scale Factor

Initially during load cell calibration, a minimum value of zero force is applied and an entry is made or an adjustment is performed to display a numerical value of zero. Next, a maximum value of load is applied or simulated and an entry is made or an adjustment is performed to display a numerical value equal to the maximum load. Occasionally during this stage of the process of calibration a value of scale factor is required to determine the maximum value displayed on a data acquisition system or computer monitor. Scale factor is a numerical value of gain that is used to amplify the output of the load cell when the maximum load is applied or simulated. Should a scale factor or gain value be required during this stage of the process a computation will need to be performed. The computation is demonstrated in the following example.

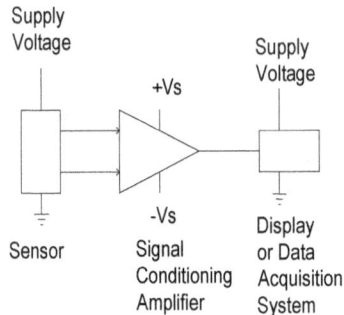

Supply
Voltage

+Vs

Supply
Voltage

-Vs

Sensor

Signal
Conditioning
Amplifier

Display
or Data
Acquisition
System

Figure 52 - A block diagram of a typical data acquisition system.

Example Problem:

Assume a 5000-pound load cell is to be connected to a data acquisition system and the system is to be calibrated. Determine the required offset and scale factor to be entered into the system during the calibration process. Assume the Rated Output of the load cell is 2.2 mV/V, the supply voltage is 10 Volts and the shunt calibration resistor simulates 72% of the 5000-pound rated load (3600-pounds).

Solution:

Once the 10 Volt supply is connected to the load cell and the load cell output is connected to the data acquisition system the system is powered-up and allowed to warm-up for a couple of minutes until the displayed value from the load cell is stable. Once the display is stable the calibration process can begin. Initially there should be no physical load applied to the load cell. The data acquisition system display should be zero since no load is applied but will probably be slightly off from zero.

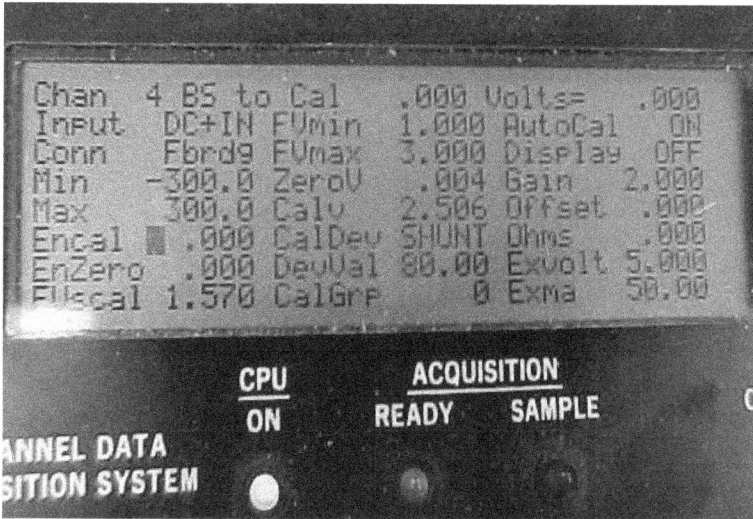

Figure 53 – The data acquisition system display. Note the indicated values, a 100-pound full bridge
load cell with a 5 Volt supply and shunt resistor simulating 80 pounds was connected to the data
acquisition unit.

The first step in the calibration process is to adjust the display to zero using the
"Offset" or "Zero" adjustment provided with the data acquisition system.
Occasionally, depending upon the type of system a value equal to the negative of
the displayed offset will need to be entered to null the display. Next, the
maximum load value or shunt calibration resistor is applied. Most data acquisition
systems currently available will provide some form of adjustment labeled "Gain"
or "Span." The adjustment should be varied until the displayed value equals the
value of the applied load or the value associated with the shunt calibration
resistor, 72% of the full rated load of 5000#, or 3600 #in this case.

However, many systems require that a scale factor be entered through a
keyboard or other data entry device to display the maximum applied or simulated
value. If a scale factor value is solicited the following procedure should be used.
First determine the millivoltage output of the load cell at the full rated load. Using

the provided specifications, the output voltage from the load cell is determined at 5000#.

$$\Delta Vout = RO * Supply\ Voltage$$
$$\Delta Vout = (2.2\ mV/V) * (10V)$$
$$\Delta Vout = 22\ mV\ at\ full\ rated\ load\ of\ 5000\#$$

Once the output voltage is determined at full rated load, the output voltage at the simulated load can be determined. Since the shunt calibration simulated load is 72% of the full rated load, the output voltage with the shunt calibration resistor applied will be 72% of the output voltage at the rated load, or,

$$Vout = [(Applied\ load \div Full\ rated\ load) * RO] \pm Offset$$

Since the "Applied load $\div$ Full rated load" equals 72% and assuming the offset has been set to zero in the previous step,

$$Vout\ at\ max\ calibration = [(72\%) * 22\ mV] \pm 0$$
$$Vout\ at\ max\ (shunt)\ calibration = 15.84\ mV$$

When the calibration resistor is applied the data acquisition display should read 3600#. The scale factor represents the gain of the amplifier in the previous figure. Since gain equals output divided by input, the desired display value of 3600 needs to be divided by the millivoltage from the load cell.

$$Scale\ Factor\ (SF) = 3600/15.84\ mV$$
$$Scale\ Factor\ (SF) = 3600/.\ 01584V$$
$$Scale\ Factor\ (SF) = 227,272.7273$$

Subsequently, a scale factor of 227,272.7273 will yield a displayed value of 3600. At this point both "Ends" of the calibration process should be re-checked and fine-tuned if required.

Figure 54 - A series of a Somat eDAQ stacked signal conditioning modules. These units offer the
flexibility of being calibrated and setup remotely via a laptop computer or a network connection.
Photo courtesy of Somat Corp.

A Scale Factor Shortcut

The previous process can be abbreviated if one determines the output voltage at
the full rated load, 5000# in this example. Knowing that the display should read
5000 at an output voltage from the load cell of 22 mV (.022V) the required scale
factor can be determined by dividing the full rated load (assuming this will be the
displayed value) by the load cell output voltage at the full rated load. The
following illustrates.

$$SF = \text{Full rated load} \div \text{full rated output voltage}$$

$$SF = 5000 \div .022V$$

$$SF = 227272.7273$$

Chapter Problems

Strain Gauges

#1 – Given the included information, determine the following values:

-Given:

Area = 1 in^2

Length = 2 inches

F_{comp} = 5000#

E = 31 E6 $_{PSI/in/in}$ (1020 steel)

μ = .288 $_{in^2/in^2/in/in}$

$R_{Strain\ Gauge}$ = 120 Ohms

$GF_{Strain\ Gauge}$ = 2.03

Vs = 5 Volts (4 element, full bridge)

-Find:

A) Stress

B) Strain (ε_L)

C) New Length

D) Strain (ε_T)

E) New Area

F) ΔRsg

G) Bridge Vout

#2 – Given the included information, determine the following values:

-Given:

Dia = 1 inch

Area = .785 in^2

Length = 12 feet (144 inches)

$F_{Tension}$ = 12000#

E = 31 E6 $_{PSI/in/in}$ (1020 steel)

μ = .288 $_{in^2/in^2/in/in}$

$R_{Strain\ Gauge}$ = 120 Ohms

$GF_{Strain\ Gauge}$ = 2.03

Vs = 5 Volts (4 element, full bridge)

-Find:

A) Stress

B) Strain (ε_L)

C) New Length

D) Strain (ε_T)

E) New Area

F) ΔRsg

G) Bridge Vout

#3 – Given the included information, determine the following values:

-Given:

Dia. = 1 in.

Length = 120 inches

$F_{Tension}$ = 10,000#

E = 10.7 E6 $_{PSI/in/in}$ (Alum)

μ = .350 $_{in^2/in^2/in/in}$

$R_{Strain\ Gauge}$ = 120 Ohms

$GF_{Strain\ Gauge}$ = 2.10

Vs = 5 Volts (4 element, full bridge)

-Find:

A) Stress

B) Strain (ε_L)

C) ΔLength

D) Strain (ε_T)

E) ΔArea

F) ΔRsg

G) Bridge Vout

#4 – Given the included information, determine the following values:

-Given:

Dia. = 1.5 in

Length = 5 feet

ΔF_{comp} = 5K, 10K, 20K lbs.

E = 31 E6 $_{PSI/in/in}$ (1020 steel)

μ = .288 $_{in^2/in^2/in/in}$

-Find:

A) Stresses

B) ΔStrains (ε_L)

C) ΔLengths

D) Strains (ε_T)

E) ΔAreas

$R_{Strain\ Gauge}$ = 350 Ohms   F) ΔRsg's
$GF_{Strain\ Gauge}$ = 2.08   G) Bridge ΔVout
Vs = 5 Volts (4 element, full bridge)

#5 – Given the included information, determine the following values:
Assume two active (strain sensing) gauges mounted on the sample and wired
into a full bridge circuit.

-Given: -Find:
    Dia. = .5 in   A) Stresses
    Length = 50 feet   B) ΔStrains (ε_L)
    ΔF_{comp} = 1K, 2K, 5K lbs.   C) ΔLengths
    E = 31E6 $_{PSI/in/in}$ (1020 steel)   D) Strains (ε_T)
    μ = .288 $_{in^2/in^2/\ in/in}$ E) ΔAreas
    $R_{Strain\ Gauge}$ = 350 Ohms   F) ΔRsg's
    $GF_{Strain\ Gauge}$ = 2.18   G) Bridge ΔVout
Note: Assume 2 active strain gauges,   H) the circuit diagram
    Vs = 5 Volts,   I) Label the active gauges in the
                                               circuit diagram

#6 – A strain-gauged column outputs 10.5 millivolts under a load of 1500 pounds,
and –1.2 millivolts when the load is removed. An instrumentation amplifier is to
be used to condition the bridge output into a 0 to 1.5 Volt (0 to 1500 pounds)
signal for display on a digital panel meter. Determine the gain and offset values
required for the amplifier output voltage span.

#7 – For each of the previous strain gauge problems assume an output of .1
millivolts per microstrain is desired. Determine the necessary circuitry and
operating values to develop the desired output. (Hint: For each problem a high
gain amplifier will be required. For each problem, determine the gain from the
strain and bridge Vout computation results.)

Shunt Calibration

#8 – A 350-Ohm bridge with a single active strain gauge (GF = 2.10) is shunt
calibrated with a 100K resistor. Determine the amount of stress represented with
a 40K shunt resistor. Hint: consider using the gauge factor equation.

#9 – A 120-Ohm bridge with a single active strain gauge (GF = 2.10) is shunt
calibrated with a 40K resistor. Determine the amount of stress represented with a
40K shunt resistor. Hint: consider using the gauge factor equation.

Load Cells

#10 – Given the tension/compression load cell specifications for the Eaton model
3169 included following, determine the requested values for the following
conditions.

A)
1000 lbs. rated load cell,
Vs = 5 Volts,
$F_{Tension}$ = 650 lbs.
- Determine the output voltage.

B)
1000 lbs. rated load cell,
- From the diagram determine the supply resistance.
- From the diagram determine the signal resistance.
- Explain the reason for any difference in resistance.

C)
1000 lbs. rated load cell,
Vs = 10 Volts,
$F_{Tension}$ = 650 lbs.
- Determine the output voltage.

D)
1000 lbs. rated load cell,
Vs = 5 Volts,
F_{Comp} = 1200 lbs.
Force is applied 30° from center,
- Determine the output voltage.

E)
- Explain the purpose and operation of a shunt calibration resistor included with the Eaton model 3169.

#11 – A load cell rated at 2000 pounds has a rated output of 1.6 mV/V and is to be used with a 5 Volt supply. Determine the output at rated load and the required scale factor to be used with a shunt calibration resistance representing 75% of the rated load. Assume a displayed value of 2000.

#12 – A load cell rated at 100 pounds has a rated output of 2.2 mV/V and is to be used with a 9 Volt battery. Determine the output voltage at rated load and the required scale factor to be used with a shunt calibration resistance representing 72% of the rated load. Assume a displayed value of 100.

MODELS 3169 AND 3169-106

Tension and compression 500 lbs.
to 3,000 lbs.

3169—English Thd.
3169-106—Metric Thd.

MODELS 3169-108
AND 3169-111

Tension and compression 500 lbs.
to 3,000 lbs.

3169-108—English Thd.
3169-111—Metric Thd.

FEATURES:

- Stainless steel
- Hermetically sealed
- Barometrically compensated
- Low deflection
- Proven design and performance
- Tension and compression
- Low sensitivity to extraneous loads

The 3169 stainless steel load cell offers sealed, barometrically compensated construction. It is ideally suited for scales and other critical weighing applications. This compact load cell's temperature compensation circuit is located in close proximity to the strain gage bridge which provides more effective compensation where the internal and external load cell temperature can vary. Precision performance as well as the structural integrity required to withstand extraneous loads are also features of this rugged load cell.

PERFORMANCE SPECS:
3169, 3169-106, 3169-108 AND 3169-111
SPECIFICATIONS

Output at rated capacity: millivolts per volt, terminal	2
Nonlinearity: of rated output	±0.05%*
Hysteresis: of rated output	±0.05%*
Repeatability: of rated output	±0.02%
Zero balance: of rated output	±5.0%
Creep: in 20 minutes of rated output	<±0.02%
Bridge resistance: ohms nominal	350
Temperature range, compensated: °F	+70 to +170
Temperature range, compensated: °C	+21 to +77
Temperature range, usable: °F	-65 to +200
Temperature range, usable: °C	-54 to +93
Temperature effect on output: of reading per °F	±0.002%
Temperature effect on output: of reading per °C	±0.0036%
Temperature effect on zero: of rated output per °F	±0.002%
Temperature effect on zero: of rated output per °C	±0.0036%
Excitation voltage, maximum: volts DC or AC rms	20
Insulation resistance, bridge/case: megohms at 50 VDC	>5,000
Number of bridges:	1

*Model 3169-106, -111: ±0.1%

LOAD CELLS

GENERAL PURPOSE: STAINLESS STEEL

442

This page retained blank intentionally.

Linear Integrated Circuits as Sensors Signal Conditioning Amplifiers:
Analog Sensors and Operational Amplifier Signal Conditioning Circuits for Data
Acquisition and Experimental Testing

Laboratory Experiments

This page retained blank intentionally.

ENGT 215 - Sensors and Analog I/O
Laboratory Experiment #1 - Introduction

Objective:
The objective of this exercise is to become familiar with the available components and to verify that all of the components are contained within your parts kit. In addition, this experiment should serve to re-acquaint the student with resistance measurements using an Ohmmeter.

Procedure:
1 - Compare the received components against the following inventory. If any of the components are found to be missing or of a substantially incorrect value, contact the author at LambertJ@bhc1.bhc.edu.

ENGT 215 Components:
    Circuit assembly bread-board (at least 830 connection points)
    Digital Multimeter (DMM)*
Resistors:
    4-100 Ohm
    1-150 Ohm
    2-1K
    1-1.2 K
    1-2.2 K
    1-2.7 K
    1-3.9 K
    1-4.7 K
    1-5.1K
    1-6.8 K
    1-8.2 K
    1-10 K
    1-20 K
    1-51 K
    4-100K
    2-200 K
    1-330 K
    1-390 K
    2-1Meg
Potentiometers (trim pots, connection pins should be breadboard compatible, such as Bourns 3352T):
    1-100 Ohm Linear
    1-1000 Ohm Linear
    1-10K Ohm Linear
    1-100K Ohm Linear
    1-200K Ohm Linear

Operational Amplifiers: three LM741 or equivalent op amps
9-Volt Batteries (2 required)*
9-Volt Battery Connectors (2 required)
Various lengths of 18-22 gauge solid wire
Various test clips or alligator clip leads
Sensors:
1 – Platinum 100 Ohm RTD element (Omega Engineering part# 1PT100GN1545)
1 – Type "T" fine wire (.036 diameter), 36" length thermocouple (Omega Engineering part# 5TC-TT-T-20-36)

*Note - These items will not be included within the parts kit and will need to be purchased by the participant. Www.MPJA.com carries most of the required components (meters, test clips and alligator clip leads) at reasonable prices.

Omega Engineering, Stamford, CT. can be reached at www.Omega.com or (203) 359-1660

2 - Assemble an Excel spreadsheet of rated and measured values. Resistor tolerance spans should be computed and indicated within the spreadsheet as well. Rated and measured battery voltages should also be included. This will aid in determining out-of-tolerance components.

3 - Using the DC Voltage measurement function on the digital multimeter (DMM), measure and record the battery voltages.

4 - Arrange and using a small piece of masking tape, label the resistors according to value. Try not to cover all of the color bands when taping. If possible, keep the components in a labeled organizer.

5 - Using the Ohmmeter function on your digital multimeter (DMM), measure and indicate on the resistor label the measured value. Include the measured results within the Excel spreadsheet. Save this data within the Excel spreadsheet, as it will become useful for troubleshooting during subsequent lab experiments. Remember to "Zero" the Ohmmeter before making a resistance measurement (touch the probes together and check for a zero indication). Also, remember not to touch the meter probes when making the resistance measurement as this will alter the resistance indication.

6 - Once all of the resistors have been measured, labeled and entered into Excel, determine if any values are outside of the computed tolerance.

Review:
Recall from your introductory circuit's course that resistors are labeled as following.

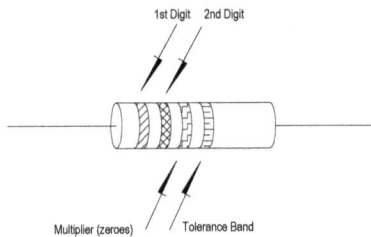

Figure 1 - Resistor color code interpretation.

Figure 2 - Resistor color code interpretation.

Following is the resistor color code and tolerance markings.

Black	0
Brown	1
Red	2
Orange	3
Yellow	4
Green	5
Blue	6
Violet	7
Grey	8
White	9

Gold	5%
Silver	10%
None	20%

Results and Conclusions:
Document any significant differences between the expected values and the measured values within Excel. Note any obvious inconsistencies in the remaining components. Save this information for later reference.

Required Submissions:
When the lab has been completed, send a copy of the Excel spreadsheet data by e-mail to LambertJ@bhc1.bhc.edu.

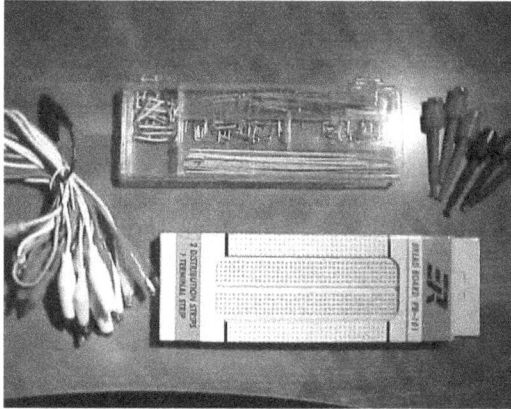

Figure 2 - The materials in this photo are representative of those used when performing the labs in ENGT 215. Included are alligator clip leads (left), the breadboard (lower right) for circuit assembly, a variety of cut and stripped 20 gauge wires (top center) and test clips. The breadboard is included with the parts kit. The remaining items can be made with materials acquired from various vendors over the web or at Radio Shack.

Objective:
The objective of this exercise is to become familiar with the circuit prototyping
board or, BreadBoard. The exercise will involve the assembly of series and
parallel resistive circuits, and measurement of the circuit's total resistance.

Introduction:
Locate and acquire the circuit prototyping board. It should appear similar to the
following photo.

Upon inspection, one will notice the construction of the board is different between
the inner (the inner columns may be labeled with the letters A through J) columns
and the outer columns. As shown in the photo, the outer 4 columns (2 columns
on each side) have points that connect vertically, or lengthwise. The inner 10
columns are connected across but NOT through the center.

The 4 outer strips are called Distribution Strips and are used to distribute power
supply, input signal, output signal or other multi-location required voltages. There

are 200 individual tie points (connection holes) in 4 strips. The strips are connected lengthwise but not across. Typically a connection is brought into the distribution strips from outside of the breadboard. Once connected, the incoming voltage will be available at 49 remaining tie points within the strip that the outside connection is made.

The 630 individual tie points (2 columns of 5 tie points across) within the center of the breadboard is called a terminal strip. Upon close inspection of the *ER* PB-101 breadboard, the outer distribution strips are actually separate from the inner terminal strip and can be replaced if desired. The terminal strips are connected 5 across but not lengthwise or through the center ridge.

Typically, circuit connections are established by inserting components into the terminal strip portion of the board. One lead of a component is inserted into an available tie point within a given row (set of 5). When an additional component leadwire is connected to a tie point within the same row, a connection is made internally within the breadboard. The following diagram illustrates.

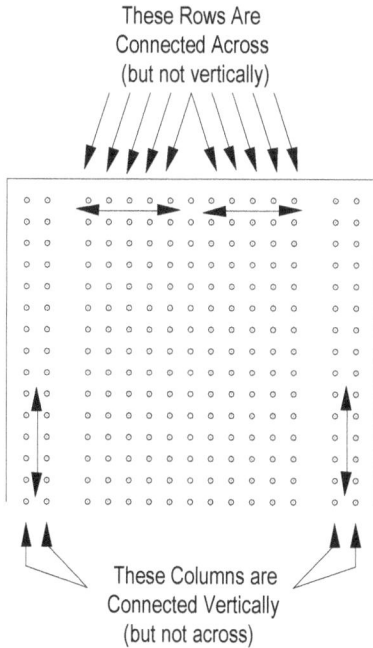

Figure 1 - Breadboard circuit connections. Note: connections exist only where indicated.

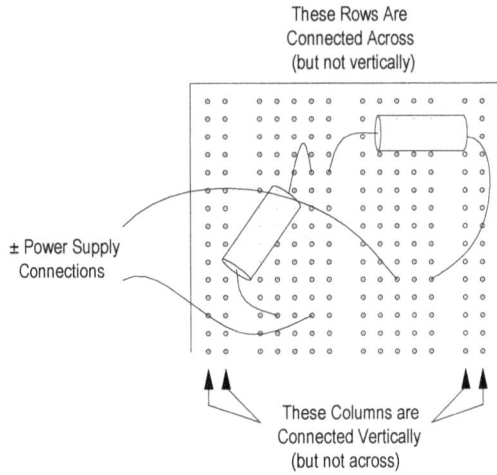

These Rows Are
Connected Across
(but not vertically)

± Power Supply
Connections

These Columns are
Connected Vertically
(but not across)

Figure 2 - An example of a two-component series circuit connection. In this example the distribution strip is not used to distribute the power supply connections.

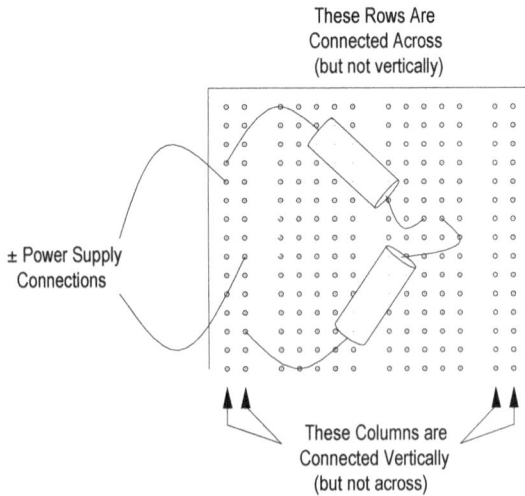

These Rows Are
Connected Across
(but not vertically)

± Power Supply
Connections

These Columns are
Connected Vertically
(but not across)

Figure 3 - An example of a two-resistor series circuit using the distribution strips for the power supply connection. The full lengths of the distribution strips will be available for additional supply connections.

451

Lab Procedure:
1 - Acquire three resistors from the parts kit. Note the values of the resistances and assemble a three-resistor series circuit using the breadboard. Computed the total resistance.

Figure 4 - Three-resistor series circuit of Lab #2.

2 - To verify proper circuit assembly, measure the total resistance across the outer two connections of the assembled circuit. Does the measured approximate the computed total? If not, re-assemble the circuit connection. Leave the circuit assembled on the breadboard once successfully completed.

3 - Acquire three additional resistors from the parts kit. Note the values of the resistances and assemble a three-resistor parallel circuit using the breadboard. Compute the total parallel resistance and assemble the circuit on the breadboard.

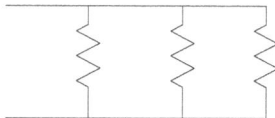

Figure 5 - Three-resistor parallel circuit of Lab #2.

4 - To verify proper circuit assembly, measure the total resistance across the assembled circuit. Does the measured approximate the computed total? If not, re-assemble the circuit connection until correct.

Results and Conclusions:
Document your results noting any significant differences between the expected values and the measured values. Save this information for later reference.

Required Submissions:
This lab was to assist the participant in becoming acquainted with circuit assembly using the PB-101 breadboard. No submitted results are required. If you should have problems assembling the circuit, leave the circuit assembled on the breadboard and contact the instructor by e-mail at LambertJ@bhc.edu. Digitally photographing or scanning the assembled circuit will be of great benefit in assisting with your assembly errors. If possible, include a photo of the circuit with your e-mail.

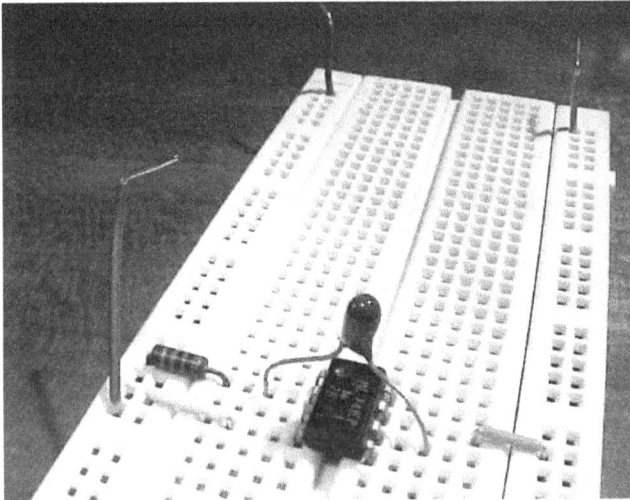

Figure 6 - This op amp circuit demonstrates the connection scheme of the breadboard. Note the supply connections from the op amp to the distribution strip (orange and yellow wires), and the resistor to capacitor (blue component) to op amp connection on the terminal strip. The capacitor is also connected to the right side of the op amp on the terminal strip. The input signal will be connected to the resistor via the distribution strip (brown wire). The two leadwires at the top of the photo are used to connected the power supply to the distribution strips.

This page retained intentionally blank.

ENGT 215 - Sensors and Analog I/O
  Laboratory Experiment #3 - Resistive Voltage Dividers

Objective:
The objective of this exercise is to review the operational concepts of the series circuit, or Voltage Divider. Specifically, the lab reviews the voltage divider concept, equation and to demonstrate the changes in voltage distribution when resistance values in a series circuit are varied. This concepts observed during the performance of this lab should serve as an introduction to resistive bridges.

Introduction:
A series circuit is an accumulation of end-to-end connected components that when connected to a voltage source result in a single current flowing, equal in amplitude through all components. The following diagram illustrates.

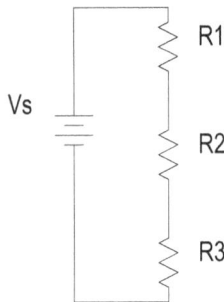

Figure 1 - A 3-component series circuit.

The current created by the DC source and the three resistors is equal to the sum of the resistances in the circuit divided into the applied voltage. As an example, assume each of the resistors is labeled as follows and 9 Volts is applied.

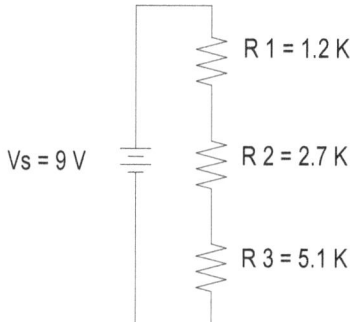

The total resistance equals the sum of all resistances.

$$R_{total} = 1.2K + 2.7K + 5.1K = 9K$$

The total resistance is 9K. The current equals 9V divided by 9K Ohms, or 1 mA.

$$I = 9V/9K = 1 \text{ mA}$$

The individual voltages can now be determined using Ohm's law.

$$V_{R1} = I * R1 = 1 \text{ mA} * 1.2K = 1.2V$$
$$V_{R2} = I * R2 = 1 \text{ mA} * 2.7K = 2.7V$$
$$V_{R3} = I * R3 = 1 \text{ mA} * 5.1K = 5.1V$$

Note the individual voltage values correspond to the individual resistance values. The larger the resistance value, the larger the voltage drops. Note also that the source voltage distribution throughout the three resistances is a proportion of any individual resistance in relation to the total circuit resistance. This can be proven using Ohm's law and results in a more rapid technique for determining the voltage drops across the individual resistors.

Assume it is desired to determine the voltage across R1.

$$V_{R1} = I * R1$$
$$I = Vs / R_{total}$$
$$\therefore VR1 = (Vs / R_{total}) * R1$$
$$\text{Or, } VR1 = Vs (R1/R_{total})$$

The previous equation demonstrates a fundamental concept in series circuits, the voltage distribution within a series circuit can be determined by the ratio of any individual resistance to the total resistance, $R1/R_{total}$, and multiplying by the applied voltage, $Vs (R1/R_{total})$. The resulting equation is known as the _Voltage Divider Equation_ and is extremely useful in determining voltage drops rapidly.

The concept of proportionality represented by the equation is equally as important and can be applied to any series connected components.

Voltage Divider Equation:

$$V_R = V_S * R/R_{total}$$

Where,
V_R is the voltage to be determined,
V_S is the applied, source voltage,
R is the resistance whose voltage is to be determined,
R_{total} is the sum of all resistances in the series circuit, also designated R_T.

It is suggested this equation and the corresponding concept be committed to memory as the author will assume student familiarization in subsequent chapters and labs.

Lab Procedure

1 - Assemble the previous circuit and verify the determined resistor voltage drops.

Be sure to measure the 9-Volt battery voltage, as batteries are typically NOT the rated value but can deviate plus or minus around the nominal rated value. Adjustments may need to be made for any deviation of the supply away from 9 Volts. As an example, if the battery measures 9.9V, the voltage is 10% higher than the rated value. Therefore, each of the measured voltage values will be 10% higher than the previously computed values.

However, keep in mind that the purpose of this lab is to reinforce the voltage divider concept and is NOT to acquire precision results. Your values need not be identical to the previously computed values. However, according to the voltage divider equation approximately 56.7% of the applied voltage should appear across the 5.1K resistor, 30% across the 2.7K and 13.3% across the 1.2K.

2 - Repeat the procedure using a larger value for R1.

3 - Document your results in Excel and save the spreadsheet as 200Lab3.

4 - Review the lab results, assemble and include a conclusion within the lab results.

5 - Send the resulting file (200lab3) to the author upon completion of the lab. Be sure to include your name on the spreadsheet.

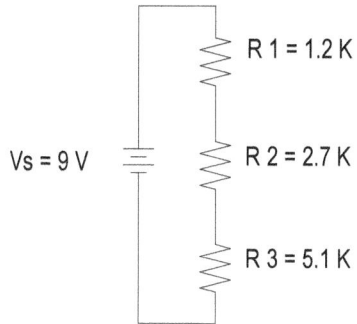

Laboratory Experiment #4 - Half and Full Bridges

Objective:
The objective of this exercise is to introduce the basic resistance bridge circuit. The bridge is a common resistive sensor (RTD's and strain gauges) signal conditioning circuit. Bridges convert sensor resistance variation into a varying voltage signal. The resulting voltage signal is amplified and mathematically scaled to represent the applied physical variable at the sensor's input.

This lab will simulate a resistive sensor to demonstrate the operation of common bridges. Resistance values of 100 and 150 Ohms will simulate an RTD (.385 Ohms/C°, 100 Ohms at 0°C) at 0° and at 130° Celsius. (The simulated RTD resistance values are left to the student to compute and verify.) Initially, a half bridge will be used to convert the "RTD" resistance into a voltage. Secondly, a full bridge will convert the same RTD simulated resistance values into voltage. The student is asked to observe and conclude on the difference between the two circuit output voltage spans.

Introduction:
A bridge circuit is a voltage divider circuit, and as mentioned in lab 3, the voltage across any resistance in a series circuit is a proportion of the resistance value to the total resistance. When a resistive sensor (R_x) is placed into a series circuit with a fixed resistor, the voltage across the resistive sensor will vary in direct proportion to the sensor's resistance value.

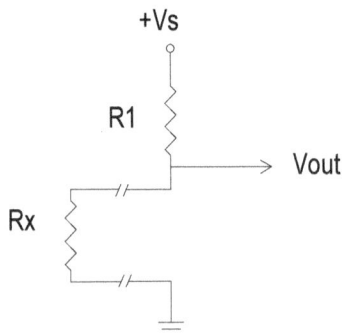

Figure 1 - A 2-component half-bridge. Rx is the sensor (simulated sensor resistance for this lab); R1 is always equal to the minimum sensor resistance value.

Half-bridges are composed of a sensor (R_X), and a fixed resistance (R_1). The fixed resistance is always equal to the minimum expected value of the sensor resistance. If the sensor resistance variation is expected to be 100 Ohms to 150 Ohms, the fixed resistance will be 100 Ohms. If the expected resistance variation is expected to be 120 Ohms to 120.05 Ohms, the fixed resistance will be 120 Ohms. Likewise, the minimum output voltage from a half-bridge is one-half of the supplied V_S. As the sensor resistance changes the output voltage from the half-bridge changes in a direct and proportional manner.

A half-bridge like figure 1 is functional but has an undesired side effect. The *resolution* of the output signal is minimal. Resolution is the ability to detect small changes clearly. Since the output voltage will be half of the supply at minimum R_X, as the sensor resistance varies the corresponding output voltage change may be small (sometimes very small, like with strain gauges) in comparison to half of the supply.

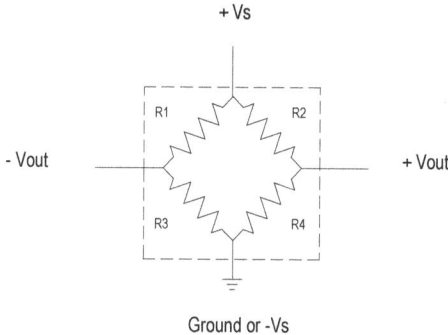

Figure 2 - A common 4-element bridge circuit, or full-bridge.

To eliminate the resolution problem another half-bridge is introduced. This results in the full-bridge circuit of figure 2. The output signal is no longer single-ended (measured Vout to ground) but requires a differential measurement between -Vout and +Vout. The resistor values composing the additional half-bridge (R1 and R3 in figure 2) will equal the fixed resistance value (minimum sensor resistance value) of figure 1 and results in an output from -Vout equal to a constant value of one-half of the supply voltage, Vs.

When the circuit is assembled and +Vout and -Vout are connected to a common meter (handheld DMM's and DVM's are differential measurement instruments since they measure the voltage difference between the probes) or differential amplifier the result will be the difference between +Vout and -Vout. Since all of the fixed resistor values equal the minimum sensor resistance (R4 in figure 2), the minimal output from the bridge will become half of the supply (+Vout) minus (differential amp or meter) half of the supply (-Vout), or zero volts. As a changing physical variable is applied to the sensor, the sensor resistance will vary and the corresponding output (+Vout) will also change. Since the output is taken between +Vout and -Vout, the output will vary from zero volts as the sensor resistance changes in accordance with the applied physical variable. The resulting output signal will vary from zero volts improving the resolution of the output signal.

The following lab experiment will demonstrate the operation of the previously mentioned half and full bridges.

Lab Procedure
1 - Using 2-100 Ohm resistors, assemble the half-bridge circuit of figure 3 following. Use a 9-V battery as Vs and measure Vout. Document the result.

Vout @ Rx =100 Ohms is _____.

2 - Locate the 150-Ohm resistor. Place the 150-Ohm resistance in the circuit in place of the resistance between +Vout and ground. (Assume the 150-Ohm Rx is simulating a Resistance Temperature Detector (RTD) type of sensor.) The circuit should appear as in the following figure. A .385 Ohm/C°, 100-Ohm Platinum RTD between the temperatures of 0° C. and approximately 130° C. will vary between 100 and 150 Ohms.) Measure and record Vout with the 150-Ohm resistor in the circuit.

Vout @ Rx =150 Ohms is _____.

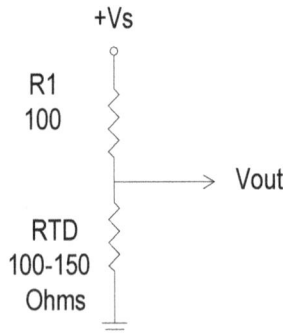

Figure 3 - Half-Bridge at minimum and maximum "RTD" resistance.

Assume the two results from step 1 and step 2 are the extreme values of the sensor's operation and note the output voltage's minimum to maximum span. Note the actual signal is changing approximately .9 volts in an environment of approximately 4.5 volts. An amplifier would have difficulty developing any significant gain with a signal of this type since the amplifier output is limited to the value of the supply voltages. In short, the 4.5V "pedestal" from which the output varies diminishes the resolution of the output voltage signal variation. Eliminating the 4.5V and referencing the signal voltage to 0V will enhance the resolution of the output voltage span. A full-bridge will be employed next to reference the output signal to zero volts.

3 - Assemble a full-bridge using four 100-Ohm resistors. Designate the resistance in the lower right corner of the bridge (+Vout) as Rx. Once completed, measure and document the output voltage. Remember that the differential output signal is now taken between the two outputs (-Vout, +Vout) and will not require a ground connection to the meter.

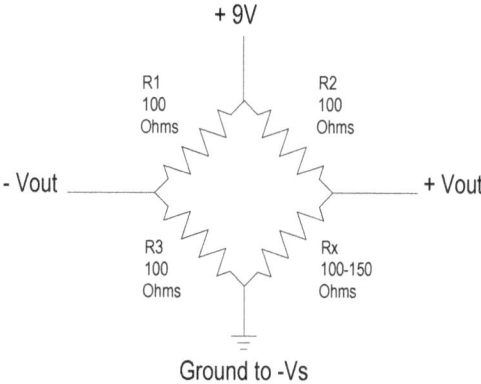

Figure 4 - Full-Bridge connection with Rx indicated.

4 - Replace the 100-Ohm designated Rx with the 150-Ohm resistance, measure and record the output again. Compare the results of the half-bridge and the full bridge at the minimum and maximum Rx values.

Vout @ Rx =100 Ohms is _____.

Vout @ Rx =150 Ohms is _____.

5 - Application Exercise:

A 100-Ohm Platinum RTD element (.385 Ohms/C°) should be included in your parts kit. Repeat this lab using the RTD in place of Rx in the half and full-bridge circuits. The RTD will vary from 100 and 138.5 Ohms when placed in a 0° C. (32° F) to 100° C (212° F) environment. Although the RTD can be submerged in water the leadwires are not insulated and will probably short given the conductive properties of water. It is suggested the RTD be taped to the side of a glass or metal vessel containing a 50/50 solution of ice and water, this should result in an approximate temperature of 0° C. (32° F). Gradually increase the temperature by applying a low heat setting until the solution is brought to a boil, or until the half and full-bridges output the expected values.

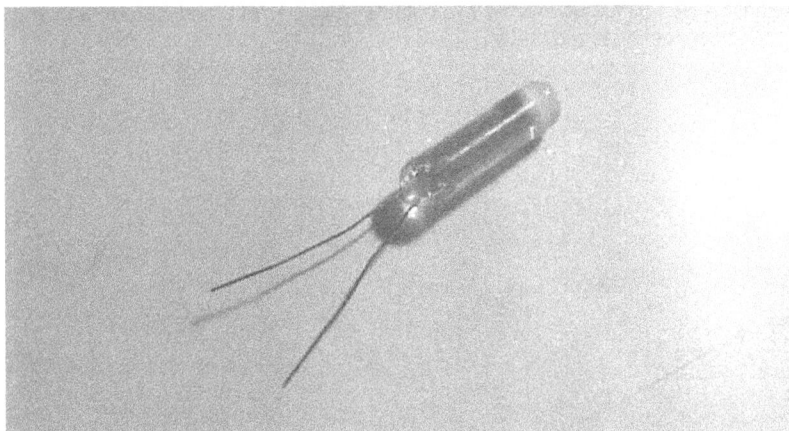

Figure 5 - The 100 Ohm Platinum RTD element.

6 - Tabulate the measured results from each bridge (half and full) at the 100 and 150-Ohm values of Rx, and/or at the values of temperature using the RTD. Assemble and include a conclusion from your measured results and submit the file (200lab4) to the author.

This page retained intentionally blank.

Objective:
The objective of this exercise is to introduce the operational amplifier and it's ability to perform mathematical operations. This lab will investigate the non-inverting amplifier, circuit analysis and input/output gain.

Introduction:
The non-inverting amplifier provides an output that follows the input in a direct, or *in-phase* manner. As the input voltage increases, the output voltage increases. All amplifiers are designed to operate over a span of input/output values. Input voltage spans are obtained from sensors, receivers, other amplifiers and any circuit requiring signal conditioning of a mathematical nature. The mathematical relationship between the input and output voltage spans is called *Voltage Gain*, and is given the symbol, Av. The non-inverting amplifier and corresponding input/output characteristic plot follows in figure 1. Note the slope of the input/output plot is the amplifier voltage gain, Av.

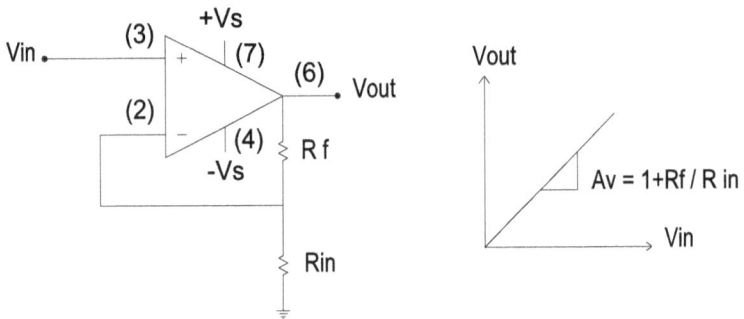

Figure 1 - A non-inverting, or direct-acting, amplifier.

Also note the amplifier requires *bipolar* supplies of plus and minus voltage applied to pins 7 and 4 respectively. (Most operational amplifiers will work with unipolar supplies but the amplifier output will assume a value of 1/2 of the available supply at an expected output of zero volts.) To accomplish the bipolar supply connection two independent DC supplies, such as 9-Volt batteries, will be required. The Batteries need to be connected as follows.

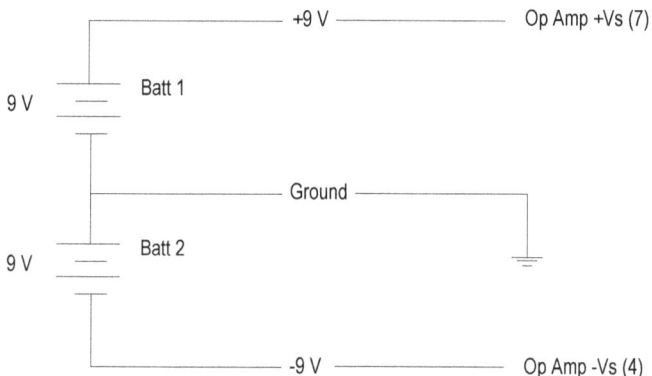

Figure 2 - Nine volt batteries arranged to provide bipolar power to operational amplifiers.

Once connected, the common connection between the batteries becomes the ground for the amplifier and any related circuitry. This circuit connection makes Batt 2 the negative voltage supply and Batt 1 the positive voltage supply. The circuit should be assembled and the outputs checked before proceeding with the lab experiment. Note that 9-Volts is a nominal battery voltage and the actual value may be slightly greater or less.

Lab Procedure
Step 1
 - Assemble the circuit of figure 2 using two 9-volt batteries or independent DC supplies if available. Measure each supply voltage and record the results below.

$+Vs = $ _____.

$-Vs = $ _____.

Step 2
- Place a 1K potentiometer across the positive supply and ground. The potentiometer output will serve as the amplifier signal voltage for the amplifier. The circuit should be connected as follows. Remember, the center connection on any potentiometer is the wiper. Small circles on figure 3 designate all three potentiometer connection points. Once the potentiometer is connected rotate the potentiometer wiper and measure the available the voltage from the wiper to ground. The approximate voltage should vary between zero and nine volts. If exactly zero and nine volts are not available do not be concerned, as the following lab will require a voltage of 1 to 3 volts from the potentiometer.

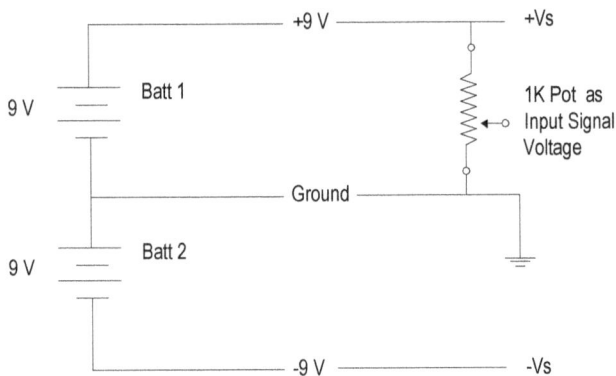

Figure 3 - Connecting a potentiometer as a variable (positive) voltage source. The voltage from the wiper to ground will serve as the amplifier input signal.

Step 3
- Assemble the circuit of figure 1. Assume an Rf of 20K and Rin of 10K. The completed circuit should appear as follows.

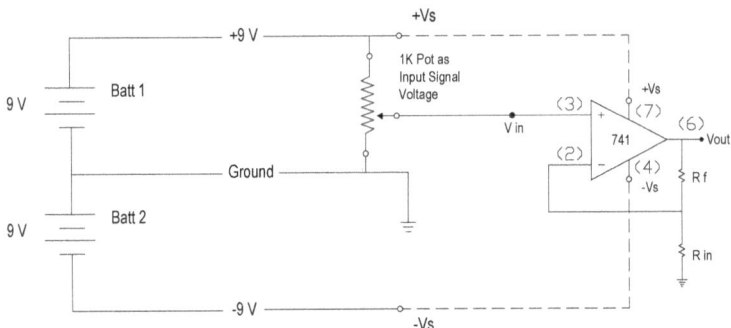

Figure 4 - The completed non-inverting amplifier circuit. Note that all voltages are measured in reference to ground and both grounds in the diagram are connected together.

Step 4
- Set the amplifier input voltage on the wiper of the potentiometer to 1 volt. Using the non-inverting amplifier circuit analysis procedure described in chapter 3 of the text, determine the resistor voltages *across* Rf and Rin at an input of 1 volt. Document the calculated voltages following.

Calculated Rf voltage = _____

Calculated Rin voltage = _____

Step 5
Measure and record the actual input and feedback resistor voltages following.

Actual Rf voltage = _____

Actual Rin voltage = _____

Step 6
- Measure and record the output voltage at an input voltage of 1 volt.

Output Voltage, Vout = _____

Do not attempt to compute the amplifier gain at this time. Since voltage gain (Av) is a slope computation requiring input and output spans (Av = ΔVout/ΔVin), Av should not be computed using a single input/output voltage points. The difference between input and output voltage extremes are required to compute voltage gain.

Step 7
- Set the amplifier input voltage on the wiper of the potentiometer to 3 volts. Again, using the non-inverting amplifier circuit analysis procedure described in chapter 3 of the text determine the resistor voltages *across* Rf and Rin at an input of 3 volts.

Calculated Rf voltage = _____

Calculated Rin voltage = _____

Step 8
- Measure and record the actual resistor voltages.

Actual Rf voltage = _____

Actual Rin voltage = _____

Step 9
- Using the non-inverting amplifier gain equation of 1+ Rf/Rin, determine the gain of the amplifier circuit. The calculated output voltage is determined by multiplying the computed gain by the input voltage. Document the calculated gain and the computed output voltage following.

Calculated Voltage Gain, Av = _____

Calculated Output Voltage, Vout = _____

Step 10
- Measure and record the actual output voltage.

Actual Output Voltage, Vout = _____

Step 11

- Verify the amplifier gain by taking the difference in output voltage values divided by the difference in input voltage values, Av = ΔVout/ΔVin. Record this value following.

Actual Voltage Gain, Av = _____

Step 12
- Determine the percentage of error between the calculated gain in step 8 and the actual gain in step 10. The percentage of error value should be within the resistor tolerance (typically ±5%) and is used to determine if the results of the lab experiment are acceptable. The percentage of error equation follows.

% Error = [(Calculated - Measured) ÷ Calculated] * 100%

Do not disassemble this amplifier circuit. The following lab (#6) can be performed with a minor connection modification of this circuit.

Results and Conclusions
Complete the lab by assembling a brief conclusion for the experiment. Include any data or procedural discrepancies, technical problems or malfunctions, inconsistencies, suggestions for improvements and the following questions in your conclusion statement.

Did the amplifier exhibit a non-inverting (direct-acting) function?
Were the calculated and measured resistor voltages comparable?

This page retained intentionally blank.

Objective:
The objective of this exercise is to introduce the operational amplifier and it's ability to perform mathematical operations. This lab will investigate the inverting amplifier, circuit analysis and input/output gain.

Introduction:
The inverting amplifier provides an output that follows the input in an inverse, or *out-of-phase* manner. As the input voltage increases, the output voltage decreases. Amplifiers are designed to operate over a span of input/output values. Input voltage spans are obtained from sensors, receiver/demodulators, other amplifiers or any circuit requiring signal conditioning of a mathematical nature. The mathematical relationship between the input and output voltage spans is called *Voltage Gain*, and is given the symbol, Av. The inverting amplifier and corresponding input/output characteristic plot follows in figure 1. Note the slope of the input/output plot is the amplifier voltage gain, Av.

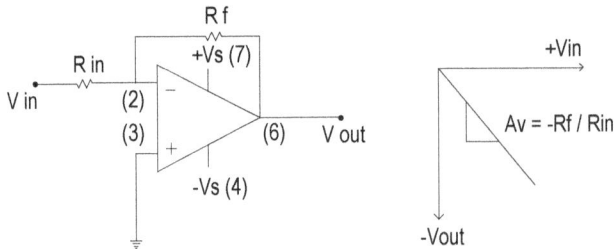

Figure 1 - An inverting, or reverse-acting, amplifier.

Also note the amplifier requires *bipolar* supplies of plus and minus voltage applied to pins 7 and 4 respectively. (Most operational amplifiers will work with unipolar supplies but the amplifier output will assume a value of 1/2 of the available supply at an expected output of zero volts.) To accomplish the bipolar supply connection two independent DC supplies, such as 9-Volt batteries, will be required. The Batteries need to be connected as follows.

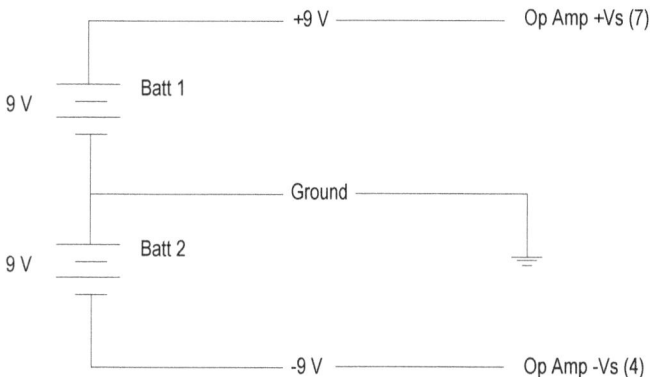

Figure 2 - Nine volt batteries arranged to provide bipolar power to operational amplifiers.

Once connected, the common connection between the batteries becomes the ground for the amplifier and any related circuitry. This circuit connection makes Batt 2 the negative voltage supply and Batt 1 the positive voltage supply. The circuit should be assembled and the outputs checked before proceeding with the lab experiment. Note that 9-Volts is a nominal battery voltage and the actual value may be slightly greater or less.

Lab Procedure
Step 1
 - Assemble the circuit of figure 2 using two 9-volt batteries or independent DC supplies if available. Measure each supply voltage in reference to ground and record the results below.

+Vs = _____

-Vs = _____

Step 2
- Place a 1K potentiometer across the positive supply and ground. The potentiometer output will serve as the amplifier signal voltage. The potentiometer circuit should be connected as follows. Remember that the center connection on any potentiometer is the wiper. Small circles on figure 3 designate all three potentiometer connection points. Once the potentiometer is connected rotate the potentiometer and measure the available the voltage from the wiper to ground. The approximate voltage should vary between zero and nine volts. If exactly zero and nine volts are not available do not be concerned, as the following lab will require a voltage of 1 to 3 volts from the potentiometer.

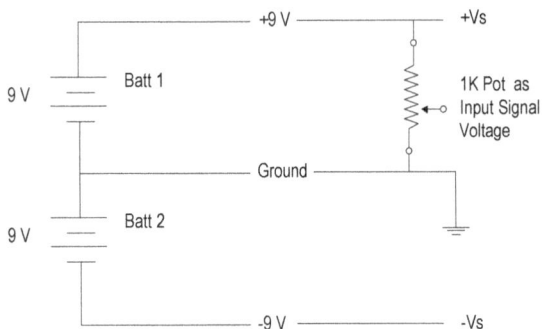

Figure 3 - Connecting a potentiometer as a variable (positive) voltage source. The voltage from the wiper to ground will serve as the amplifier input signal.

Step 3
- Assemble the following circuit of figure 4; swapping the amplifier ground and input connections as in figure 4 can modify the previous circuit of lab 5 for use in this lab. Rf will again equal 20K and Rin will equal 10K. Note: do not swap the connections by switching the connections to pins 2 and 3 as this will not result in the required negative feedback. The completed circuit should appear as follows.

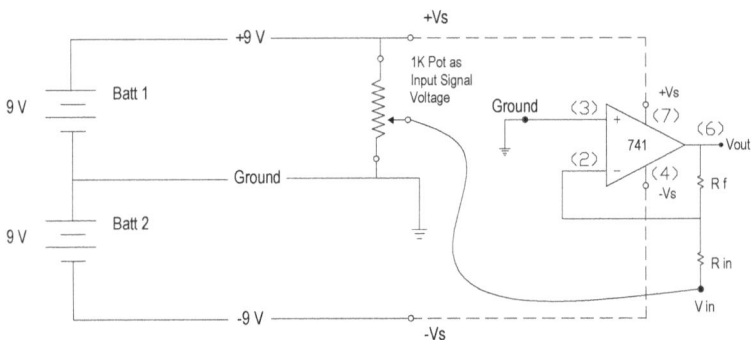

Figure 4 - The completed inverting amplifier circuit. Note that all voltages are measured in reference to ground and both grounds in the diagram are connected together.

Step 4
- Set the amplifier input voltage on the wiper of the potentiometer to 1 volt. Using the inverting amplifier circuit analysis procedure described in chapter 3 of the text, determine the resistor voltages *across* Rf and Rin at an input of 1 volt. Document the calculated voltages following.

Calculated Rf voltage = _____

Calculated Rin voltage = _____

Step 5
Measure and record the actual input and feedback resistor voltages following.

Actual Rf voltage = _____

Actual Rin voltage = _____

Step 6
- Measure and record the output voltage at an input voltage of 1 volt.

Output Voltage, Vout = _____

Note: Do not attempt to compute the amplifier gain at this time. Since voltage gain (Av) is a slope computation requiring input and output spans (Av = ΔVout/ΔVin), Av should not be computed using a single input/output point. The difference between input and output voltage extremes are required to compute voltage gain.

Step 7
- Set the amplifier input voltage on the wiper of the potentiometer to 3 volts. Again, using the inverting amplifier circuit analysis procedure described in chapter 3 of the text, determine the resistor voltages *across* Rf and Rin at an input of 3 volts. Measure and record the voltages following.

Calculated Rf voltage = _____

Calculated Rin voltage = _____

Actual Rf voltage = _____

Actual Rin voltage = _____

Step 8
- Using the inverting amplifier gain equation of - Rf/Rin, determine the calculated gain of the amplifier circuit. The calculated output voltage is determined by multiplying the computed gain by the 3V input voltage. Document the calculated gain and output voltage following.

Calculated Voltage Gain, Av = _____

Calculated Output Voltage, Vout = _____

Step 9
- Measure and record the actual output voltage.

Actual Output Voltage, Vout = _____

Step 10

- Verify the amplifier gain by taking the difference in output voltage values divided by the difference in input voltage values, Av = ΔVout/ΔVin. Record this value following.

Actual Voltage Gain, Av = _____

Step 11
- Determine the percentage of error between the calculated gain in step 8 and the actual gain in step 10. The percentage of error value should be within the resistor tolerance (typically ±5%). Percentage of error computations is used to determine if the results of the lab experiment are acceptable. The percentage of error equation follows.

% Error = [(Calculated - Measured) ÷ Calculated] * 100%

Results and Conclusions
Complete the lab by assembling a brief conclusion for the experiment. Include any data or procedural discrepancies, technical problems or malfunctions, inconsistencies, suggestions for improvements and the following questions in your conclusion statement.

Did the amplifier exhibit an inverting (reverse-acting) function?
Were the computed and actual resistor voltage values comparable?

ENGT 215 - Sensors and Analog I/O
Laboratory Experiment #7 - Biased Non-Inverting Amplifiers

Objective:
The objective of this exercise is to introduce the operational amplifier as a signal scaling circuit. This lab will investigate the biased non-inverting amplifier, circuit analysis, input/output gain and output offset.

Introduction:
The biased non-inverting (or direct-acting) amplifier provides an output that follows the input in a direct, or *in-phase* manner. As the input voltage increases, the output voltage increases. In addition, the output is biased, or offset from zero volts when the input is zero volts.

All amplifiers are designed to operate over a span of input and output values. Input voltage spans are obtained from sensors, receivers, other amplifiers and any circuit requiring signal conditioning of a mathematical nature. The dynamic relationship between the input and output voltage spans is called *Voltage Gain*, and is given the symbol, Av. The output offset voltage is given the symbol "V_b" and corresponds to the Y-axis intercept value in the Y=mX+b equation. In order to accomplish output offset, a constant DC bias voltage will be introduced to the op amp input that is normally grounded.

The biased non-inverting amplifier schematic and corresponding input/output characteristic plot follows in figure 1. Note the slope of the input/output plot is the amplifier voltage gain, Av, and the output offset is designated "Vb" in the characteristic diagram.

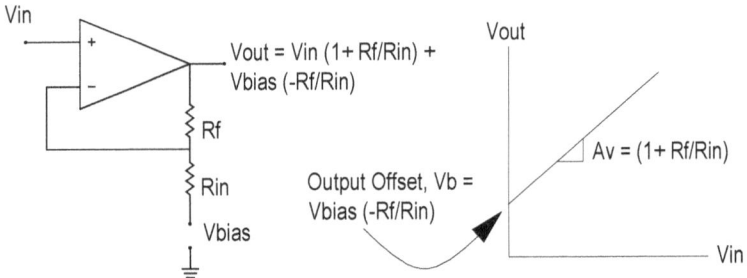

Figure 1 - A biased non-inverting, or direct-acting, amplifier. Gain and offset (intercept) are indicated on the amplifier characteristic plot.

To differentiate between the input (signal) voltage and the bias voltage, the input signal will be a span of values such as 1V to 5V, or -1V to +1V, and the bias voltage will be a constant value. Bias voltages are usually acquired from a DC supply or resistive network connected to a supply. A more complete diagram of a biased non-inverting amplifier follows.

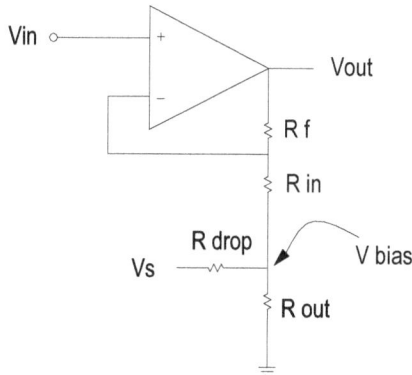

Figure 2 - Vbias is usually applied using a voltage divider connected to an available supply.

The biasing voltage circuit below Rin in figure 2 is composed of two resistors, Rdrop and Rout. (In the future these designations will be replaced with R1 and R2 or Ra and Rb.) Rout is responsible for providing the necessary bias voltage to the op amp and Rdrop will "drop" the difference between the supply voltage and the required bias voltage.

The resulting output voltage can be determined by the equation provided in figure 1 and following, and is the result of summing the individual inputs multiplied by their respective gains. Vbias is determined using Vs and the voltage divider equation provided following.

$$Vout = Vin \ (1+Rf/Rin) + Vbias \ (-Rf/Rin)$$
$$Vbias = Vs \ [Rout/(Rdrop+Rout)]$$

Lab Procedure
Step 1 - Assembly
 - Assemble the following biased non-inverting amplifier. After connecting the circuit measure each supply voltage and record the results below. This is done to assure that neither of the supplies is over-loaded, resulting in a substantial voltage difference between the two supplies. Excessively mismatched supplies ($\geq$ 2 Volts) will result in an output voltage span with significant error. If an excessive difference exists measure each battery unloaded (removed from the circuit) as one of the batteries may have become weak. If both measure about equal, check the circuit connections.

As with the previous lab, a 1K pot will provide the input signal voltage of 1-3 Volts.

Battery Voltages:

+Vs = _____:

-Vs = _____:

+Vs

+9 V

9 V — Batt 1

Input Signal
Voltage =
1V to 3V

1K Pot

(3) +Vs
(7)
V in
741 (6)
Vout
(2) − (4) Rf =
-Vs 200K

Ground

Batt 2

9 V

R in =
100K

R drop
= 4.7K

-9 V

-Vs

R out =
2.2 K

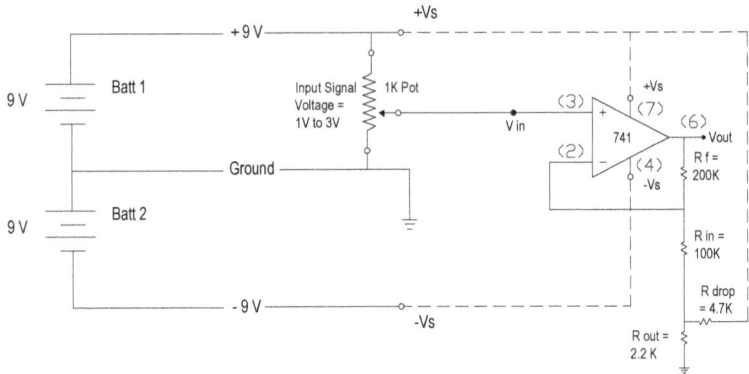

Step 2 - Calculations: Gain and Offset
- Determine the input signal voltage gain (A_V) and output offset (Vb) values from the previous circuit. If assistance is required, use the equations provided in figure 1 for slope (A_V) and intercept (Vb).

Calculated Circuit Characteristics:

Calculated Gain, Av = _____.

Calculated Output Offset, Vb = _____.

Step 3 - Calculations: Output Voltage Span
- Assuming an input of 1V in the previous circuit, compute the minimum output voltage. Then, using an input of 3V, determine the maximum output voltage.

Calculated Output Voltages:

Vout @ 1V of Vin = _____.

Vout @ 3V of Vin = _____.

Step 4 - Calculations: Resistor Voltages
- Using the non-inverting amplifier circuit analysis procedure described in chapter 4 of the text, determine the resistor voltages across Rf and Rin at an input of 1 volt. Document the calculated voltages following.

Calculated Rf and Rin Voltages:

Calculated Rf voltage = _____

Calculated Rin voltage = _____

Again, using the non-inverting amplifier circuit analysis procedure described in chapter 4 of the text determine the resistor voltages *across* Rf and Rin at an input of 3 volts.

Calculated Rf voltage = _____

Calculated Rin voltage = _____

Step 5 - Measurements
- Set the amplifier input voltage on the wiper of the potentiometer to 1 volt. Measure and record the actual resistor and output voltages following.

Actual Rf voltage @ 1V = _____

Actual Rin voltage @ 1V = _____

Actual Output Voltage @ 1V, Vout = _____

Step 6 - Measurements
- Set the amplifier input voltage on the wiper of the potentiometer to 3 volts. Measure and record the actual resistor and output voltages following.

Actual Rf voltage @ 3V = _____

Actual Rin voltage @ 3V = _____

Actual Output Voltage @ 3V, Vout = _____

Step 7 - Spreadsheet Calculations
- Acquire a number of random but corresponding input and output voltage measurements between the input voltage extremes of 1V and 3V. Enter the values into Excel and using the slope and intercept mathematical functions within Excel determine the actual gain (slope) and offset (intercept) values for the amplifier. Enter the values following.

Actual Amplifier Characteristics

Actual Voltage Gain, Av = _____

Actual Output Offset, Vb = _____

Step 8
- Determine the percentage of error between the calculated gain in step 2 and the actual gain in step 7. The percentage of error value should be within the resistor tolerance (typically ±5%) and is used to determine if the results of the lab experiment are acceptable. The percentage of error equation follows.

% Error = [(Calculated - Measured) ÷ Calculated] * 100%

Error Calculations:

Gain Error % = _____.

Output Offset Error % = _____.

Results and Conclusions
Complete the lab by assembling a brief conclusion for the experiment. Include any data or procedural discrepancies, technical problems or malfunctions, inconsistencies, suggestions for improvements and the following questions in your conclusion statement.

Did the amplifier exhibit a biased non-inverting (direct-acting) function?
Were the calculated and measured gain values comparable?
Were the calculated and measured offset values comparable?
Were the calculated and measured resistor voltages comparable?

Do not disassemble this amplifier circuit. The following lab (#8) can be performed with a connection modification of this circuit.

This page retained blank intentionally.

Objective:
The objective of this exercise is to introduce the operational amplifier as a signal scaling circuit. This lab will investigate the biased inverting amplifier, circuit analysis, voltage gain and output offset.

Introduction:
The biased inverting (or reverse-acting) amplifier provides an output that follows the input in an indirect, or out-of-phase manner. As the input voltage increases, the output voltage decreases. In addition, the output is biased, or offset from zero volts when the input is zero volts.

All amplifiers are designed to operate over a span of input and output values. Input voltage spans are obtained from sensors, receivers, other amplifiers and any circuit requiring signal conditioning of a mathematical nature. The dynamic relationship between the input and output voltage spans is called Voltage Gain, and is given the symbol, Av. The output offset voltage is given the symbol "V_b" and corresponds to the Y-axis intercept value in the Y=mX+b equation. In order to accomplish output offset, a constant DC bias voltage will be introduced to the op amp input that is normally grounded.

The biased inverting amplifier schematic and corresponding input/output characteristic plot follows in figure 1. Note the slope of the input/output plot is the amplifier voltage gain, Av, and the output offset is designated "Vb" in the characteristic diagram.

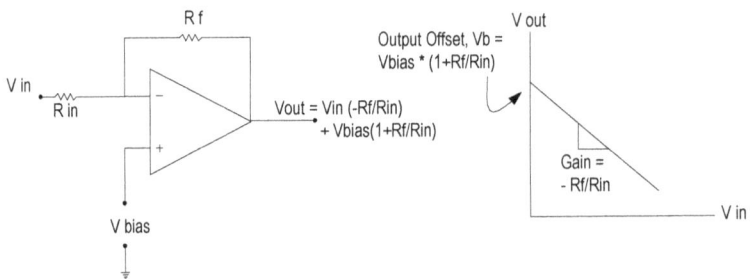

Figure 1 - A biased inverting, or reverse-acting, amplifier. Gain (slope) and offset (intercept) are indicated on the amplifier characteristic plot.

In order to differentiate between the input (signal) voltage and the bias voltage, the input signal will be a span of values such as 1V to 5V, or -1V to +1V, and the bias voltage will be a constant value. Bias voltages are usually acquired from a DC supply or resistive network connected to a supply. A more complete diagram of a biased inverting amplifier follows.

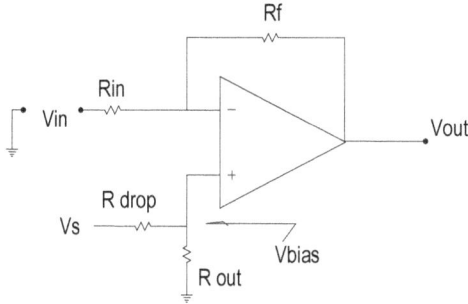

Figure 2 - Vbias is usually applied using a voltage divider connected to an available supply.

The biasing voltage circuit at the non-inverting input in figure 2 is composed of two resistors, Rdrop and Rout. (In the future these designations will be replaced with R1 and R2 or Ra and Rb.) Rout is responsible for providing the necessary bias voltage to the op amp and Rdrop will "drop" the difference between the supply voltage and the required bias voltage.

The resulting output voltage can be determined by the equation provided in figure 1 and following, and is the result of summing the individual inputs multiplied by their respective gains. Vbias is determined using Vs and the voltage divider equation provided following.

$$Vout = Vin\ (-Rf/Rin) + Vbias\ (1+Rf/Rin)$$
$$Vbias = Vs\ [Rout/(Rdrop+Rout)]$$

Lab Procedure
Step 1 - Assembly
 - Assemble the following biased non-inverting amplifier. After connecting the circuit measure each supply voltage and record the results below. This is done to assure that neither of the supplies is over-loaded, resulting in a substantial voltage difference between the two supplies. Excessively mismatched supplies ($\geq$ 2 Volts) will result in an output voltage span with significant error. If an excessive difference exists measure each battery unloaded (removed from the circuit) as one of the batteries may have become weak. If both measure about equal, check the circuit connections.

As with the previous lab, a 1K pot will provide the input signal voltage of 1-3 Volts.

Battery Voltages:

+Vs = _____.

-Vs = _____.

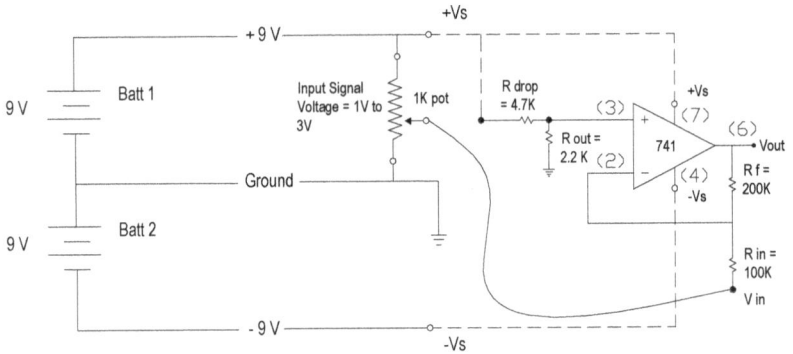

Step 2 - Calculations: Gain and Offset

- Determine the input signal voltage gain (A_V) and output offset (Vb) values from the previous circuit. If assistance is required, use the equations provided in figure 1 for slope (A_V) and intercept (Vb).

Calculated Circuit Characteristics:

Calculated Gain, Av = _____ .

Calculated Output Offset, Vb = _____ .

Step 3 - Calculations: Output Voltage Span

- Assuming an input of 1V in the previous circuit, compute the corresponding output voltage. Then, using an input of 3V, determine the minimum output voltage.

Calculated Output Voltages:

Vout @ 1V of Vin = _____ .

Vout @ 3V of Vin = _____ .

Step 4 - Calculations: Resistor Voltages

- Using the inverting amplifier circuit analysis procedure described in chapter 4 of the text, determine the resistor voltages across Rf and Rin at an input of 1 volt. Document the calculated voltages following.

Calculated Rf and Rin Voltages:

Calculated Rf voltage = _____

Calculated Rin voltage = _____

Again, using the inverting amplifier circuit analysis procedure described in chapter 4 of the text determine the resistor voltages across Rf and Rin at an input of 3 volts.

Calculated Rf voltage = _____

Calculated Rin voltage = _____

Step 5 - Measurements
- Set the amplifier input voltage on the wiper of the potentiometer to 1 volt. Measure and record the actual resistor and output voltages following.

Actual Rf voltage @ 1V = _____

Actual Rin voltage @ 1V = _____

Actual Output Voltage @ 1V, Vout = _____

Step 6 - Measurements
- Set the amplifier input voltage on the wiper of the potentiometer to 3 volts. Measure and record the actual resistor and output voltages following.

Actual Rf voltage @ 3V = _____

Actual Rin voltage @ 3V = _____

Actual Output Voltage @ 3V, Vout = _____

Step 7 - Spreadsheet Calculations
- Acquire a number of random but corresponding input and output voltage measurements between the input voltage extremes of 1V and 3V. Enter the values into Excel and using the mathematical functions within Excel determine the actual gain (slope) and offset (intercept) values for the amplifier. Enter the values following.

Actual Amplifier Characteristics

Actual Voltage Gain, Av = _____

Actual Output Offset, Vb = _____

Step 8
- Determine the percentage of error between the calculated gain in step 2 and the actual gain in step 7. The percentage of error value should be within the resistor tolerance (typically ±5%) and is used to determine if the results of the lab experiment are acceptable. The percentage of error equation follows.

% Error = [(Calculated - Measured) ÷ Calculated] * 100%

Error Calculations:

Gain Error % = _____.

Output Offset Error % = _____.

Results and Conclusions
Complete the lab by assembling a brief conclusion for the experiment. Include any data or procedural discrepancies, technical problems or malfunctions, inconsistencies, suggestions for improvements and the following questions in your conclusion statement.

Did the amplifier exhibit a biased inverting (reverse-acting) function?
Were the calculated and measured gain values comparable?
Were the calculated and measured offset values comparable?
Were the calculated and measured resistor voltages comparable?

This page retained blank intentionally.

ENGT 215 - Sensors and Analog I/O
Laboratory Experiment #9 - Biased Non- Inverting Amplifier Design

Objective:
The objective of this exercise is to introduce design of biased non-inverting operational amplifiers. This lab will investigate design of a biased non-inverting amplifier and it's ability to perform direct-acting Y=MX+B mathematical operations.

Introduction:
Biased operational amplifiers utilize a constant DC voltage to *offset* the output voltage from zero volts when the input is zero volts. The effect of adding the bias voltage is called *output offset* and is represented by the y-axis intercept "b" quantity in the Y=mX+b equation. Many sensor-based circuits and systems utilize a "Live" zero where the output is a non-zero quantity when the input is zero. Common examples are found in the standard signal spans found in the industrial instrumentation and process control fields. Signals are commonly transmitted between control loop components as 4 to 20 milliamps or 3 to 15 psi as a process varies between 0% and 100%. Live zero spans are beneficial in determining when a signal to a control loop component is lost as the transmitted signal will become zero should a signal loss occur.

Another example of output offset is when a signal-conditioning amplifier is designed to support a sensor whose output is not zero at an input condition of zero, such as most temperature sensors. Temperature sensing circuits commonly output a non-zero resistance, voltage or current when the applied temperature is 0° Centigrade. In order to condition a signal from a sensor of this type, the output voltage is set to zero using a bias voltage since the input from the sensor will be a non-zero quantity.

Independent of the application and focusing solely on amplifier output characteristics, three types of amplifier signal spans exist. A *true zero* output span assumes an output voltage of zero volts when the input is zero volts and output offsetting bias voltages are not required for this type of amplifier. A *suppressed zero* voltage span requires a biased amplifier to generate an output voltage that contains all positive voltage values where zero is suppressed below the output voltage span. An *elevated zero* output voltage span either contains zero within the range of output values or consists of entirely negative output voltage values. The three types of spans are shown following.

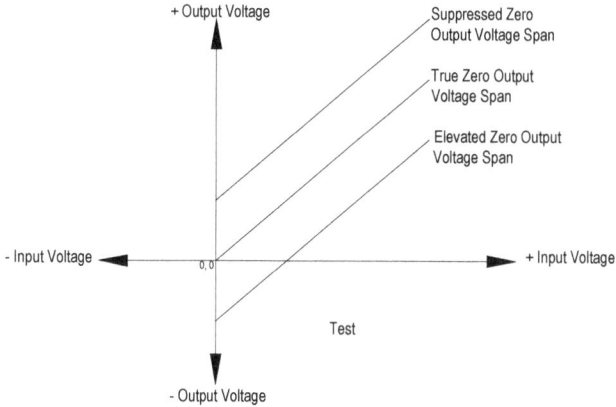

Figure 1 - Suppressed, True and Elevated zero output voltage spans. Suppressed and Elevated spans require a bias voltage to establish an output offset.

Biased amplifier design is a straightforward and mechanical process if one understands the purpose and operation of the biased operational amplifier. In fact, biased amp design is understood to be less demanding than the analysis of the same circuit. If amplifier operation is well understood, the design process can be reduced to a few rapid steps. If amplifier operation is not well understood, it may be best to review the text material and previous lab results until becoming familiar with the application and circuit analysis of the biased operational amplifier.

Biased Amplifier Design Steps:
1 - Determine the specific values of the input and output voltage spans, and graph the resulting input/output characteristic.

2 - From the previous input and output voltage spans determine the gain (slope) and offset (intercept) values. The amplifier type can now be determined from observation of the slope of the input/output characteristic. A positive slope will require a non-inverting amplifier. A negative slope will require an inverting amp.

Biased Non-Inverting Amplifier

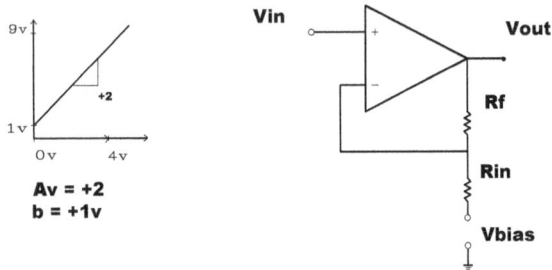

Av = +2
b = +1v

Figure 2 - A biased non-inverting amp and corresponding input/output characteristic plot.

3 - Establish the gain determining resistances by assuming an input resistance (Rin) value of 100K Ohms. Using the appropriate amplifier gain equation (1+Rf/Rin for non-inverting and -Rf/Rin for inverting), rearrange the gain equation and solve for the required value of Rf.

4 - Assemble a circuit diagram of the amplifier including values for Rin and Rf. Determine and label the locations for Vin and Vbias. Looking into the amplifier from the location of Vbias, determine the voltage gain "seen" by Vbias. For a non-inverting amp (A_{Vin}), Vbias will see an inverting gain (-Rf/Rin). For an inverting amp (A_{Vin}), Vbias will see a non-inverting gain. The gain seen by Vbias is termed A_{Vbias}.

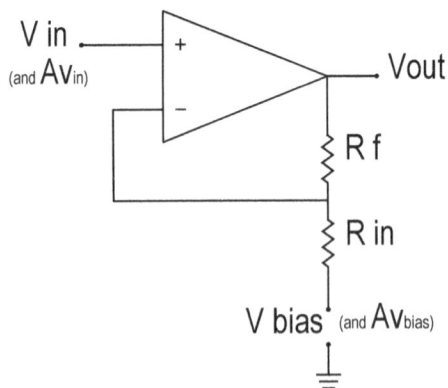

Figure 3 - A biased non-inverting amplifier. Vin "sees" the signal voltage gain (Avin), and Vbias "sees" the bias voltage gain (Avbias).

5 - Once A_{Vbias} is determined, Vbias can be determined by dividing the output-offset voltage (intercept) found in step 2 by A_{Vbias}. The resulting quantity will be the amount of bias voltage required to offset the output by the desired amount.

6 - Assemble a voltage divider, potentiometer or other circuit to provide the correct amount of Vbias.

Lab Procedure
Step 1
- Following each of the previous steps, design an amplifier to convert an input voltage span of 1V to 3V into an out put voltage span of 1V to 7V. Document the computed values following.

Av = _____

b = _____

Rin = _____100K_____

Rf = _____

Av bias = _____

Vbias = _____

Step 2
- Place a 1K potentiometer across the positive supply and ground. The potentiometer output will serve as the amplifier signal voltage. The potentiometer circuit should be connected as follows. Remember that the center connection on any potentiometer is the wiper. Small circles on figure 4 designate all three potentiometer connection points.

- Once the potentiometer is connected rotate the potentiometer and measure the available the voltage from the wiper to ground. The approximate voltage should vary between zero and nine volts. If exactly zero and nine volts are not available do not be concerned, as the following lab will require a voltage of 1 to 3 volts from the potentiometer.

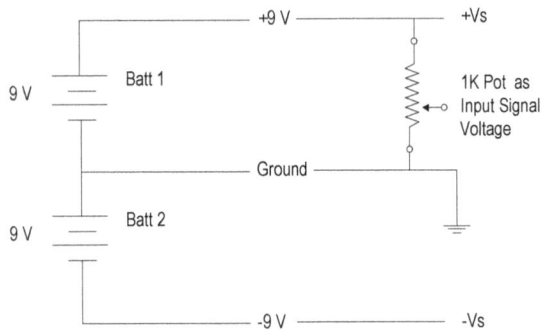

Figure 4 - Connecting a potentiometer as a variable (positive) voltage source. The voltage from the wiper to ground will serve as the amplifier input signal.

Step 3

- An additional circuit is required to accommodate Vbias. A simple voltage divider as Rout and Rdrop in the following circuit will work well. Rdrop will drop the difference between the supply voltage and the desired Vbias. Rout is responsible for establishing Vbias at the inverting input of the amplifier. Once Vbias is determined, the voltage across Rout should equal Vbias and the voltage across Rdrop will equal the difference between Vbias and the supply. Once the voltages across Rdrop and Rout are known, the resistor values can be assigned according to the voltages. As an example, if Rdrop is to drop 5 volts and Rout requires 4V, size Rdrop at 5K and Rout at 4K.

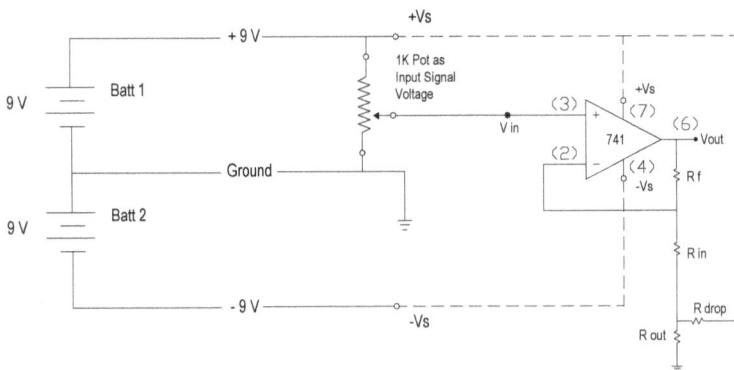

Figure 5 - Bipolar supplies with amplifier and biasing resistor connections. Note: In this circuit a positive Vbias is required. A negative Vbias would require a connection to the negative supply.

Step 4
- Once the circuit is assembled, it should be tested for proper operation. An appropriate test is to apply the minimum and maximum input voltage values and check for the correct minimum and maximum output values.

- When proper operation is confirmed, apply numerous input voltage values between 1V and 3V and document the corresponding output voltage values. Document the values of the resulting input/output voltages in Excel or other spreadsheet and determine the resulting gain (slope) and offset (intercept) values by utilizing the mathematical functions within the spreadsheet software. Document the resulting gain and offset following.

Actual Gain (slope from Excel) = _____

Actual Offset (intercept from Excel) = _____

Step 5
- Determine the percentage of error between the desired gain and offset and the actual gain in the preceding step. The percentage of error value should be within the resistor tolerance (typically ±5%). Percentage of error computations is used to determine if the results of the lab experiment are acceptable. The percentage of error equation follows.

% Error = [(Calculated - Measured) ÷ Calculated] * 100%

% Gain Error (slope) = _____

% Offset Error (intercept) = _____

Note: Do not attempt to reconstruct or "fine-tune" your lab results, even if you find major errors and/or discrepancies. Submitting the actual results will assist in improving the lab experiments.

Results and Conclusions
Complete the lab by assembling a brief conclusion for the experiment. Include any data or procedural discrepancies, technical problems or malfunctions, inconsistencies, suggestions for improvements and the following questions in your conclusion statement.

Did the amplifier exhibit the desired biased non-inverting function?
Were the computed and actual gain and offset values comparable?

ENGT 215 - Sensors and Analog I/O
  Laboratory Experiment #10 - Biased Inverting Amplifier Design

Objective:
The objective of this exercise is to introduce design of biased inverting
operational amplifiers. This lab will investigate design of a biased inverting
amplifier and it's ability to perform reverse-acting Y=MX+B mathematical
operations.

Introduction:
Biased operational amplifiers utilize a constant DC voltage to *offset* the output
voltage from zero volts when the input is zero volts. The effect of adding the bias
voltage is called *output offset* and is represented by the y-axis intercept "b"
quantity in the Y=mX+b equation. Many sensor-based circuits and systems utilize
a "Live" zero where the output is a non-zero quantity when the input is zero.
Common examples are found in the standard signal spans found in the industrial
instrumentation and process control fields. Signals are commonly transmitted
between control loop components as 4 to 20 milliamps or 3 to 15 psi as a
process varies between 0% and 100%. Live zero spans are beneficial in
determining when a signal to a control loop component is lost as the transmitted
signal will become zero should a signal loss occur.

Another example of output offset is when a signal-conditioning amplifier is
designed to support a sensor whose output is not zero at an input condition of
zero, such as most temperature sensors. Temperature sensing circuits
commonly output a non-zero resistance, voltage or current when the applied
temperature is 0° Centigrade. In order to condition a signal from a sensor of this
type, the output voltage is set to zero using a bias voltage since the input from
the sensor will be a non-zero quantity.

Independent of the application and focusing solely on amplifier output
characteristics, three types of amplifier signal spans exist. A *true zero* output
span assumes an output voltage of zero volts when the input is zero volts and
output offsetting bias voltages are not required for this type of amplifier. A
suppressed zero voltage span requires a biased amplifier to generate an output
voltage that contains all positive voltage values where zero is suppressed below
the output voltage span. An *elevated zero* output voltage span either contains
zero within the range of output values or consists of entirely negative output
voltage values. The three types of spans are shown following.

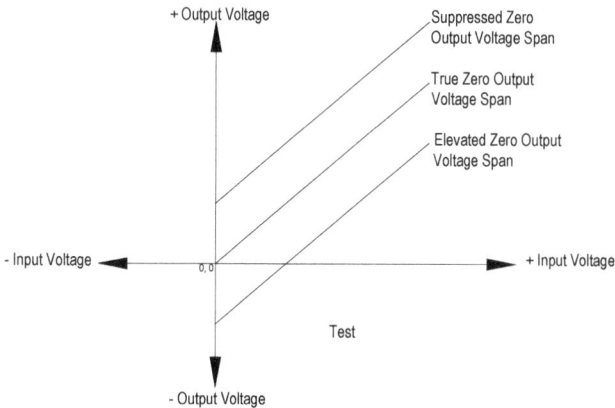

Figure 1 - Suppressed, True and Elevated zero output voltage spans. Suppressed and Elevated spans require a bias voltage to establish an output offset. Note - the positive slope is representative of a non-inverting amplifier. Biased inverting amplifiers produce an input/output characteristic with a negative slope.

Biased amplifier design is a straightforward and mechanical process if one understands the purpose and operation of the biased operational amplifier. In fact, biased amp design is understood to be less demanding than the analysis of the same circuit. If amplifier operation is well understood, the design process can be reduced to a few rapid steps. If amplifier operation is not well understood, it may be best to review the text material and previous lab results until becoming familiar with the application and circuit analysis of the biased operational amplifier.

Biased Amplifier Design Steps:
1 - Determine the specific values of the input and output voltage spans, and graph the resulting input/output characteristic.

2 - From the previous input and output voltage spans determine the gain (slope) and offset (intercept) values. The amplifier type can now be determined from observation of the slope of the input/output characteristic. A positive slope will require a non-inverting amplifier. A negative slope will require an inverting amp.

BIASED INVERTING AMPLIFIER

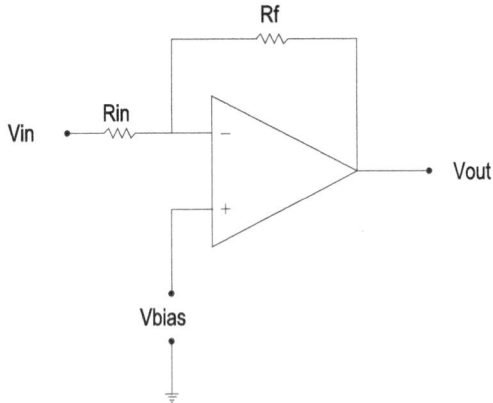

Rf

Rin

Vin

Vout

Vbias

$$Vout = Vin\ (-Rf/Rin) + Vbias\ (1+Rf/Rin)$$

Figure 2 - A biased inverting amp and corresponding input/output characteristic plot.

3 - Establish the gain determining resistances by assuming an input resistance (Rin) value of 100K Ohms. Using the appropriate amplifier gain equation (1+Rf/Rin for non-inverting and -Rf/Rin for inverting), rearrange the gain equation and solve for the required value of Rf.

4 - Assemble a circuit diagram of the amplifier including values for Rin and Rf. Determine and label the locations for Vin and Vbias. Looking into the amplifier from the location of Vbias, determine the voltage gain "seen" by Vbias. For a non-inverting amp (A_{Vin}), Vbias will see an inverting gain (-Rf/Rin). For an inverting amp (A_{Vin}), Vbias will see a non-inverting gain. The gain seen by Vbias is termed A_{Vbias}.

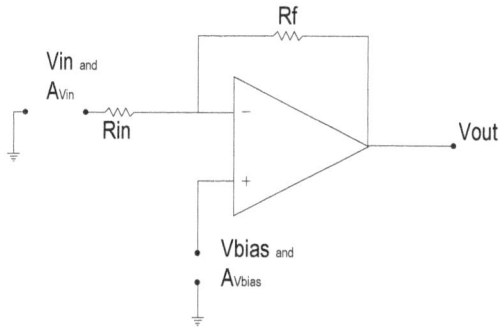

Figure 3 - A biased inverting amplifier. Vin "sees" the signal voltage gain (Avin), and Vbias "sees" the bias voltage gain (Avbias).

5 - Once A_{Vbias} is determined, Vbias can be determined by dividing the output-offset voltage (intercept) found in step 2 by A_{Vbias}. The resulting quantity will be the amount of bias voltage required to offset the output by the desired amount.

6 - Assemble a voltage divider, potentiometer or other circuit to provide the correct amount of Vbias.

Lab Procedure
Step 1
 - Following each of the previous steps, design an amplifier to convert an input voltage span of 1V to 3V into an out put voltage span of 7V to 1V. Document the computed values following.

$$Av = \underline{\hspace{4cm}}$$

$$b = \underline{\hspace{4cm}}$$

$$Rin = \underline{\hspace{1.5cm}100K\hspace{1.5cm}}$$

$$Rf = \underline{\hspace{4cm}}$$

$$Av\ bias = \underline{\hspace{4cm}}$$

$$Vbias = \underline{\hspace{4cm}}$$

Step 2
- Place a 1K potentiometer across the positive supply and ground. The potentiometer output will serve as the amplifier signal voltage. The potentiometer

circuit should be connected as follows. Remember that the center connection on any potentiometer is the wiper. Small circles on figure 4 designate all three potentiometer connection points.

- Once the potentiometer is connected rotate the potentiometer and measure the available the voltage from the wiper to ground. The approximate voltage should vary between zero and nine volts. If exactly zero and nine volts are not available do not be concerned, as the following lab will require a voltage of 1 to 3 volts from the potentiometer.

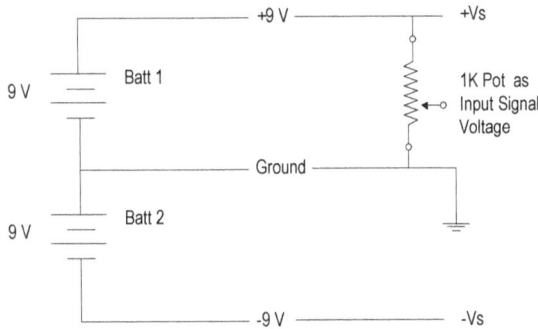

Figure 4 - Connecting a potentiometer as a variable (positive) voltage source. The voltage from the wiper to ground will serve as the amplifier input signal.

Step 3

- An additional circuit is required to accommodate Vbias. A simple voltage divider as Rout and Rdrop in the following circuit will work well. Rdrop will drop the difference between the supply voltage and the desired Vbias. Rout is responsible for establishing Vbias at the inverting input of the amplifier. Once Vbias is determined, the voltage across Rout should equal Vbias and the voltage across Rdrop will equal the difference between Vbias and the supply. Once the voltages across Rdrop and Rout are known, the resistor values can be assigned according to the voltages. As an example, if Rdrop is to drop 5 volts and Rout requires 4V, size Rdrop at 5K and Rout at 4K.

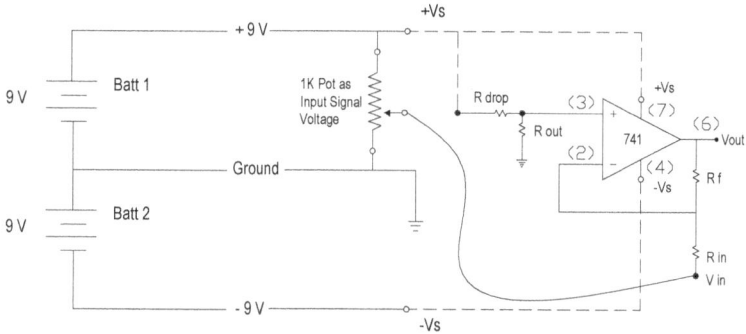

Figure 5 - Bipolar supplies with amplifier and biasing resistor connections. Note: In this circuit a positive Vbias is required. A negative Vbias would require a connection to the negative supply.

Step 4
- Once the circuit is assembled, it should be tested for proper operation. An appropriate test is to apply the minimum and maximum input voltage values and check for the correct minimum and maximum output values.

- When proper operation is confirmed, apply numerous input voltage values between 1V and 3V and document the corresponding output voltage values. Document the values of the resulting input/output voltages in Excel or other spreadsheet and determine the resulting gain (slope) and offset (intercept) values by utilizing the mathematical functions within the spreadsheet software. Document the resulting gain and offset following.

Actual Gain (slope from Excel) = _____

Actual Offset (intercept from Excel) = _____

Step 5
- Determine the percentage of error between the desired gain and offset and the actual gain in the preceding step. The percentage of error value should be within the resistor tolerance (typically ±5%). Percentage of error computations is used to determine if the results of the lab experiment are acceptable. The percentage of error equation follows.

% Error = [(Calculated - Measured) ÷ Calculated] * 100%

% Gain Error (slope) = _____

% Offset Error (intercept) = _____

Note: Do not attempt to reconstruct or "fine-tune" your lab results, even if you find major errors and/or discrepancies. Submitting the actual results will assist in improving the lab experiments.

Results and Conclusions
Complete the lab by assembling a brief conclusion for the experiment. Include any data or procedural discrepancies, technical problems or malfunctions, inconsistencies, suggestions for improvements and the following questions in your conclusion statement.

Did the amplifier exhibit the desired biased inverting function?
Were the computed and actual gain and offset values comparable?

This page retained intentionally blank.

ENGT 215 - Sensors and Analog I/O
Laboratory Experiment #11 - Biased Non- Inverting Amplifier Design with
Variable Gain and Variable Offset

Objective:
The objective of this exercise is to introduce design of biased non-inverting
operational amplifiers with variable gain and variable offset. This lab will
investigate design of a biased non-inverting amplifier and the procedure for
variable resistance value determination when variable gain and variable offset
adjustments are required.

Introduction:
Operational amplifier circuits are frequently used with signals where the voltage
span available to the amplifier input operates within a tolerance. A good example
is the Sensym SX series of pressure sensors. The SX30 generates a typical
output voltage span of −20 to +90 millivolts when operated from a 5-volt supply.
However the SX30's output signal varies *ratio metrically* to supply voltage
variations. (This is referred to as *ratiometricity* and represents the ability of a
circuit to vary its operating characteristics in a proportional manner with the
supply voltage. Ratiometricity is always associated with resistive bridge circuits.)

In addition to ratiometricity, the SX series specifications contain columns of
minimum, typical and maximum offset and gain (sensitivity in the SX
specifications, the SX series of pressure sensor specifications are available at
http://www.pressure.invensys.com/pdf/sx.pdf). Such variations in sensitivity
(gain) and offset characteristics will translate into widely varying output voltage
spans. Not surprising considering the SX series of sensor contains
semiconductor strain gauges as the active sensing components and sells for
about $20 each. Inexpensive sensors commonly require calibration to firmly
establish the specific input/output values and the resulting error or deviation from
norm.

The SX series of pressure sensor is rugged, repeatable and linear but it is not
very replaceable. Should one require replacing, the new sensor will require a
complete re-calibration of the associated signal-conditioning amplifier since the
SX's specifications vary widely from unit to unit. For this reason it is often
desirable to make the gain and offset of signal conditioning amplifiers variable −
to compensate for variations in the input signal due to variations in the signal
source circuit's operating specifications.

Designing amplifiers with variable gain and variable offset is not difficult if one
considers the inverting and non-inverting circuits and their corresponding gain
and output offset circuitry.

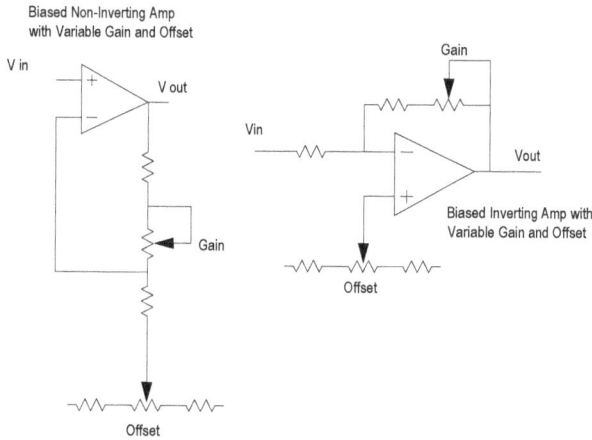

Figure 1 – Biased Inverting (right) and Non-inverting (left) amplifiers with variable gain and variable offset circuitry.

As can be seen in the previous diagram, additional resistors and potentiometers will need to be added to the biased amplifier circuits studied in previous labs. The first step is to determine the extent of the variation for the gain and offset circuits. This will depend upon the extent of the variation of the input signal. The overall design of the amplifier is the same with the exception of adding plus and minus values to the computed gain and offset. The following example demonstrates.

Non-inverting Amplifier Example -
Assume an input span (defined as the difference in input signal end points) is expected to vary by plus or minus 10%. This means an input signal span of 2 volts can vary by plus or minus .2 volts, from 1.8 to 2.2 volts. A nominal signal span of 1V to 3V could become 1V to 2.8V or 1V to 3.2V assuming the input starting value of 1V can be assured. In most applications this is not the case however, and compensation for variations in the minimum input value (input signal offset) must also be included. Likewise, the amplifier gain must accommodate input signal span variations and the amplifier bias circuit should accommodate input signal offset variations. In addition, it is suggested the amplifier gain and bias variation be made slightly larger than the expected signal variations. In this case, amplifier gain and offset bias variations of plus and minus 25% should adequately cover the anticipated ±10% input signal variations. The additional variation will also allow for easier determination of all amplifier resistance values.

Assume an amplifier is to convert the previous input signal of 1-3 volts into an output signal of 1-5 volts. Further assume the amplifier gain and offset is to be made variable by plus and minus 25%.

Amplifier Gain Variation
The ideal amplifier gain, assuming the input is exactly 1-3 volts, becomes,

$$\text{Ideal Amplifier Voltage Gain, } Av = \Delta Vout / \Delta Vin = 4V / 2V = +2$$

The amplifier gain with ±25% variation becomes,
$$\Delta Av = 1.5 \text{ to } 2.5$$

Given the non-inverting amplifier gain equation is 1+ Rf/Rin, and assuming Rin is 100K Ohms, the Rf variation for Av = 2 ± 25% computes to,

$$\text{Rf at minimum gain of } 1.5 = 50K$$
$$\text{Rf at maximum gain of } 2.5 = 150K$$

The feedback resistance will need to vary from 50K to 150K Ohms. This will require a fixed resistance of 50K and a variable resistance of 100K.

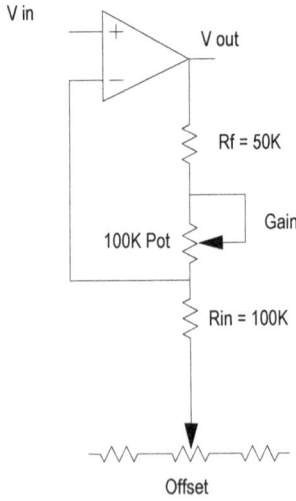

Figure 2 - Non-inverting amplifier with variable gain components.

In this case the gain determining resistances computed to commonly available standard resistance values. This may not always happen. Since potentiometer values are usually available in multiples of 1, 2 and 5 (such as 1K, 2K, 5K, 10K,

20K, 50K, 100K, 200K, 500K, etc.), the values of Rin and Rf minimum (the fixed feedback resistor) may need to be scaled to accommodate the available potentiometer values. Should this be the case, determine the gain resistances as in the previous example and scale (multiply or divide) the resulting potentiometer computation to the available potentiometer value. Then scale Rin and Rf minimum by the same factor. As an example, assume the previous potentiometer computation yielded a value of 75K and the available potentiometer value was 100K. Rin and Rf minimum will need to be increased by the ratio of the available potentiometer to the computed value, or 1.333. Scaling the fixed resistor values as described previously may result in unavailable fixed resistor values. Use the closest available values. This should result in acceptable amplifier operation since the percentage of variation was made larger than required.

Amplifier Offset Variation
The ideal amplifier output offset, assuming the input offset is exactly 1 volt, becomes,

$$\text{Ideal Amplifier Output Offset Voltage, } b = Vout - (Av * Vin) = 5V - (+2 * 3V)$$
$$b = -1V$$

The amplifier output offset with ±25% variation becomes,
$$\Delta b = -.75V \text{ to } -1.25V$$

Determining the required Vbias variation to accommodate the desired output offset variation is complicated by A_{Vbias} now that the gain has been made variable. In order to determine A_{Vbias} assume the gain potentiometer is centered resulting in the value of ideal gain when viewed from the input. With the gain potentiometer adjusted to center in the previous diagram, the potentiometer will exhibit 50K of resistance resulting in a total feedback resistance of 100K. The bias voltage gain, A_{Vbias}, becomes,

$$A_{Vbias} = -Rf / Rin = -100K / 100K = -1 \text{ with gain pot centered}$$

To determine the bias voltage variation given $\Delta b = -.75V$ to $-1.25V$ and $A_{Vbias} = -1$, the bias voltage variation becomes,

$$\Delta Vbias = \Delta b / A_{Vbias} = -.75V \text{ to } -1.25V / -1$$

$$\Delta Vbias = +.75V \text{ to } +1.25V$$

The voltage on the wiper of the offset potentiometer needs to be .75V to 1.25V. Determining the resulting resistance values is identical to the procedure provided previously for fixed bias amplifier circuits, the resistance values are sized according to the desired voltage drop. Assuming available supplies of plus and minus 9V, the plus 9V supply will be used to provide the output offset bias voltage. The resulting voltage drops (circled) and subsequent resistance values are indicated in the following diagrams.

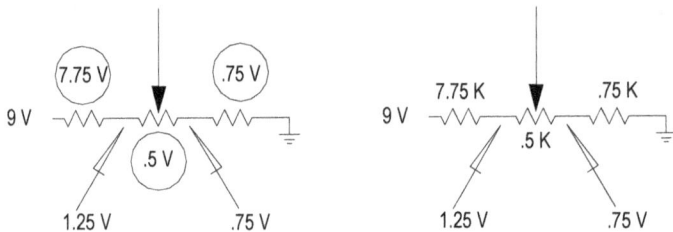

Figure 3 – Distribution of voltages around the offset biasing circuit and the resulting resistance values.

As mentioned previously the potentiometer resistance may not always compute to commonly available values, scaling the fixed resistances around the available potentiometer may be required. The procedure described previously can be applied here as well. A word of caution – the resistance values used to bias the amplifier also influence the gain determining resistor (Rin) on the non-inverting amp. Bias resistance values greater than 1's of K's will cause Rin to appear larger and will result in a decrease in the expected gain. As a rule of thumb use gain resistors in the 100's of Kilohms range and biasing resistors of 1's of Kilohms.

The resulting amplifier appears as follows.

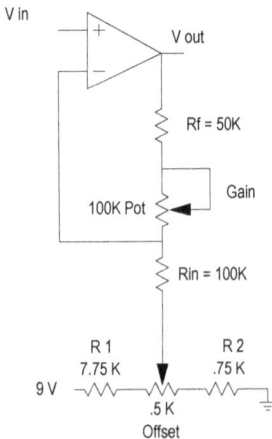

Figure 4 – The completed non-inverting amplifier with variable gain and offset.

Amplifier Calibration

The gain and offset potentiometers are set in a manner consistent with the parameters they represent on an input/output characteristic plot. The offset adjustment will determine the Y-intercept value of the amplifier's input/output graph and the gain will determine the slope of the plot. Specifically, with the input at the minimum value, the amplifier offset should be adjusted until the desired minimum output voltage is acquired. This may need to be performed again after setting the gain pot since both pots will be at random values after the amplifier is assembled. The gain should be set with the input at the maximum value. Rotate the gain potentiometer to the required corresponding output voltage. After setting the gain pot check the output at the minimum input and fine-tune if necessary.

Lab Procedure

Design and assemble an amplifier to convert an input voltage span of 1V to 3V into an output voltage span of –5 to +5 volts. Assume gain and bias variations of ±25%. The following circuit can be used as a template for the design. Submit the computed and actual resistance values for evaluation. Include a brief description of your lab results.

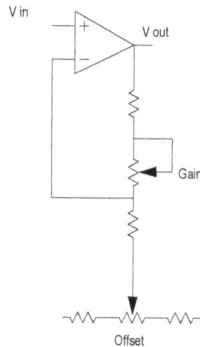

Computed resistance values, note the actually used resistance values in parenthesis –
Rin = 100K

Rf = _____

Gain Pot = _____

Offset Pot = _____

R1 = _____

R2 = _____

Bias Supply Voltage = _____

Objective:
The objective of this exercise is to introduce design of biased inverting operational amplifiers with variable gain and variable offset. This lab will investigate design of a biased non-inverting amplifier and the procedure for variable resistance value determination when variable gain and variable offset adjustments are required. A procedure for setting the adjustments is also provided.

Introduction:
Operational amplifier circuits are frequently used with signals where the voltage span available to the amplifier input operates within a tolerance. A good example is the Sensym SX series of pressure sensors. The SX30 generates a typical output voltage span of −20 to +90 millivolts when operated from a 5-volt supply. However the SX30's output signal varies *ratio metrically* to supply voltage variations. (This is referred to as *ratiometricity* and represents the ability of a circuit to vary its operating characteristics in a proportional manner with the supply voltage. Ratiometricity is always associated with resistive bridge circuits.)

In addition to ratiometricity, the SX series specifications contain columns of minimum, typical and maximum offset and gain (sensitivity in the SX specifications, the SX series of pressure sensor specifications are available at http://www.pressure.invensys.com/pdf/sx.pdf). Such variations in sensitivity (gain) and offset characteristics will translate into widely varying output voltage spans. Not surprising considering the SX series of sensor contains semiconductor strain gauges as the active sensing components and sells for about $20 each. Inexpensive sensors commonly require calibration to firmly establish the specific input/output values and the resulting error or deviation from norm.

The SX series of pressure sensor is rugged, repeatable and linear but it is not very replaceable. Should one require replacing, the new sensor will require a complete re-calibration of the associated signal-conditioning amplifier since the SX's specifications vary widely from unit to unit. For this reason it is often desirable to make the gain and offset of signal conditioning amplifiers variable – to compensate for variations in the input signal due to variations in the signal source circuit's operating specifications.

Designing amplifiers with variable gain and variable offset is not difficult if one considers the inverting and non-inverting circuits and their corresponding gain and output offset circuitry.

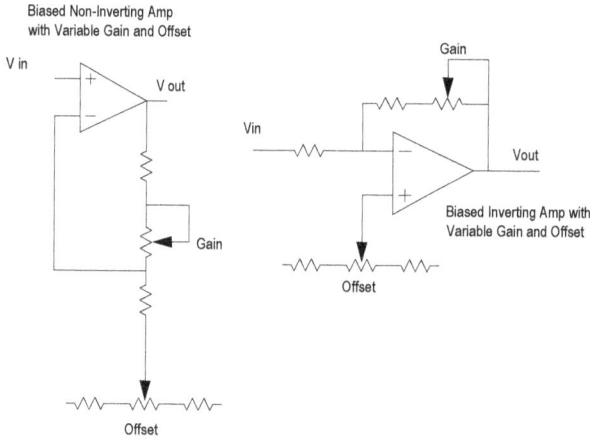

Figure 1 – Biased Inverting and Non-inverting amplifiers with variable gain and variable offset circuitry.

As can be seen in the previous diagram, additional resistors and potentiometers will need to be added to the biased amplifier circuits studied in previous labs. The first step is to determine the extent of the variation for the gain and offset circuits. This will depend upon the extent of the variation of the input signal. The overall design of the amplifier is the same with the exception of adding plus and minus values to the computed gain and offset. The following example demonstrates.

Inverting Amplifier Example -
Assume an input span (defined as the difference in input signal end points) is expected to vary by plus or minus 10%. This means an input signal span of 2 volts can vary by plus or minus .2 volts, from 1.8 to 2.2 volts. A nominal signal span of 1V to 3V could become 1V to 2.8V or 1V to 3.2V assuming the input starting value of 1V can be assured. In most applications this is not the case however, and compensation for variations in the minimum input value (input signal offset) must also be included. Likewise, the amplifier gain must accommodate input signal span variations and the amplifier bias circuit should accommodate input signal offset variations. In addition, it is suggested the amplifier gain and bias variation be made slightly larger than the expected signal variations. In this case, amplifier gain and offset bias variations of plus and minus 25% should adequately cover the anticipated ±10% input signal variations. The

additional variation will also allow for easier determination of all amplifier resistance values.

Assume an amplifier is to convert the previous input signal of 1-3 volts into a reverse-acting output signal of 5-1 volts. Further assume the amplifier gain and offset is to be made variable by plus and minus 25%.

Amplifier Gain Variation
The ideal amplifier gain, assuming the input is exactly 1-3 volts, becomes,

$$\text{Ideal Amplifier Voltage Gain, } Av = \Delta Vout / \Delta Vin = -4V / 2V = -2$$

The amplifier gain with ±25% variation becomes,
$$\Delta Av = -1.5 \text{ to } -2.5$$

Given the inverting amplifier gain equation is -Rf/Rin, and assuming Rin is 100K Ohms, the Rf variation for Av = 2 ± 25% computes to,

$$Rf \text{ at minimum gain of } -1.5 = 150K$$
$$Rf \text{ at maximum gain of } -2.5 = 250K$$

The feedback resistance will need to vary from 150K to 250K Ohms. This will require a fixed resistance of 150K and a variable resistance of 100K.

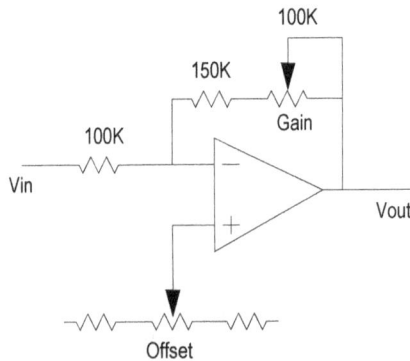

Figure 2 - Non-inverting amplifier with variable gain components.

In this case the gain determining resistances computed to commonly available standard resistance values. This may not always happen. Since potentiometer values are usually available in multiples of 1, 2 and 5 (such as 1K, 2K, 5K, 10K, 20K, 50K, 100K, 200K, 500K, etc.), the values of Rin and Rf minimum (the fixed feedback resistor) may need to be scaled to accommodate the available potentiometer values. Should this be the case, determine the gain resistances as in the previous example and scale (multiply or divide) the resulting potentiometer computation to the available potentiometer value. Then scale Rin and Rf minimum by the same factor. As an example, assume the previous potentiometer computation yielded a value of 75K and the available potentiometer value was 100K. Rin and Rf minimum will need to be increased by the ratio of the available potentiometer to the computed value, or 1.333. Scaling the fixed resistor values as described previously may result in unavailable fixed resistor values. Use the closest available values. This should result in acceptable amplifier operation since the percentage of variation was made larger than required.

Amplifier Offset Variation
The ideal amplifier output offset, assuming the input signal offset is exactly 1 volt, becomes,

$$\text{Ideal Amplifier Output Offset Voltage, b} = Vout - (Av * Vin) = 5V - (-2 * 1V)$$
$$b = +7V$$

The amplifier output offset with ±25% variation becomes,
$$\Delta b = 5.25V \text{ to } 8.75V$$

Determining the required Vbias variation to accommodate the desired output offset variation is complicated by A_{Vbias} now that the gain has been made variable. In order to determine A_{Vbias} assume the gain potentiometer is centered resulting in the value of ideal gain when viewed from the input. With the gain potentiometer adjusted to center in the previous diagram, the potentiometer will exhibit 150K of resistance resulting in a total feedback resistance of 100K. The bias voltage gain, A_{Vbias}, becomes,

$$A_{Vbias} = -Rf / Rin = 1 + 200K / 100K = +3 \text{ with gain pot centered}$$

To determine the bias voltage variation given Δb = 5.25V to 8.75V and A_{Vbias} = +3, the bias voltage variation becomes,

$$\Delta Vbias = \Delta b / A_{Vbias} = 5.25V \text{ to } 8.75V / +3$$

$$\Delta Vbias = +1.75V \text{ to } 2.92V$$

The voltage on the wiper of the offset potentiometer needs to be 1.75V to 2.92V. For reasons of convenience the maximum bias voltage will be rounded-up to 3V. This results in a bias voltage span of 1.75V to 3V. Determining the resulting resistance values is identical to the procedure provided in previous labs for fixed bias amplifier circuits, the resistance values are sized according to the desired voltage drop. Assuming available supplies of plus and minus 9V, the plus 9V supply will be used to provide the output offset bias voltage. The resulting voltage drops (circled) and subsequent resistance values are indicated in the following diagrams.

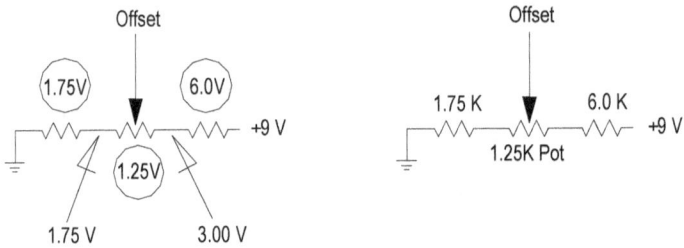

Figure 3 – Distribution of voltages around the offset biasing circuit and the resulting resistance values.

Scaling of the resistor values will be required since a 1.25K-Ohm potentiometer is not usually available. Reducing the size of the potentiometer from 1.25K to a standard and commonly available value of 1K.is a change of 20%, therefore the adjoining dropping resistors should also be decreased by 20%. This results in the biasing network of the following figure.

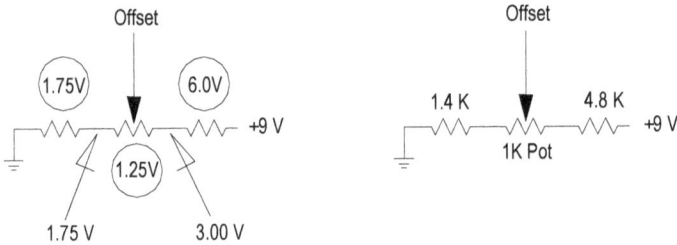

The resulting amplifier appears as follows.

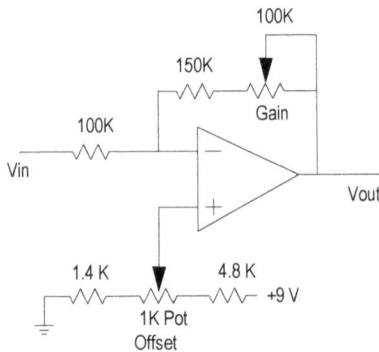

Figure 4 – The completed inverting amplifier with variable gain and offset.

Amplifier Calibration

The gain and offset potentiometers are set in a manner consistent with the parameters they represent on an input/output characteristic plot. The offset adjustment will determine the Y-intercept value of the amplifier's input/output graph and the gain will determine the slope of the plot. Specifically, with the input at the minimum value, the amplifier offset should be adjusted until the desired maximum output voltage is acquired. This may need to be performed again after setting the gain pot since both pots will be at random values after the amplifier is assembled. The gain should be set with the input at the maximum value. Rotate the gain potentiometer to the required corresponding output voltage. After setting the gain pot check the output at the minimum input and fine-tune if necessary.

Lab Procedure
Design and assemble an inverting amplifier to convert an input voltage span of 1V to 3V into a reverse-acting output voltage span of +5 to -5 volts. Assume gain and bias variations of $\pm 25\%$. The following circuit can be used as a template for the design. Submit the computed and actual resistance values used for evaluation. Include a brief description of your lab results.

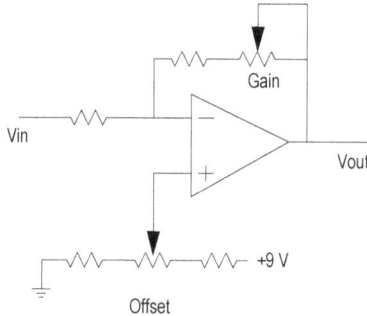

Computed resistance values, note the actually used resistance values in parenthesis –
Rin = 100K

Rf = _____

Gain Pot = _____

Offset Pot = _____

R1 = _____

R2 = _____

Bias Supply Voltage = _____

Objective:
The objective of this exercise is to introduce and test a differential amplifier for Common-Mode Rejection. In addition, output offset voltage and offset voltage compensation will be investigated.

Introduction:
The signals utilized in the previous labs have all been single-ended. Single-ended signals require only a single leadwire and a ground connection to carry information from stage to stage. Within electronic circuits and systems single-ended signals and circuits are very desirable as signal processing is easily afforded when the signal voltage is carried through a single circuit connection. However, when signals are generated outside of, or away from electronic equipment the possibility of AC noise being induced into the signal exists.

Induced AC noise can occur wherever magnetic fields exist. To induce noise into a signal line the line must be within a magnetic field and the field must be changing. These criteria are no problem for an AC magnetic field; constantly expanding/contracting AC fields are generated from motors, relays, overhead lights and all other AC current carrying loads and their associated conductors. Signal lines from low-level sensors (strain gauges, load cells, pressure sensors, etc.) are particularly sensitive to AC noise induction and need to be defended against noise and any other potential source of error since low-level sensors generate signal voltages on the order of microvolts to millivolts.

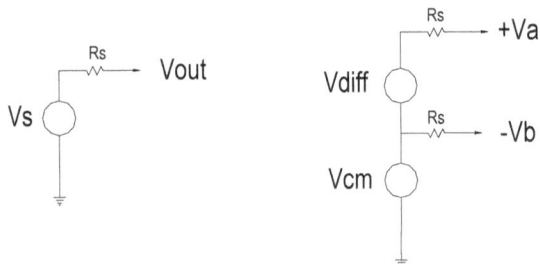

Figure 1 - Equivalent circuit diagrams of single-ended (left) and differential signal (right) sources. The "Vcm" designation represents noise or any other undesired voltage common to both output leads and is termed common-mode voltage.

Note the polarities of the differential signal source outputs in the previous diagram. Differential signals are generated out-of-phase, as one changes in a positive direction the other changes in a negative direction. The polarities represent the direction the voltages will change with an increase in signal. Similarly, an amplifier will be required to subtract the individual Va and Vb output signal voltages as the output signal is contained "between" the output lead voltages. In this respect the differential amplifier will amplify the difference between the positive and negative outputs, effectively subtracting the voltages at the two outputs.

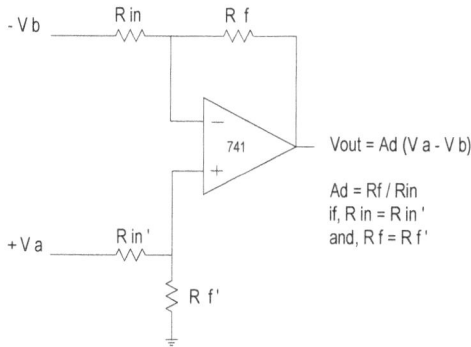

Figure 2 – A differential amplifier and associated equations. The output voltage is the amplified difference between the inputs Va and Vb. The designation Ad represents differential gain.

Induced AC noise is minimized by the use of differential signal sources and differential amplification. As an example, assume a full-bridge type sensor such as a strain gauge or load cell is applied to an experimental test fixture. Lengthy leadwires are run from the sensor to a remotely located differential amplifier. All four of the bridge leadwires are run between the sensor and amplifier in the same cable and are approximately the same length. Should the connecting cable between the sensor and differential amplifier pass through a varying magnetic field a voltage will be induced into the leadwires.

Contrary to popular belief, shielding the cable will not reduce the induced noise significantly since cable shielding attempts to minimize the radiation of magnetic fields emanating from the conductors within the cable. Any AC noise induced into the connecting leadwires will be of approximately equal amplitudes since the individual signal carrying wires are of the same size and length, and the signal leadwires are in close proximity of each other. By subtracting the voltage available at each of the bridge sensor signal leads, the differential amplifier output will eliminate the induced noise since the noise voltage will be equal on each input.

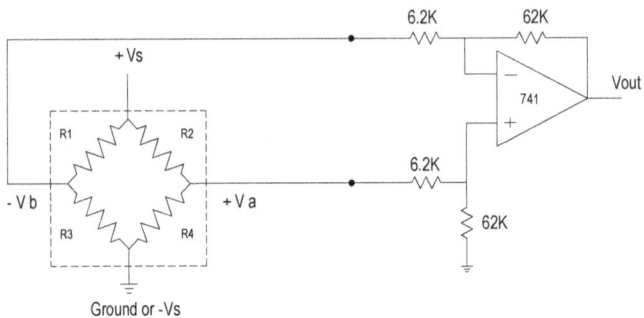

Figure 3 - A full-bridge sensor and differential amplifier. With the amplifier differential gain set to 10 (Rf/Rin), the amplifier output will be ten times the bridge output voltage (Va-Vb).

The ability of the differential amplifier to eliminate signals common to both inputs is referred to as Common-Mode Rejection, or CMR. CMRR (Common Mode Rejection Ratio) is used to quantify the ability of a differential amplifier to eliminate common-mode signals. CMRR is the ratio of differential gain to common-mode gain.

$$CMRR_{db} = 20 \log_{10} (Ad/Acm)$$

In the CMRR equation, Rf/Rin determines differential gain (Ad) and although the common-mode gain (Acm) can be approximated mathematically, best results are obtained experimentally. Decibel units of ratio are normally used to express CMRR.

Lab Procedure A
Offset Error Quantification and Compensation

Upon inspection, the Common Mode Rejection Ratio equation is actually a measurement of how well a differential amplifier can amplify in relation to how well it rejects common-mode signals. Differential gain is established by the ratio of Rf to Rin, assuming Rin = Rin' and Rf = Rf'. The resulting differential gain can be verified using the following circuit.

Before measuring the differential gain it is beneficial to determine the amount of *output offset voltage, Vos*. Vos is the amount of output voltage that exists at an input voltage of zero volts and can distort CMR test results if not quantified beforehand. Theoretically an output of zero volts should exist when an input of zero volts is applied. Unfortunately, when an op amp is used with a high gain such as 1000, internal errors within the IC become magnified and appear as a DC error voltage in the output. Once Vos is known, the amount of output variation created by an applied non-zero input can be accurately determined. Output offset voltage can be determined by connecting both inputs together at ground and measuring Vout as in the following diagram.

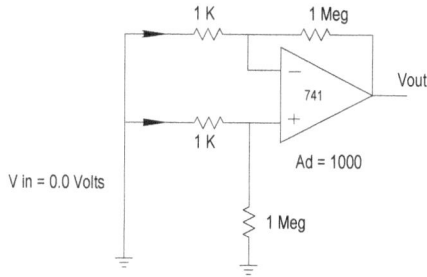

Figure 4 – Offset Voltage (Vos) test circuit. The amount of output voltage that exists at high gain with the input set to zero volts is the offset error voltage.

Output offset errors are normally ignored with low gain (<100) amplifiers and with biased amplifiers but should be considered and compensated against at gains of 100 or greater as the output error voltage grows in proportion to the amplifier gain. Output offset errors can cause AC signal clipping which results in distorted AC outputs and cause excessive errors in DC signal conditioning amplifiers. Compensation can be provided by connecting a 10K pot between pins #1 and #5 (with the wiper connected to –Vs) of a conventional 741 as in the following circuit diagram. If additional information and specifications are desired, the National Semiconductor 741 op amp datasheet can be found at http://www.national.com/ads-cgi/viewer.pl/ds/LM/LM741.pdf.

Step 1 – Vos measurement
Connect the circuit of figure 4 (don't forget the supply voltage connections, they've been omitted in the diagram) and measure the offset error at the amplifier output. Document the error following.

Vos = _____

To determine if the error is within spec, divide the measured output offset voltage (Vos) by the amplifier gain (1000) and check with the National Semiconductor spec sheet. Look for a specification titled Input Offset Voltage. Offset voltage is normally referred to the amplifier input (often abbreviated RTI on spec sheets) and represents the typical amount of voltage that would need to be applied to the input to drive the output offset error to zero. Depending upon the specific amplifier type under test (741A, 741, 741C) the Input Offset Voltage should range between 1 and 6 millivolts. Document the Input Offset Voltage results following.

Input Offset Voltage = _____

Step 2 – Vos Compensation

If desired, install the Vos compensation pot (also called Offset Null compensation) as pictured in figure 5, and set the output to zero volts. This will be a touchy adjustment since the amplifier gain is large. If a null (zero) setting is unachievable, remove the compensation pot, as an output of exactly 0.000 volts is not required for successful lab results.

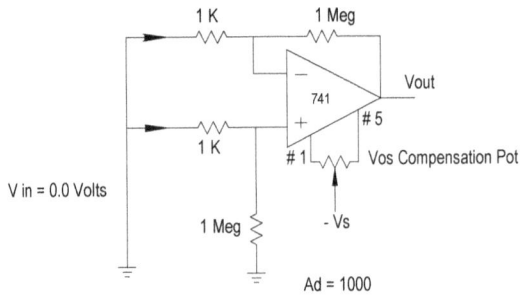

Figure 5 - Offset compensation is afforded by a 10K pot connected to pins #1 and #5, and the negative supply. Always check the op amp's specifications before connecting a compensation network.

Lab Procedure B
CMRR$_{db}$ Measurement

Step 3 – Differential Gain Verification

Assemble the following circuit. Note that only the input circuitry need be introduced to the previous circuit. Before measuring the amplifier output, compute the amplifier differential input by subtracting Vb from Va.

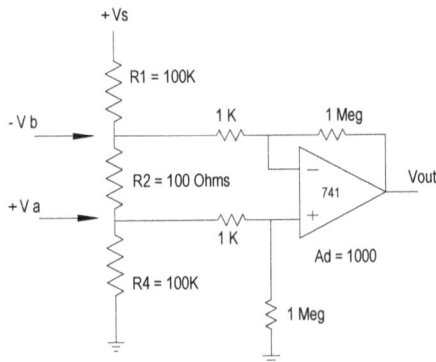

Figure 6 - Differential amplifier with gain of 1000 and a differential input voltage of - .0045 volts. The theoretical output voltage is 1000 times greater, or –4.45 volts.

The differential input voltage will be negative since Va is less than Vb (an unintentional error by the author, if a positive input is desired simply switch the connections to the voltage divider circuit at Va and Vb). Also compute the expected output voltage. Vout should approximate 1000 * (Va – Vb) and should also be a negative value due to the values of Va and Vb. Once the measured values are obtained, determine the differential gain by dividing the output by the input. Document the results following.

Computed Differential Input Voltage = _____

Computed Output Voltage = _____

Measured Differential Input Voltage* = _____

Measured Output Voltage = _____

Actual Differential Gain, Ad = _____

*Differential voltages are measured by applying a voltmeter across the amplifier inputs. In this case the positive voltmeter lead will be applied to Va and the negative probe to Vb. Differential signals are never measured by placing one of the voltmeter leads to ground as this, by definition is a single-ended measurement.

Step 4 – Common-Mode Input and Output Voltages
By this step, a review is in order. This lab is testing a differential amplifier for Common Mode Rejection (CMR). Common-Mode Rejection is a measure of a differential amplifier's ability to amplify differential input signals while rejecting any voltages common to both inputs. Common Mode Rejection Ratio (CMRR) is measured in decibels and is determined by dividing a differential amplifier's differential gain (Ad) by the amplifier's common-mode gain. Differential gain can be determined by dividing Rf/Rin. Common-mode gain will be determined experimentally.

$$CMRR_{db} = 20 \log_{10} (Ad/Acm)$$

To determine the common-mode gain, a common-mode signal must be applied to both differential inputs. Assemble the following circuit by removing the 100-Ohm resistor and connecting both diff. amp inputs between the 100K resistors. Note that the signal applied to each amplifier input is the same (approximately 4.5V). Measure and document the Common-Mode input voltage following.

Measured Common-Mode Input Voltage = _____

Once the circuit is assembled measure the resulting output voltage. Don't be surprised if the output voltage approaches zero. Since the function of a differential amp is to eliminate common mode signals, an output voltage near zero is to be expected. Document the resulting output voltage following.

Measured Common-Mode Output Voltage = _____

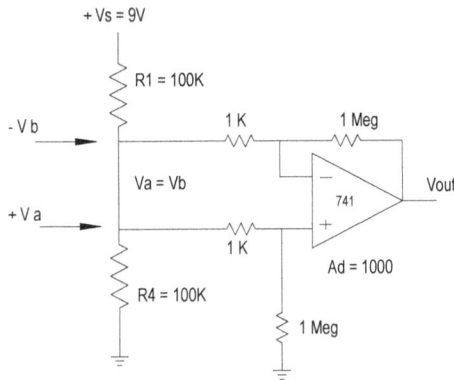

Figure 7 - Common-mode gain test circuit. This test is usually performed with a low frequency, high amplitude (10 Vpp) AC signal generator applied to both differential inputs simultaneously.

Step 5 - Common-Mode Gain, Acm
Once the common-mode input and output voltages have been measured, determine the common-mode gain (Acm) by dividing the output voltage by the input voltage. The value of Acm should compute to well below unity (typically below .01). Note – If offset voltage compensation has not been employed during this part of the lab the amount of output offset voltage measured in step 1 should be subtracted from the common-mode output voltage measured in step 4.

Common-Mode Gain, Acm = _____

The Acm portion of the lab is best obtained with an AC signal generator at the input. AC is used to make measuring the resulting output easier since the resulting output signal will be about a thousandth of the input if the amplifier is operating within specifications. When DC voltages are used for CMR determination the measured common-mode output cannot be distinguished from Vos. Likewise, the Vos was subtracted from the common-mode output voltage in the previous step. An AC signal would be easier to observe at low levels in the diff. amp output and would not require subtracting Vos. For this reason an AC sinusoidal signal source set to around 100 Hz is normally utilized as the common-mode input signal.

Step 6 - $CMRR_{db}$
Once the common-mode (Acm) and the differential (Ad) gains are known, CMRR can be computed and compared against the 741 specifications.

$$CMRR_{db} = 20 \log_{10} (Ad/Acm)$$

For a general purpose 741 op amp, the ratio of Ad/Acm usually approaches 1000/.001 or 1,000,000. This implies that if two signals, one noise and one informational, of equal amplitudes appear at the inputs of a differential amplifier simultaneously, the output signal will be a million times larger than the noise. Document the computed $CMRR_{db}$ following.

$$CMRR_{db} = \underline{\hspace{3cm}}$$

Many students find the appearance of the decibel confusing. A decibel is merely an expression of ratio. Decibels are normally found where relationships between the input and output are logarithmic or extremely large, such as in this case where the ratio is typically a million to one. Expressing the ratio as 100 to 120 db is a convenience.

Once $CMRR_{db}$ has been determined, locate the National Semiconductor LM741 Common Mode Rejection Ratio spec on the datasheet (page 3) and perform a comparison with the results obtained in step 6.

Results and Conclusions
This lab was more complicated and demanding than the previous labs. If you find that you did not follow the procedure or concepts presented don't be concerned. Of greatest importance is to recognize that differential amplifiers amplify the difference between inputs and when the inputs are equal, the output approximates zero volts. Of equal importance is to recognize when one is working with differential signals and/or single-ended signals since data acquisition, process control and PLC systems require one to setup the I/O accordingly.

To complete this lab, submit the following items and a brief description of your overall understanding and results, include;
-   The measured Output Offset Voltage, Vos,
-   Were you able to compensate for Vos?
-   What was the actual Ad? (The theoretical Ad was 1000)
-   What was the measured common-mode output voltage?
-   What was the Acm?
-   What was the $CMRR_{db}$?
-   Was $CMRR_{db}$ in spec?

ENGT 215 - Sensors and Analog I/O
Laboratory Experiment #14 – Differential Signals and Amplification

Objective:
The objective of this exercise is to introduce a differential signal source and to demonstrate the operation and application of a differential amplifier.

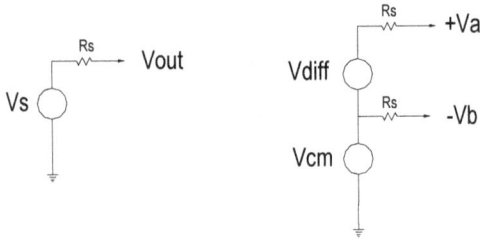

Figure 1 - Equivalent circuit diagrams of single-ended (left) and differential signal (right) sources. The "Vcm" designation represents noise or any other undesired voltage common to both output leads and is termed common-mode voltage.

Introduction:
Differential signal sources are commonly found in commercial audio amplification and experimental applications where signals with low noise content are required. Sensors utilized in experimental testing applications commonly output signals in the microvolts to millivolts range and with lengthy signal lines being common, the sensor output signals are susceptible to induced AC noise. To minimize the effects of induced noise strain gauges, load cells, pressure sensors, RTD's and other resistive sensors commonly employ a differential signal circuit configuration such as a full bridge.

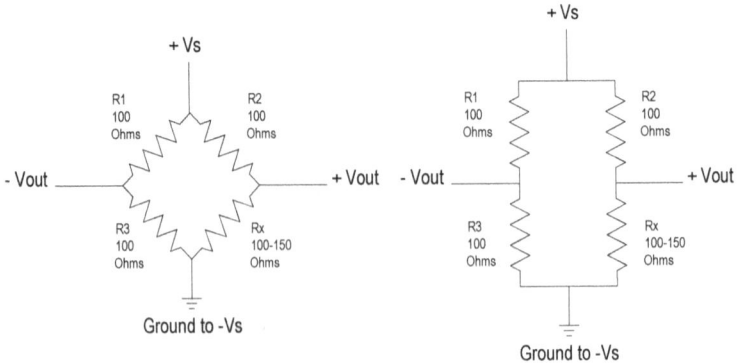

Figure 2 - A resistive bridge circuit drawn two ways to better demonstrate the individual
voltage divider circuits that compose the bridge. The sensor is located at +Vout in the
lower right leg of the bridge.

Note the polarities of the bridge outputs in the previous diagram. Differential
signals are generated out-of-phase, as one changes in a positive direction the
other changes in a negative direction (or remains constant). The polarities
represent the direction the voltages will change with an increase in signal. in the
previous diagram the actual output signal is taken from between the outputs,
meaning the individual outputs of –Vout and +Vout must be subtracted. A
voltmeter placed across the outputs (+Vout and –Vout) will easily measure the
output signal. Measurement of a differential signal should never require a ground
connection.

Converting a differential output signal into a single-ended signal is performed
through the use of a differential amplifier. Differential amplifiers amplify the
difference in the two voltages (+Vout and –Vout) applied to each of the amplifier's
inputs. For this reason the amplifier's inputs are polarized (+Va and –Vb in the
following diagram). Diff. amps can also amplify the differential input voltage if
desired. A diagram of a differential amplifier follows.

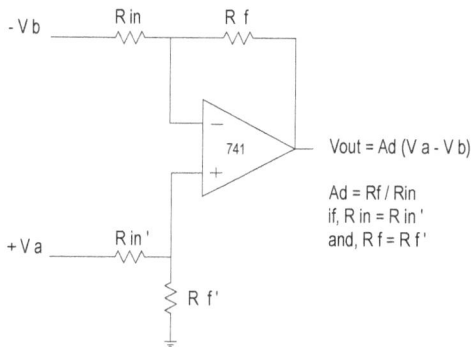

Figure 3 – A differential amplifier and associated equations. The output voltage is the
amplified difference between the inputs Va and Vb. The designation Ad represents
differential gain.

Induced AC noise is minimized by the use of differential signal sources and
differential amplification. As an example, assume a full-bridge type sensor such
as a strain gauge or load cell is applied to an experimental test fixture. Lengthy
leadwires are run from the sensor to a remotely located differential amplifier. All
four of the bridge leadwires are run between the sensor and amplifier in the same
cable and are approximately the same length. Should the connecting cable
between the sensor and differential amplifier pass through a varying magnetic
field a voltage will be induced into the leadwires.

Contrary to popular belief, shielding the cable will not reduce the induced noise significantly since cable shielding attempts to minimize the radiation of magnetic fields emanating from the conductors within the cable. Any AC noise induced into the connecting leadwires will be of approximately equal amplitudes since the individual signal carrying wires are of the same size and length, and the signal leadwires are in close proximity of each other. By subtracting the voltage available at each of the bridge sensor signal leads, the differential amplifier output will eliminate the induced noise since the noise voltage will be equal on each input.

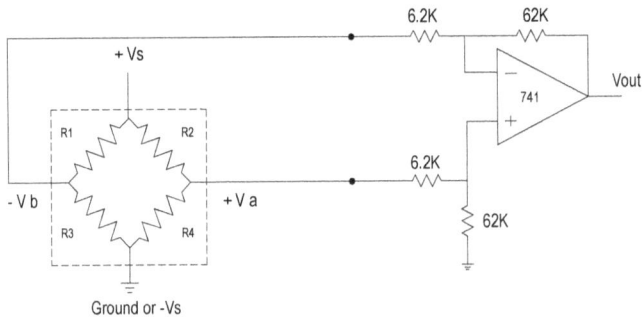

Figure 4 - A full-bridge sensor and differential amplifier. With the amplifier differential gain set to 10 (Rf/Rin), the amplifier output will be ten times the bridge output voltage (Va-Vb).

Lab Experiment
Differential Signal Sources, Differential Amplifiers and Signal Conditioning

The following lab will use a Resistive Temperature Detector (RTD) to demonstrate differential signals and the operation of a bridge circuit. The lab will contain three procedural sections. Procedure "A" will assemble a full-bridge circuit and measure the output voltage between upper and lower temperature extremes. Procedure "B" will convert the differential signal to an amplified single-ended signal using a differential amplifier. Procedure "C" will condition the differential output signal to a voltage span representative of temperature.

Lab Procedure A - RTD and Bridge
Locate the Resistive Temperature Detector among the components in the parts kit. If the parts kit was not acquired, inexpensive (and expensive) RTD's can be ordered from Omega (www.Omega.com), look for the temperature products and search for RTD. RTD's are constructed by placing a pure metal wire within a sheath (probe). RTD's in stainless steel sheaths cost between $75 and $100, however less expensive units are available under $30 as cement-on patches and non-submersible RTD element assemblies.

If you desire not to purchase an RTD, a variable resistor or combination of fixed resistors may be substituted in place of the RTD in the following lab. The substitute resistor(s) will need to be varied between the values of 100 and approximately 110 Ohms to simulate the RTD resistance at minimum temperature (100 Ohms at 0°C) and maximum temperature (109 Ohms at 23°C).

Resistive Temperature Detectors (RTD) -
RTD's are understood to be composed of a pure metal, are linear (although not perfectly linear at temperatures above 100°C), and exhibit a positive resistance/temperature characteristic. A positive resistance/temperature coefficient means the resistance will increase with a temperature increase. Note – don't confuse the RTD with a *thermistor,* a thermistor is a resistive temperature sensor that exhibits characteristics opposing the RTD in almost every sense. Thermistors are highly non-linear, produce a negative resistance/temperature characteristic and are composed of semiconductor (metal alloy) materials. Thermistors work well for temperature compensating electronic components and circuits over limited temperature spans but are not generally accepted as "industrial strength." Likewise, thermistors are rarely used in industrial temperature sensing applications. Thermistors can be purchased at Radio Shack, RTD's cannot. However, since the thermistor is a resistive temperature sensor the signal conditioning requirements are similar to the RTD requiring at least a bridge and differential amplifier.

The most common RTD is composed of Platinum and exhibits Alpha (gain) coefficients of .00385 Ohms/Ohm/C° and .003902 Ohms/Ohm/C°. (Other RTD's are available, such as Copper and Nickel, but are not as popular currently as the Platinum type.) The gain coefficient of .00385 Ohms/Ohm/C° is termed the European Alpha and the .003902 Ohms/Ohm/C° the American Alpha coefficient. Oddly, the European Alpha (.00385 Ohms/Ohm/C°) is the most commonly used in the USA.

The Alpha (gain) coefficient of .00385 Ohms/Ohm/C° appears intimidating but when broken-down becomes easy to interpret. First, one must know that most RTD's exhibit 100 Ohms at 0 degrees Centigrade. Other values are available but the "100-Ohm RTD" is the most popular. 100 Ohms at 0° C. is the offset or Y-axis intercept value for the sensor. Given this, one must read the Alpha coefficient as the amount of resistance change exhibited by the RTD for every Ohm of resistance at 0° C. Multiplying the Alpha coefficient by the resistance exhibited at 0° C yields the amount of resistance change per degree Centigrade. The following demonstrates.

Alpha (gain) coefficient = .00385 Ohms/Ohm/C°
Multiplying the Alpha coefficient by the RTD resistance at 0°C (100 Ohms) yields the resistance/temperature gain,
Gain in Ohms/C° = .00385 Ohms/Ohm/C° * 100 Ohms = .385 Ohms/C°

In summarizing, the RTD Alpha coefficient is normally expressed as the amount of resistance change for every Ohm of resistance exhibited at zero degrees, per degree Centigrade. Confusing, but with time and exposure makes sense.

Step 1 – RTD Resistance Determination
Given the RTD offset of 100 Ohms at zero degrees and the resulting gain of .385 Ohms per degree Centigrade change, determine the RTD resistance at the temperature extremes of 0 and 100 degrees Centigrade. Since the RTD is linear, a Y=mX+b approach is required.

Resistance at 0° = _____ Ohms

Resistance at 100° = _____ Ohms

The computations should have resulted in RTD resistance values of 100 Ohms and 138.5 Ohms at the temperature extremes. If not, use the following equation and re-compute.

Resistance = (Gain*Temperature) + Offset

R = (.385 Ohms/C° * Temperature) + 100 Ohms

Step 2 – Temperature Application
If an RTD is available in the parts kit it probably is of the non-submersible type, meaning it shouldn't be immersed in liquids. Likewise, two other temperature extremes and a slightly different approach (compared to what is performed in the lab at the college) are suggested.

First, measure the RTD resistance at room temperature. Indicate the measured resistance following.

Measured resistance at room temperature = _____ Ohms

From the measured room temperature value determine the RTD temperature. Don't be surprised if the temperature is found to be different than expected as many variables are at work on the RTD. This temperature will be required in later computations, document the temperature following. The re-arranged RTD equation follows to assist in determining the temperature.

Temperature = (Measured resistance – 100 Ohms) ÷ .385 Ohms/C°

Computed room temperature = _____ C°

Next, a 50/50 combination of ice and water will create a solution close to 0 degrees Centigrade. A thermometer will be required to verify this temperature but if none is available, trust me and assume the solution is close to 0 degrees

Centigrade. Tape the RTD element to the vessel (preferably glass, metal may create ground problems and plastic is not a good thermal conductor) containing the solution and after a minute or two, measure the resistance of the RTD. Document the resistance following.

Measured Resistance at 0° = _____ Ohms

The resistance should be close to 100 Ohms. Leave the RTD on the vessel containing the minimum temperature solution.

Step 3 – RTD Bridge
Bridges are used to convert a resistance change into a voltage change. Full bridges are used to establish a minimum differential output of zero volts. Assemble the following bridge circuit. The output connected to the RTD will be the positive leadwire and the connection opposite the bridge will be the negative output. When employing bridge circuits, all other bridge resistors will be equal to the minimum sensor resistance. In this manner the bridge output at the minimum applied temperature will be zero volts. For this reason the remaining bridge resistances should be 100 Ohms.

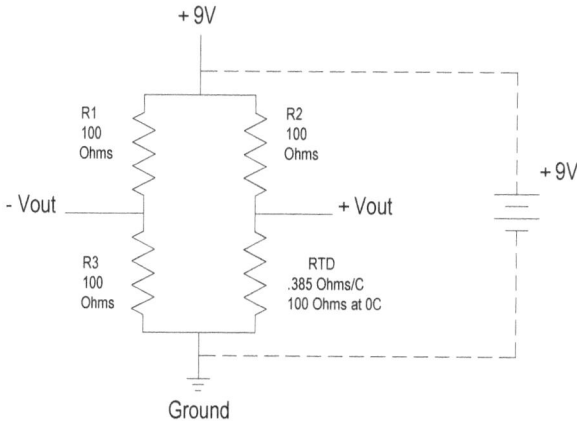

With the circuit assembled and the RTD attached to the vessel containing the 50/50 solution, measure and record the bridge output. Remember to measure the output as a differential signal by placing the voltmeter between the positive and negative output leads. The measured voltage should be close to zero.

Measured bridge output at approximately 0° = _____ Volts

At this point it should be mentioned that bridges frequently have "zeroing" circuits attached. Bridge zero or "Nulling" circuits allow the output to be manually adjusted to zero volts when the minimum sensor input is applied. This type of adjustment is usually set when the sensor is initially applied and/or when factory calibration is performed. The bridge null adjustment is rarely located on the front panel of an instrument or data acquisition system. Front panel zero and gain adjustments are normally associated with the scaling amplifier stage. The scaling amplifier and associated calibration adjustments will be addressed in lab procedure "C".

After measuring the bridge output at the minimum temperature, carefully remove the RTD from the vessel, as it will need to be attached again later. Let the RTD set at room temperature for a few minutes and document the bridge output following.

Measured bridge output at room temperature = _____ Volts

Lab Procedure "A" Review –
During this portion of the lab the RTD resistance and bridge output voltage spans were determined at minimum and maximum applied temperatures. In the next section a differential amplifier will be connected to the bridge to convert the differential signal into a single-ended signal and to provide a small amount of amplification. The amplifier output span at the two temperature extremes will be measured and in the final section a scaling amplifier will be designed to output a voltage span representative of the applied temperature extremes.

Lab Procedure B - Differential Amplifier
Step 4 – Differential Amplifier
Assemble the following circuit. Assure that the positive and negative outputs from the bridge are connected to the positive and negative amplifier inputs respectively (if the connections are reversed, the amplifier output will decrease when the input increases). Don't forget the ±V supplies connections to the op amp, they're not indicated in the diagram.

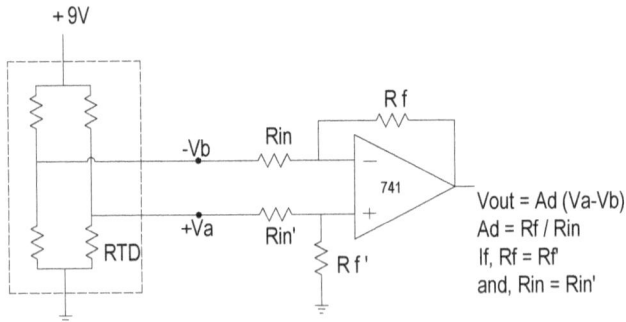

+9V

Rf

Rin

-Vb

741

+Va   Rin'

RTD

Rf'

Vout = Ad (Va-Vb)
Ad = Rf / Rin
If, Rf = Rf'
and, Rin = Rin'

Consider making Rin and Rin' values of 10K and Rf and Rf' 20K. This will result in a differential gain of 2. When circuit assembly is complete measure and record the output voltage at room temperature.

Reapply the RTD to the minimum temperature vessel and measure the output voltage at the minimum temperature. Remember to give the RTD a couple of minutes to settle at the new temperature. Also, add ice as needed to maintain an approximate 50/50 ratio and to keep the minimum temperature constant. Document the voltages following.

Diff. Amp output at room temperature = _____ Volts

Diff. Amp output at minimum temperature = _____ Volts

Lab Procedure "B" Review –
If your results are comparable to those acquired in the lab at the college, you should have measured a bridge voltage of approximately 0.0 with the RTD on the vessel to 0.18 volts at a room temperature of 22° C. (72° F). The amplifier output should be very close to twice these values, or approximately 0.0 to .36 volts. If your values are comparable proceed to the next section, if not, attempt to locate the discrepancy. Feel free to contact the author with your results or for assistance but be prepared to make additional circuit measurements if necessary.

Lab Procedure C –Scaling Amplifier
As a final step in the differential amplifier lab, a scaling amp will be designed and applied to the output of the differential amplifier. The scaling amplifier will provide an output voltage span that will be representative of the measured temperature while also providing calibration adjustments for accuracy.

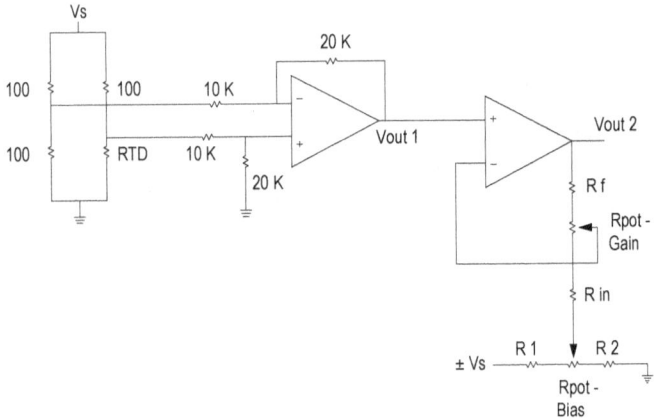

The completed circuit will appear as in the previous diagram. Using the following criteria, design and assemble the final stage scaling amplifier.

Assume the scaling amplifier output stage voltage will be .32 volts with the RTD applied to the vessel and .72 volts with the RTD at room temperature. Make the gain and offset (bias) of the scaling amp variable by ± 25% or some other reasonable percentage. Provide the indicated values following.

Computed scaling amp gain = _____

Computed scaling amp offset = _____

Rin = _____

Rf = _____

Rpot - gain value = _____

Biasing resistor R1 = _____

Biasing resistor R2 = _____

Rpot - Bias value = _____

Calibrate the scaling amplifier by placing the RTD on the minimum temperature vessel and set the bias pot for an output of .32V. Then place the RTD at room temperature and set the gain pot to .72V. Check the minimum output again by placing the RTD on the minimum temperature vessel, fine-tune the bias pot if necessary. Fine-tune the gain by setting the RTD at room temperature again. This process can be minimized by setting each pot to about half of its adjustment range before beginning the calibration procedure but double checking and fine-tuning is to be expected when first calibrating a newly designed and assembled sensor amplifier.

Once calibration is complete, the RTD can be placed at other locations and temperatures. The resulting output voltage from the scaling amplifier should provide a fairly accurate indication of temperature at .01V per degree Fahrenheit.

Results and Conclusions –
Submit the results of this lab to the author by writing a brief description of your results. Include the measured results of each section and any inconsistencies and discrepancies.

Objective:
The objective of this laboratory exercise is to design, assemble and calibrate an
instrumentation amplifier. The lab will involve acquiring a signal from a low-level
sensor (thermocouple) and conditioning the sensor output into a voltage
representative of the physical variable being measured (temperature).

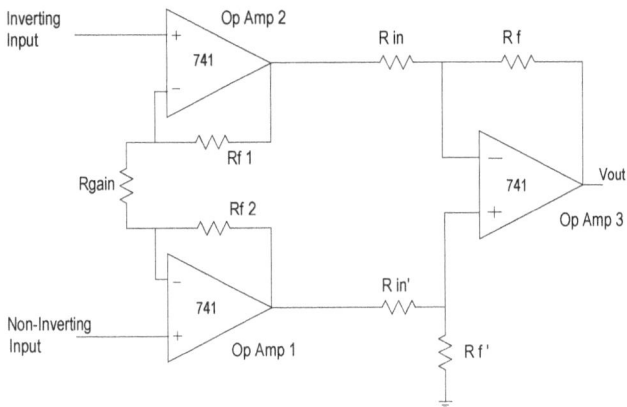

Figure 1 – The "Classic" three op amp instrumentation amplifier.

Introduction:
This lab will require the student to apply all of the principles presented in the
ENGT 215 course. Using the material presented in chapter 8 and the Type "T"
(copper-constantan) thermocouple included in the parts kit, the student is
expected to determine the thermocouple output millivoltage span, determine the
instrumentation amplifier output voltage gain and output offset voltage, determine
the amplifier components, assemble the amplifier circuit and calibrate the
amplifier. The results of this lab will be submitted for evaluation.

Sensor:
Thermocouples are the most common form of industrial temperature sensor.
Thermocouple operation is understood to be empirical, or based upon scientific
observation, rather than predicted or proven with mathematics. Established
millivoltage vs. Temperature reference tables are used to determine the
thermocouple output voltage for a corresponding temperature.

Figure 2 - Two common thermocouples, a type "K" adhesive patch (left) and a type "T" exposed junction (right). The thermocouple junction on the right demonstrates the joined dissimilar metals forming the "Hot" junction.

A millivoltage is achieved at the thermocouple output by coupling (connecting) two dissimilar metals together and applying the joined end (called the "Hot" junction) to the temperature to be measured. The temperature difference between the 'hot' junction at the process and the 'cold' or reference junction (where the thermocouple output is located) generates a small, but very predictable milli-voltage. The generated millivoltage emf is the result of the temperature difference between the two ends of the thermocouple, at the process and at the measurement apparatus or transmitter. The greater the temperature difference, the greater the millivoltage generated - the smaller the temperature difference, the smaller the millivoltage. Should the two temperatures be equal with no temperature difference, the resulting millivoltage would be zero.

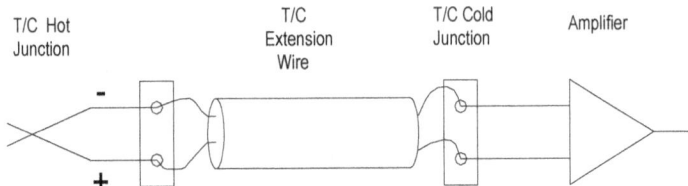

-The Cold Junction is located where the T/C
extension wires connect to a different material type.

-To find the millivoltage Vout of Thermocouple:
Determine millivoltage at temperature extremes
from millivoltage tables, reference junction convert by
subtracting millivoltage equivalent of actual reference
junction temperature.

Figure 3 - An operational diagram of the thermocouple. The "Cold" or reference junction for this lab will be located where the thermocouple connects to the amplifier. The negative leadwire of a thermocouple is always colored red.

Thermocouple tables clearly indicate the generated millivoltage in relation to the applied temperature for the various types of sensor composition. Type 'J' (iron/constantan) and type 'T' (copper/constantan) thermocouple millivoltage tables have been included in the appendix of the manuscript. If the tables are not included, the thermocouple millivoltage vs. temperature tables can also be located at Omega Engineering, Inc.'s web site (http://www.omega.com/temperature/Z/zsection.asp), look for the section titled "Thermocouple Reference Data."

To use the information provided in the tables one must know the type of thermocouple used (types T, J or K are included in the parts kit), the temperature span to be measured (32° to 212° F) and the expected temperature of the reference junction (or "Cold" junction). Given that you'll most likely being assembling and operating the lab circuit indoors (and assuming you don't live in an igloo), the assumed reference junction temperature will be 72 degrees Fahrenheit (22.22° C.). Once these items are known, one can determine the expected thermocouple millivoltage from the thermocouple tables at the previously mentioned Omega site.

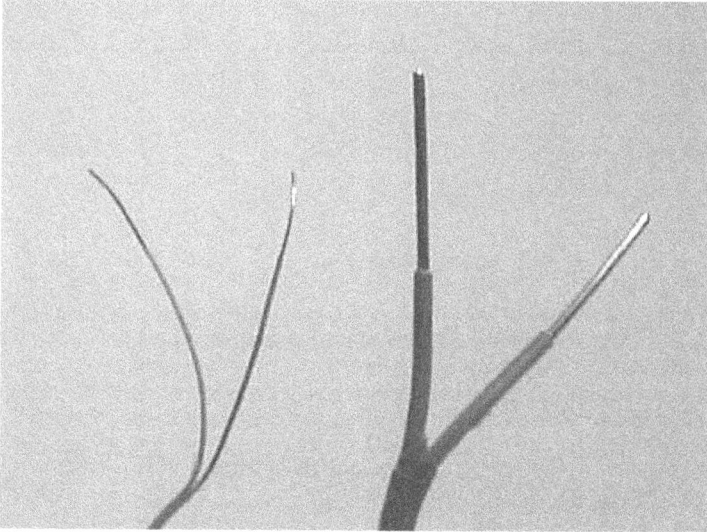

Figure 4 – Two possible thermocouple reference junctions. Thermocouples are distinguished by a characteristic color code. In this case the yellow and red leads designate a type "K" (chromel-alumel) and the blue and red leads represent a type "T" (copper-constantan) thermocouple. The red lead is always negative on thermocouples.

Thermocouple EMF:
At the end of this lab is a table of temperature vs. millivoltage (also referred to as electromotive force, or EMF) for a type "K" (chromel-alumel) thermocouple. Using the values provided in the table the amount of millivoltage available at the thermocouple output can be determined. Note the reference junction temperature is assumed to be 32° F. This is common for thermocouple data tables even though most reference junctions will not be at a temperature of 32° F. Centigrade tables use 0° C. as a reference junction temperature.

To determine the expected thermocouple output millivoltage, momentarily assume the reference junction temperature for the circuit is as specified on the table. Also assume the desired hot junction temperature span to be measured will be from 32 to 212 degrees Fahrenheit. The table lists individual temperatures from 0 to 10 across the top, and groups of temperature in 10-degree increments vertically along the left column. Given that the desired temperature to be measured begins at 32° F, locate "30" along the left column and move three positions to the right to find the 32 degree millivoltage. Explained in another way,

°F	0	1	2	3	4	5	6	7	8	9	10°
-10	-1.114	-1.094	-1.073	-1.052	-1.031	-1.010	-0.989	-0.968	-0.947	-0.926	-0.905
0	-0.905	-0.883	-0.862	-0.841	-0.820	-0.799	-0.778	-0.756	-0.735	-0.714	-0.692
10	-0.478	-0.457	-0.435	-0.413	-0.392	-0.370	-0.349	-0.327	-0.305	-0.284	-0.262
20	-0.262	-0.240	-0.218	-0.197	-0.175	-0.153	-0.131	-0.109	-0.088	-0.066	-0.044
30	-0.044	-0.022	0.000	0.022	0.044	0.066	0.088	0.110	0.132	0.154	0.176
40	0.176	0.198	0.220	0.242	0.264	0.286	0.308	0.330	0.353	0.375	0.397

Locate the intersection of "30" in the left column and "2" across the top of the table. The result will be the 0.000 millivoltage value. Since the hot junction and reference junction are at the same temperature, no millivoltage can be expected at the thermocouple output. Thermocouple output voltage varies in proportion to the temperature *difference* between the hot junction and the cold (reference) junction. In addition, the millivoltage output will be *positive* if the hot junction is hotter than the cold junction, and *negative* if the hot junction is colder than the reference junction.

Next locate the millivoltage value for a reference junction of 32° F and a hot junction of 212° F by finding "210" in the left column and dropping down from the value of "2" across the top. The result should yield 4.096 millivolts as shown following.

°F	0	1	2	3	4	5	6	7	8	9	10°
200	3.820	3.843	3.866	3.889	3.912	3.935	3.958	3.981	4.004	4.027	4.050
210	4.050	4.073	4.096	4.119	4.142	4.165	4.188	4.211	4.234	4.257	4.280
220	4.280	4.303	4.326	4.349	4.372	4.395	4.417	4.440	4.463	4.486	4.509

The resulting output millivoltage span for an applied temperature of 32° to 212° F with a reference junction temperature of 32° F is 0.000 millivolts to 4.096 millivolts. However, the actual reference junction temperature will most probably be closer to a room temperature of 72° F.

Reference Junction Conversion:
The process of converting the previously acquired millivoltage data to values corresponding to a reference junction temperature of room temperature (72° F) or any other temperature is called *Reference Junction Conversion* and is a relatively simple process. To begin, attempt to determine the reference junction temperature where the thermocouple connects to the amplifier circuit. If a thermometer is not available assume a reference junction temperature of 72° F. Next, locate the millivoltage corresponding to a temperature of 72° F. From the table a value of .888 millivolts is found.

°F	0	1	2	3	4	5	6	7	8	9	10°
60	0.619	0.642	0.664	0.686	0.709	0.731	0.753	0.776	0.798	0.821	0.843
70	0.843	0.865	0.888	0.910	0.933	0.955	0.978	1.000	1.023	1.045	1.068
80	1.068	1.090	1.113	1.136	1.158	1.181	1.203	1.226	1.249	1.271	1.294

To perform the reference junction conversion, subtract the value corresponding to 72° F from each of the previously obtained millivoltages at 32° F (0.000 mV) and 212° F (4.096 mV). The resulting millivoltage span for an applied hot junction temperature span of 32° F to 212° F is -.888 mV to +3.208 mV respectively.

The concept behind reference junction conversion follows the operational principle of the thermocouple. With a hot junction temperature of 32° F and a reference junction temperature of 72° F, the generated millivoltage will be negative since the hot junction is colder than the reference junction. When the hot junction temperature increases from 32° F to 72° F, the millivoltage will move in a positive direction from -.888 mV towards 0 mV since the hot junction and reference junction will be approaching a room temperature of 72° F. As the hot junction temperature increases beyond 72° F the millivoltage generated will move into positive values since the hot junction is becoming hotter than the reference junction. When compared to a reference junction temperature of 32° F, the resulting 212° F millivoltage can be expected to be less since the temperature difference between the hot and reference junction temperatures is less with a reference junction temperature is at 72° F.

Lab Experiment
Thermocouple Temperature Sensing, Instrumentation Amplifier Signal Conditioning and Scaling Amplifier Calibration

Step 1 – Millivoltage Determination
The thermocouple provided in the parts kit will be of a different type than the "K" type described previously. Locate the thermocouple in the parts kit and determine the expected millivoltage output for an applied temperature of 32° F to 212° F, and a reference junction temperature of 72° F. Feel free to use a reference junction temperature other than 72° F if necessary. The thermocouple reference tables are located at
http://www.omega.com/temperature/Z/zsection.asp. Note the expected millivoltages before and after reference junction conversion following.

Before Reference Junction Conversion:
Millivoltage at 32° F = _____ millivolts

Millivoltage at 72° F = _____ millivolts

Millivoltage at 212° F = _____ millivolts

After Reference Junction Conversion:
Millivoltage at 32° F = _____ millivolts

Millivoltage at 72° F = _____ millivolts

Millivoltage at 212° F = _____ millivolts

Feel free to contact the ENGT 215 faculty member for verification of the millivoltage results before proceeding.

Step 2 – Amplifier Design
Assume the desired amplifier output voltage is to represent the applied hot junction temperature in degrees Fahrenheit. Likewise, an amplifier output voltage span of .32V to 2.12V is appropriate.

Given the expected output voltage span of .32V to 2.12V and the sensor input voltage from Step #1, determine the required amplifier gain and offset values.

Computed amp gain = _____

Computed amp offset = _____

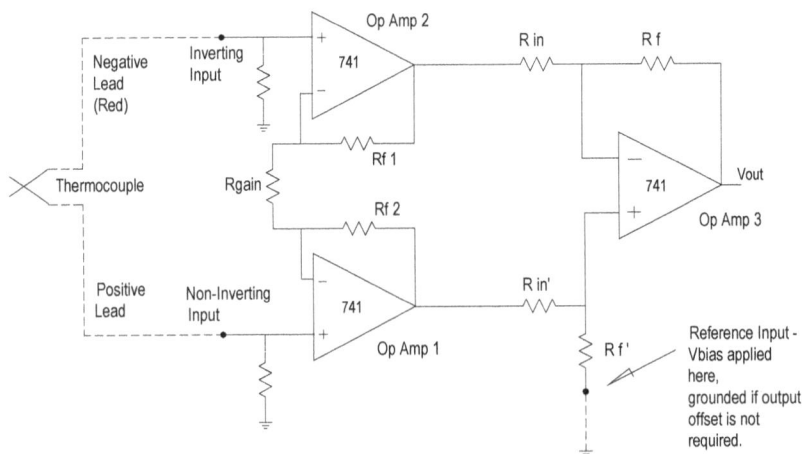

Figure 5 - Thermocouple and instrumentation amplifier of lab #15. Note – the resistors to ground from each of the op amp inputs (op amp1, op amp 2) are required to maximize

CMR and to provide a path to ground for bias currents. If omitted, the negative lead of the thermocouple and the corresponding input will need to be connected to ground.

From the text material on instrumentation amplifiers, determine the required resistor and bias voltage values. Note: you will be required to determine the values of all other instrumentation amplifier resistors. It is suggested the second stage differential amplifier have a gain of unity (Rin, Rin', Rf and Rf' are all equal) and Rf1 and Rf2 in the first stage be greater than 10K. Note the values chosen following.

First stage resistances:

Rf1 = _____

Rf2 = _____

Rgain = _____

Second stage resistances:

Rin = _____

Rin' = _____

Rf = _____

Rf' = _____

Second stage Vbias value

Vbias = _____

Step #3 – Gain and Bias Variation
For calibration purposes the gain and bias values are to be made variable. It is suggested that the gain be made variable by plus and minus 50%, and the bias voltage be made variable by plus and minus 100%. Modify the circuit values to accommodate the variation. The following circuit can be used as a model.

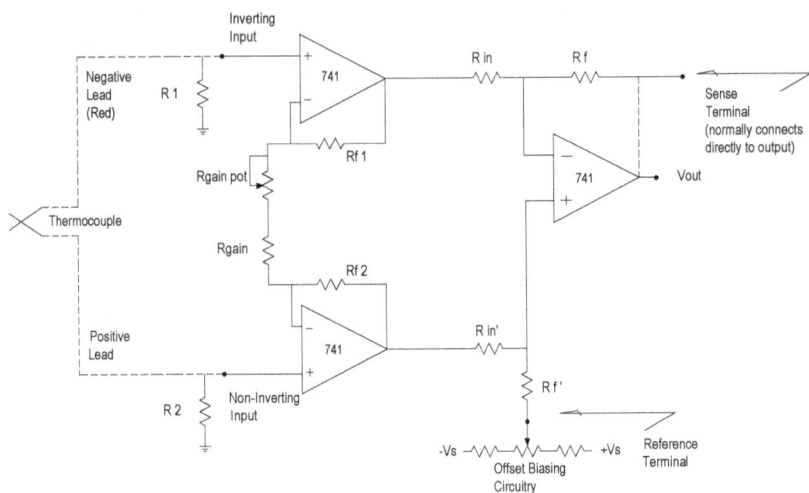

Figure 6 - The thermocouple and instrumentation amplifier schematic complete with gain and output offset variation. R1 and R2 can be any value over 10K, preferably 100K.

Include the resistance values for the gain and bias variation in the following table.

Gain resistances:

Rgain pot = _____

Rgain = _____

Bias voltage resistances:

Rpot = _____

Rdrop 1 = _____

Rdrop 2 = _____

Step #4 – Calibration
Once the circuit is assembled and determined to be functional, the gain and offset (bias) potentiometers will require setting. The process of setting the potentiometers is referred to as calibration and is best performed by substituting the thermocouple with an accurately known millivoltage source. The thermocouple is removed and a voltage divider, designed to provide the minimum to maximum millivoltage span, is placed across the inputs to the instrumentation amplifier. As an example and using the previously determined millivoltage values for the type "K" thermocouple, a voltage divider would be assembled to output -.888 mV to +3.208 mV. The circuit need not provide exactly

-.888 mV to +3.208 mV but should a millivoltage span that contains these extreme values. In this manner the thermocouple voltage is substituted with the millivoltage from the voltage divider. The bias and gain potentiometers can then be set accordingly for an amplifier output voltage of .32V to 2.12V. An example of the voltage divider substitution circuit would appear as follows.

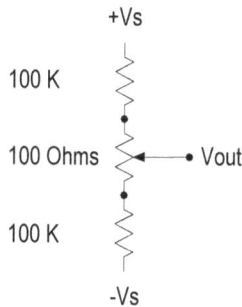

```
                    +Vs
                     |

        100 K         ⊰

        100 Ohms      ⊰─────• Vout

        100 K         ⊰

                     |
                    -Vs
```

Figure 7 – Divider circuit to simulate a thermocouple input voltage. Using the values indicated and ±9 Volt supplies the output voltage will be approximately -9 mV to +9 mV. With the positive lead of the thermocouple disconnected, the output from the wiper connects to the non-inverting amplifier input. The amplifier's inverting input should be grounded. A multi-turn potentiometer works best for this application.

Before beginning the calibration effort, disconnect the positive thermocouple lead from the non-inverting amplifier input. Connect the wiper of the divider to the non-inverting amplifier input. Connect the inverting amplifier input to ground. The following diagram illustrates.

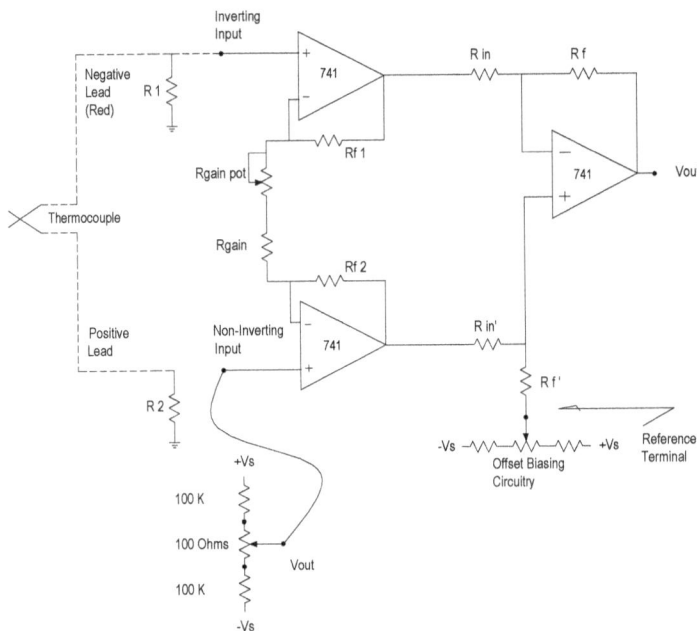

Figure 8 - Thermocouple simulation circuit applied to the instrumentation amplifier to facilitate calibration.

Calibrate the amplifier by placing the output from the divider at the minimum expected thermocouple output millivoltage. Set the bias pot for an output of .32V. Then place the divider to the maximum expected thermocouple millivoltage and set the gain pot for an amplifier output of 2.12V. Check the minimum output again by setting the divider to the minimum expected thermocouple millivoltage; fine-tune the bias pot if necessary. Fine-tune the gain by setting the divider to the maximum millivoltage again. This process can be minimized by setting each pot to about half of its adjustment range before beginning the calibration procedure but double checking and fine-tuning is to be expected when first calibrating a newly designed and assembled sensor amplifier.

Once calibration is complete, the thermocouple can be placed at other locations and temperatures. The resulting output voltage from the scaling amplifier should provide a fairly accurate indication of temperature at .01V per degree Fahrenheit.

Step #5 – Application
If a thermometer with a wide measurement span (at least 32° to 212° F) is available, consider taping the thermocouple to a glass (metal will also work with a

gas stove but may create problems with an electric stove) vessel and filling the vessel with a 50/50 solution of water and ice. This will result in a temperature of approximately 32°. Apply heat to the vessel by placing it on a stove and setting the burner to low. Use caution to avoid damaging the thermocouple, extension leadwires or the amplifier circuit. Also, <u>keep the reference junction of the thermocouple at a remote distance from the heat source</u> or the expected millivoltage will be greatly reduced. Periodically compare the output of the amplifier with the indicated value of the thermometer. To minimize the effects of thermal gradients within the ice/water solution, gently stir before taking a reading. Attempt to take the water temperature to boiling (212° F). Tabulate your results and submit them for assessment.

Results and Conclusions –
This lab will be assessed for completion and understanding. Submit the results to the author by writing a brief description of your experience. Include the computed and measured results of each section as well as any inconsistencies and/or discrepancies. If the lab was not completed submit evidence in support of your progress prior to termination.

Type "K" Thermocouple – Temperature vs. Millivoltage Table
Nickel-Chromium vs. Nickel-Aluminum
(Chromel – Alumel)
Reference Junction Temperature = 32° F.

°F	0	1	2	3	4	5	6	7	8	9	10°
-10	-1.114	-1.094	-1.073	-1.052	-1.031	-1.010	-0.989	-0.968	-0.947	-0.926	-0.905
0	-0.905	-0.883	-0.862	-0.841	-0.820	-0.799	-0.778	-0.756	-0.735	-0.714	-0.692
10	-0.478	-0.457	-0.435	-0.413	-0.392	-0.370	-0.349	-0.327	-0.305	-0.284	-0.262
20	-0.262	-0.240	-0.218	-0.197	-0.175	-0.153	-0.131	-0.109	-0.088	-0.066	-0.044
30	-0.044	-0.022	0.000	0.022	0.044	0.066	0.088	0.110	0.132	0.154	0.176
40	0.176	0.198	0.220	0.242	0.264	0.286	0.308	0.330	0.353	0.375	0.397
50	0.397	0.419	0.441	0.463	0.486	0.508	0.530	0.552	0.575	0.597	0.619
60	0.619	0.642	0.664	0.686	0.709	0.731	0.753	0.776	0.798	0.821	0.843
70	0.843	0.865	0.888	0.910	0.933	0.955	0.978	1.000	1.023	1.045	1.068
80	1.068	1.090	1.113	1.136	1.158	1.181	1.203	1.226	1.249	1.271	1.294
90	1.294	1.316	1.339	1.362	1.384	1.407	1.430	1.453	1.475	1.498	1.521
100	1.521	1.543	1.566	1.589	1.612	1.635	1.657	1.680	1.703	1.726	1.749
110	1.749	1.771	1.794	1.817	1.840	1.863	1.886	1.909	1.931	1.954	1.977

120	1.977	2.000	2.023	2.046	2.069	2.092	2.115	2.138	2.161	2.184	2.207
130	2.207	2.230	2.253	2.276	2.298	2.321	2.344	2.367	2.390	2.413	2.436
140	2.436	2.459	2.483	2.506	2.529	2.552	2.575	2.598	2.621	2.644	2.667
150	2.667	2.690	2.713	2.736	2.759	2.782	2.805	2.828	2.851	2.874	2.897
160	2.897	2.920	2.944	2.967	2.990	3.013	3.036	3.059	3.082	3.105	3.128
170	3.128	3.151	3.174	3.197	3.220	3.244	3.267	3.290	3.313	3.336	3.359
180	3.359	3.382	3.405	3.428	3.451	3.474	3.497	3.520	3.544	3.567	3.590
190	3.590	3.613	3.636	3.659	3.682	3.705	3.728	3.751	3.774	3.797	3.820
200	3.820	3.843	3.866	3.889	3.912	3.935	3.958	3.981	4.004	4.027	4.050
210	4.050	4.073	4.096	4.119	4.142	4.165	4.188	4.211	4.234	4.257	4.280
220	4.280	4.303	4.326	4.349	4.372	4.395	4.417	4.440	4.463	4.486	4.509

This page retained blank intentionally.

About the author –

Jonathan Lambert is a professor emeritus of engineering technology at Black Hawk College in Moline, Illinois. As a senior member of the International Society of Automation (ISA) and a thirty-year member of the American Society for Engineering Education (ASEE), Jon holds an AAS degree in Electrical/Instrumentation Engineering Technology from Black Hawk College, a BS degree in Electrical Engineering Technology (EET) from Bradley University and an MS degree in Industrial Engineering from the University of Iowa. He began a career in electronics, sensors and controls repairing traffic signals for the City of Quincy, Illinois in 1970 and has held positions as a test engineer for Motorola Corporation, a broadcast chief engineer for Black Hawk College Educational Television/WQPT Quad-Cities Public Television and served for 35 years as a professor, lead instructor and department chair at Black Hawk College. He has taught engineering and technology courses at Western Illinois University, Purdue University and Universiti Teknologi Malaysia (UTM) and has provided technical assistance, engineering consulting and product development testing services for numerous Quad-Cities regional employers including the Harvester and Seeding divisions of John Deere Worldwide Product Development, Roth Pump, Small Newspapers, Chrome Locomotive, National Railway, the Moline Dispatch, Martin Engineering, Chemplex Corporation, US Army/Edgewood Chem/Bio Center, River Stone Group, Moline Consumers Company, Ipsco Steel, Kewanee Boiler Manufacturing and numerous other regional firms.

Jon is currently providing contract engineering services and adjunct engineering technology instruction at Black Hawk College in Moline, Illinois. He can be reached at LambertJ@bhc.edu.

www.ingramcontent.com/pod-product-compliance
Lightning Source LLC
Chambersburg PA
CBHW021023210326
41598CB00016B/893